Springer-Verlag Berlin Heidelberg GmbH

Gordon Filby (Ed.)

Spreadsheets in Science and Engineering

With 123 Figures and 24 Tables

 Springer

Dr. Gordon Filby
Forschungszentrum Karlsruhe GmbH
76021 Karlsruhe
Germany

Cataloging-in-Publication Data applied for

Die Deutsche Bibliothek – CIP-Einheitsaufnahme
Spreadsheets in science and engineering : with 24 tables / Gordon Filby (ed.). - Berlin ; Heidelberg ; New York ; Barcelona ; Budapest ; Hong Kong ; London ; Milan ; Paris ; Santa Clara ; Singapore ; Tokyo : Springer 1997
   ISBN 978-3-642-80251-5   ISBN 978-3-642-80249-2 (eBook)
   DOI 10.1007/978-3-642-80249-2

Additional material to this book can be downloaded from http://extras.springer.com.

© Springer-Verlag Berlin Heidelberg 1998
Softcover reprint of the hardcover 1st edition 1998

Cover design: Struve & Partner, Heidelberg
Typesetting: Data conversion by MEDIO, Berlin

SPIN: 10532203   52/3020 - 5 4 3 2 1 0 – Printed on acid-free paper.

Preface

Time was, and it was not so long after the "spreadsheets...they're only for accountants" era, that one could scarcely open a scientific journal without spotting one or more articles with titles like "using spreadsheets to..." or "applying spreadsheets in the...". Sometimes it appeared that the only applications for spreadsheets still missing were "... to hang up your washing", "...cure baldness" or "...get rid of your mother-in-law". Times change and now there is little doubt that these "financial" tools are so all pervading that even titles like these would hardly attract special attention. So why bother preaching to the converted? Why another book on spreadsheets in something or other? Wellllll...the answer to that is simple. This book is unique; it sets out to do two main things that most other spreadsheet books do not even try. Let me explain what I mean.

Previous books of the kind "Spreadsheets in Science and Technology" are in the main single author works. They reflect the knowledge, interests and ability of the author in the, by definition, very broad fields involved. However, it is clear that no single author can possess the wherewithal to cope equally well with all applications of spreadsheet software in the modern scientific world. This book is different. At the outset, the invited authors – all experts in their fields – were asked to provide examples and quality literature references on the use of spreadsheets in their special branch of expertise. This way, it was hoped, the reader could easily see what had already been achieved in their speciality. The idea was to provide help in answering questions like "what's in it for me?" and "is it worth the trouble to learn?".

The second major goal of the book was to help those readers already on their way after having answered "...apparently quite a lot" and "yes" to the prior questions. Namely to provide potential users with expert tuition on how to perform useful work with a spreadsheet program. With this approach the reader gains a much broader look at various styles of doing things in the spreadsheet world than from a single author demonstrating his favourite tips and tricks. He can copy these styles or, if he prefers features of another authors style, mix them up to find his own way of working.

This book describes the fundamentals of spreadsheet use together with typical examples taken from the several sectors of contemporary science and engineering.

Starting with focusing the readers attention on the basic mechanics of spreadsheet handling the first chapter is a simple "getting started" type chapter. Here the reader will see how to move, copy, format his worksheet etc against the background of a couple of simple chemical examples. The barebones of spreadsheeting can be found here. Chapters 2 and 3 enlarge on this by providing step-by-step details of spreadsheet problem solving in Physics, Electronic Engineering and Statistics. Look out here for elegant models of Brownian motion and the statistics of budgeting for old-age pensions. By the end of these chapters all the basics should have been mastered. Chapter 4 looks at geological applications. At this stage the reader is expected to be able to reproduce examples from nearly empty skeleton worksheets provided for practice. The two following chapters are for the chemically-minded. In Chapter 5 we deal with aquatic chemistry and in Chapter 6 with chemical engineering problems. Both of these chapters use the advanced equation-solving feature of Excel known as Solver. Chapter 7 deals with what was for me the most astonishing application of spreadsheet software. Their use in molecular biology. Here you'll see how the software helps, among other things, in solving nucleic acid-sequencing problems and the analysis of protein properties. The last chapter in the book concerns itself with Materials Science. The emphasis here is the Visual Basic for Applications (VBA) programming language. In this chapter it is used to produce turnkey applications requiring little knowledge of Excel from the reader. VBA was supplied as part of the Excel package as of Version 5.

The book closes with three appendices. Appendix A provides a brief but thorough overview of VBA with the stress on the current version (Excel 97). A disk example demonstrates the principles at work. Appendix B contains a comprehensive list of newer literature on spreadsheets (mainly Excel). Books (general and specialised scientific ones) are listed as well as references from the primary research literature in fields such as chemistry and the geosciences. This appendix closes with a list of World Wide Web (www) sites with extensive material on spreadsheets. Appendix C is provided for readers either considering upgrading to one of the recent Office packages like MS Office 97 or those having just received one. It describes the features added to MS Excel 97. Examples are provided to smooth the transition.

The general aim of the book is to provide interested scientists, teachers, students and professionals with the know-how to use these increasingly important tools in their own disciplines. We think it has succeeded. We hope that you, the reader, do too and wish you, as they say in German "viel Spaß" (something like Enjoy!) on your way.

Of course, in the successful preparation of a book like this, many people are involved. First and foremost the chapter contributors, without whose work this book would not exist. Thanks everybody, especially those who got their chapters in on time. In close second place must come the publisher, in the person of Mr Peter Enders for his quiet support. The assistance of his staff is also not forgotten.

The editors son, Rafael was again instrumental in removing some of his fathers sillier ideas.

The editor also wishes to express his gratitude to Microsoft and Lotus Development for providing him with gratis copies of their main Office suites.

Last not least, thanks to the editors employers, the Research Center of Karlsruhe (Forschungszentrum Karlsruhe, formerly Kernforschungszentrum Karlsruhe). Especial gratitude for their permission to participate in the project goes to Prof. Dr. M. Popp and Dr. F. Horsch.

Karlsruhe, Summer 1997 W.G. Filby

Table of Contents

The Authors and Their Chapters

Chapter 1. Spreadsheet Basics
Gordon Filby is a senior scientist in the Forschungszentrum Karlsruhe, Germany (Karlsruhe Research Center). He also compiled the Appendices B and C in the book. His previous books are Turbo Pascal for Chemists (VCH 1993) and Spreadsheets for Chemists (VCH 1995)

Chapter 2. Applying Spreadsheets in Physics and Electronic Engineering
William Jay Orvis received his BS and MS degrees in Physics at the University of Denver in 1973 and 1976 respectively. Since then, he has built nuclear reactor instrumentation at the Idaho National Engineering Laboratory and has designed and modelled solid state devices at the Lawrence Livermore National Laboratory. He is currently a member of CIAC, the US Department of Energy's computer incident response team. He is the author of more than 10 books on computer programming, including several on the use of spreadsheets in scientific applications.

Chapter 3. Spreadsheets as Tools in Mathematical Modelling and Numerical Mathematics
Erich Neuwirth is the head of the Computer Supported Didactics Working Group at the University of Vienna, and one of his areas of interest is using computers to enhance learning and teaching in mathematics, statistics and science. He has been visiting professor at universities in the United States, and he has been invited to give presentations about spreadsheets in education in many different countries. He also is active in many international scientific bodies concerned with computers in education, and he is an acknowledged expert in using computers for creating interactive learning environments. In 1996 he won the European Academic Software Award for a multimedia program about the mathematical foundations of musical tuning systems.

You can find out more from his homepage at *http://sunsite.univie.ac.at/ Spreadsite*.

Chapter 4. Applications of Spreadsheets in the Earth Sciences
J. P. Le Roux received an M.Sc. from the University of Stellenbosch in 1977 and a Ph.D. from the University of Port Elizabeth in 1985. He worked for the Atomic Energy Corporation of South Africa from 1978–1985 and has been lecturing in physical geology, sedimentology and stratigraphy at the University of Stellen-

bosch since 1985. Research projects in South Africa, Israel, Australia and Chile have culminated in about 100 publications in journals, books and conference proceedings.

R. D. O'Brien graduated from the University of Stellenbosch in 1991 with a B. Sc. (Hons) and is currently completing a M. Sc. Degree. He has been working for the University of Stellenbosch since 1992 as a technical officer.

Chapter 5. Spreadsheet Applications in Aquatic Chemistry

Stephen Leharne is Reader in Environmental Chemistry at the University of Greenwich, England. His major teaching interests are focused upon contaminated land assessment and remediation, environmental modelling and environmental risk assessment. He has major research interests in scanning microcalorimetry, the physical chemistry of biopolymer and surfactant systems and contaminated land remediation. He has extensive experience of the use of spreadsheets in these various subject areas and has published several papers on the subject.

Chapter 6. Using Spreadsheets in Chemical Engineering Problems

Frank M. Julian recently retired after 42 years with DuPont, serving at sites throughout the United States and Europe. Most recently he was a Principal Consultant in DuPont's Central Engineering Department. He received a BS in Chemical Engineering from Rice University and an MS from the University of Houston. He is co-teacher of the continuing education course titled "Spreadsheet Power!" offered by the American Institute of Chemical Engineers.

Chapter 7. Spreadsheets in Molecular Biology

Gerry Shaw was born in Nottingham, England and obtained a B.Sc. in zoology at University College London in 1975. He then moved to Dennis Bray's laboratory at King's College London and was awarded a Ph.D. degree there in 1980. He then moved to Göttingen, Germany to the laboratory of Professor Klaus Weber were he worked first as a Post-Doctoral fellow then as a staff member. In 1986 he moved to Gainesville, Florida where he is now a Professor of Neuroscience. Dr. Shaw's major research interests are the cytoskeletal and signalling molecules of the nervous system, which he studies using a variety of biochemical, immunological, molecular biological and computational techniques.

His home page is at URL *http://www.ufbi.ufl.edu/people/shaw/shawhp.htm*

Chapter 8. Spreadsheet Applications in Materials Science

Antonio Augusto Gorni studied material science at the Federal University of São Carlos and the Polytechnic of São Paulo, Brazil. Currently he is conducting research into hot rolling at the Companhia Siderúrgica Paulista, Cubatão. He also lectures in polymer science at the Faculty of Industrial Engineering, São Bernardo do Campo. He has published more than 60 papers in the fields of hot rolling and heat treatment, with the emphasis on mathematical modelling. His home page is at *http://www.geocities.com/SiliconValley/5978*.

Appendix A: The Visual Basic for Applications Programming Language (VBA)
John Walkenbach has been involved with spreadsheets since the early days of Visi-Calc. He holds a Ph.D. in experimental psychology, and is the author of two dozen spreadsheet books and approximately 250 articles and reviews in various computer publications. In addition, he maintains several sites on the World Wide Web (http://www.j-walk.com). He lives in San Diego, California.

File Directory

Readme.txt

This CD-ROM contains example files from book:

Spreadsheets in Science and Engineering

Edited by Gordon Filby

Copyright Springer-Verlag Berlin Heidelberg 1998

Published by
Springer-Verlag
Postfach 10 52 80
D-69121 Heidelberg
Germany

Readme.txt contains:

- Reading and Printing this Document
- System Requirements
- CD-ROM Contents
- What to do if you have problems

Reading and Printing this Document

To view Readme.txt on screen in Notepad, maximize the Notepad window.

To print Readme.txt, open it in Notepad or another word processor, then use the Print command on the File menu.

System Requirements

IBM or compatible PC, 80486 processor or higher, 8 MB RAM, VGA color monitor, CD-ROM drive, Windows 95, Excel 97. Whereas the majority of the example files presented in the chapters will run with earlier versions of Excel (5.0 or higher) we strongly recommend the use of an English version of Excel 97. The CD-ROM has not been tested for Excel under Windows NT or older Windows versions (e.g. V 3.11). Microsoft currently (September 1997) offers a free download of the Excel Viewer 97, please visit http://www.microsoft.com/excel.

http://www.microsoft.com/Excel

The CD has not been tested using MS Excel running under Windows NT (any version).

A HTML browser is needed to open the file Pense.HTM in Chapter 3. We assume that the user has a browser, otherwise please visit

http://www.microsoft.com   or   http://www.netscape.com

What to do if you have Problems

1. Contact the editor at:
   Dr. W. G. Filby
   Forschungszentrum Karlsruhe
   Postfach 3640
   D-76021 Karlsruhe
   Germany

   Fax: 0 72 47 - 82 - 39 29
   E-mail: Gordon.Filby@pef.fzk.de

2. Contact the contributing author directly
   See the section "The Authors and Their Chapters" for some e-mail addresses and Web homepage sites. Otherwise 1.

The following tables contain a listing of all example files contained on the electronic data medium provided with this book.

Chapter 1. Getting Started with Spreadsheets

ActEner.XLS	Calculates reaction rate parameters. A walk-through of the main features of spreadsheet software
OrgSolvts.XLS	Demonstrates the database (list) feature of spreadsheets
Chp1.XLS	Six tagged worksheets for practising the basic mechanics of spreadsheets

Chapter 2. Applying Spreadsheets in Physics and Electronic Engineering

Cantilev.XLS	Uses the Euler equation to describe oscillations of a cantilever beam
Drift.XLS	Models mobility and drift velocity of electrons in GaAs

Euse.XLS	Statistics and data fitting of electric power usage
Henon.XLS	Reader is invited to prepare a worksheet for the Henon function provided
Iter.XLS	Iterative solution of coupled equations
Matrix.XLS	Solving linear equations with a matrix
Planck.XLS	Calculates and plots the spectral emission of a blackbody
Poisson.XLS	Calculates electric fields within an electron accelerator using the Poisson equation

Chapter 3. Using Spreadsheets as Tools in Mathematical Modeling and Numerical Mathematics

Brown.XLS	Simulates Brownian motion. Covered in the section Simple Stochastic Methods of the text
CoinFlip.XLS	Simulates throwing a coin using the RAND function. Covered in the section Simple Stochastic Methods of the text
MathMod.XLS	A multiple tabbed workbook covering all the simple models in the earlier part of the chapter.
Logistic.XLS	The Feigenbaum diagram as a worksheet
Pense.XLS, Pense.DOC, Pense.HTM	Models effects of demographic structures on pension budgets. Documentation in Word and HTML format

Chapter 4. Applications of Spreadsheets in the Earth Sciences

| ErthSci.XLS | Tabbed worksheet containing all exercises in the chapter. Includes examples from financial analysis, geostatistics and stratigraphy |

Tabs prefixed with D are empty except for information (column headers etc.) and are meant for practice. P-prefixed files are the complete worked through examples as described in the chapter.

Chapter 5. Spreadsheet Applications in Aquatic Chemistry

AcidSpec.XLS	Visual analysis of chemical equilibria
AcidTitr.XLS	User-defined function (Newton-Raphson) used to follow of acid-base titration
AlSpec.XLS	Fish mortality explained! Speciation of Al as a function of pH
MgSpec.XLS	Uses Solver to describe Mg – carbonate speciation

Chapter 6. Using Spreadsheets in Chemical Engineering Problems

ChemEng.XLS   A process flowsheet is developed stepwise in a multiply tabbed workbook

Chapter 7. Spreadsheets in Molecular Biology

Aacomp.XLS   Counts the number of occurrences of an amino acid in a protein sequence

Charged.XLS   Plots an estimate of the local charge along a protein molecule

Chofas.XLS   Predicts 3-D structure of proteins using sequence data

Coilcoil.XLS   Predicts regions likely to form α-helical coiled coils in coil proteins

DNA.XLS   Quantitative analysis of nucleic acid sequences

DotPlot.XLS   Shows a dot plot of protein sequences on screen

Memb.XLS   Searches for membrane-spanning domains along protein chains

Spacer.TXT, Spacer.EXE   Two utility programs to remove unwanted characters from protein sequences

Chapter 8. Spreadsheet Applications in Materials Science

Each worksheet described in this section is provided with an empty version for practice. These files are those without the ending -ex in their filenames.

Diffracex.XLS   Identification of cubic crystalline systems using X-ray diffraction

Grsizeex.XLS   Determination of the grain size of a microstructure

Nbtiex.XLS   A large scale macro example to determine the microalloying solubility of elements in austenite as a function of temperature. An alternative version NbTi2Ex.XLS is also provided.

Toughex.XLS   Material toughness determination using the Charpy impact test

Visex.XLS   Determination of the viscometric molecular weight of polymers

Volfracex.XLS   Simulates determination of the fraction of second phase in a microstructure using the point-count method

Appendix A. The VBA Programming Language – a Conceptual Overview

VBA.XLS   Two short VBA macros for practice

Appendix B. Further Sources of Information on Spreadsheets

This section contains no examples on disk.

Appendix C. The Major New Features of Excel

FeatXL97.XLS   The new Natural language and formatting features
                                are shown in a three-part worksheet

Spreadsheet Basics

W.G. Filby

What is a Spreadsheet?

The main object of any spreadsheet program is to manipulate and present data (numbers, text) found in tabular form. Now, as most bare tables of anything will not mean very much to the lay observer, they will usually (*always* in this book!) need to be accompanied by a few other important things such as:

- Titles and other explanatory information explaining the nature of the data and how and when and why it was acquired.
- Labels attached to individual numbers or sequences of numbers explaining to what the numbers refer.
- Formatting of the numbers themselves to make their meaning as self-explanatory as possible.
- Formatting behind the numbers to draw the readers attention to the important parts of the table and highlight significant data.

Like most other things electronic spreadsheets have evolved over the years and the above definition has become rather oversimplified. For, as we'll see here, the working data need no longer be "just" accountancy figures but also amino-acid sequences, experimental data taken from meteorology, physics, earth sciences etc. are also possible. In fact, there seem to be only few fields concerned with data reduction which could not profit from the application of spreadsheets to them. Furthermore, powerful built-in so-called **formulas** can be caused to act on any range (cell, block of cells) of the worksheet allowing quite complex calculations to be carried out without, at least to begin with, any special programming being necessary. Of that more later.

In recent years Microsoft Excel in all its versions has become, with little doubt, the most popular spreadsheet package on the market. It is this on which we shall concentrate in this book. At the time of writing (mid-1997) competitor packages, notably Lotus 1-2-3 (Lotus Development) and Quattro Pro (Corel), are still putting up a fight. The rise of Excel to a monopolistic position in the field seems however to be unstoppable. We shall not discuss these products, except in passing, in this book. Over wide areas all (Windows-based) spreadsheets are strongly similar and in many cases a certain standardization has taken place. This has lead to the situation that most files prepared in one spreadsheet program, say 1-2-3 can be read and if required, converted to the format of another spreadsheet program like Quattro Pro or Excel.

Our Approach

In this opening chapter you'll learn step-by-step the basic mechanics of using the Microsoft Excel spreadsheet program. For reasons of topicality this will be the most recent version at the time of writing, Excel 97. However, all examples provided with this chapter will, as long as no specifically Excel 97 advanced features are involved, also run in earlier versions.

Our intention is to give the beginning user a basic working knowledge of the generic spreadsheet commands and features. By giving the reader plenty to play with in the form of the practice worksheets supplied with the book he or she will rapidly get up to speed to deal with the later specialized scientific chapters. As Excel is a very extensive software packet, it is probably neither possible nor even desirable to attempt to learn "everything" about it all at once. In fact, in view of the multitude of possibilities available we shall scarcely scratch the surface. The remainder of the book will fill in more than enough details. Above all at this stage, experiment, experiment, experiment.

We'll be adopting a tutorial step-by-step approach involving the solution of two simple scientific problems. These have been chosen to illustrate as many everyday features of Excel as reasonably possible. As you follow through the examples you will learn most of the basic features of Excel and thereby most other Windows-based spreadsheet programs. Unfortunately, in order to keep this introduction down to a tolerable length we have had to make some assumptions about our readers. In essence these are as follows:

- That the reader possesses a basic knowledge of MS Windows methods, nomenclature and usage. Essentially this means knowing what a mouse and cursor are and do, knowing what is meant by left-click, right click, double click, what a dialog box is etc. Knowledge of another Windows product like MS-Word will more than stand you in stead.
- That both MS Windows (version 3.11 or higher) and MS Excel (version 5 or higher) are correctly installed on the users machine

In addition to the two scientific workbooks we provide another, tagged one, for practising the basic mechanics of spreadsheets. With these you can train your worksheet navigation, formatting and function-calling skills etc. as you go. They can found in the appropriate worksheets of the **Chp1.XLS** workbook.

The Disc-Based Examples

In the following two examples we'll be walking the reader through two simple chemical examples chosen to illustrate the most useful features of spreadsheet software. Although the language is, for reasons pointed out above, that of Excel, the examples will run in at least one of the competitor commercial spreadsheets.

In the first, more comprehensive example we'll be working through the simple data reduction involved in evaluating chemical kinetic data. The second example shows how Excel's listing features can be used to manage a small database of organic solvents.

1. Processing Kinetic Data

In our first example we place ourselves in the position of a researcher just completing a successful week of kinetics experiments. He has carried out a series of five repeat runs of a gas-phase reaction rate constant at eight different temperatures between 700 and 1000 K. The raw data is given in Table 1.

Table 1. Raw data from experiments

700	0.011	0.011	0.012	0.013	0.0114
730	0.035	0.035	0.032	0.0356	0.0349
760	0.105	0.105	0.1055	0.104	0.1045
790	0.343	0.343	0.345	0.342	0.342
810	0.789	0.789	0.787	0.784	0.79
840	2.17	2.17	2.168	2.18	2.19
910	20	20	20.05	20.1	20.09
1000	145	145	146	146	145

He now wishes to calculate the pre-exponential factor (A) and the Arrhenius activation energy (E_a) for this reaction. The Arrhenius equation is familiar:

$$\ln k = \ln A - E_a / RT$$

so that plotting ln k vs 1/T should yield a straight line of slope $-E_a/R$ and intercept ln A.

As we shall show, using Microsoft Excel this is very simple. The workbook **ActEner.XLS** on the CD-ROM accompanying this book contains the ten steps of a step-by-step guide made up of separate tagged worksheets. Let's get started.

Step 1: Entering the Data

From Excel's main menu we use *File* | *Open...* (or Cntl-O) and click on the file name. To save you typing in everything, nearly all the work has already been done. After opening the file and left-clicking on the so-called *tagged* worksheet named **Step 1. Raw Data** the screen is shown in Fig. 1.

In Windows-based spreadsheet programs like Excel data are entered by simply moving the cell highlight to the cell to contain data, typing in the desired numbers or text and either confirming with the Enter key (↵) or selecting another cell using the mouse or keyboard.

Actually it is a little more complicated since Excel needs to know what kind of data you mean. That is, whether numerical or character data are intended. Also you have to decide how you want your data to look on screen. Do you want four

	A	B	C	D	E	F	G	H
1								
2								
3								
4								
5			700	0.011	0.011	0.012	0.013	0.0114
6			730	0.035	0.035	0.032	0.0356	0.0349
7			760	0.105	0.105	0.1055	0.104	0.1045
8			790	0.343	0.343	0.345	0.342	0.342
9			810	0.789	0.789	0.787	0.784	0.79
10			840	2.17	2.17	2.168	2.18	2.19
11			910	20	20	20.05	20.1	20.09
12			1000	145	145	146	146	145

Fig. 1. The ActEner.XLS worksheet

following decimal points or will two do? Would you like the text strings under-
lined or in italic script? This is known as *formatting* the worksheet and we will
see more of it in the following sections. Use the file **Chp1.XLS** for practice.

Note a couple of interesting properties of the worksheet tags.

1. They can be renamed. For this move to the tag you wish to rename and select
   it by left-clicking with the mouse. Right-click, select *Rename* from the pop-up
   menu and start to type over the existing name. It will disappear and the typed
   characters substituted. Press the Enter key or the left mouse key to complete
   the operation. On the same pop-up other options are found for deleting, moving and copying and selecting all tags.
2. They can be moved by the standard Windows drag and drop methods. Hold
   down the left mouse key and drag. A small black arrow appears showing where
   you can deposit your worksheet.
3. New ones can be inserted anywhere at the users whim. Just select *Insert Worksheet* and a new one will be added to the existing list. New sheets are automatically tagged with the name Sheet n+1, where n is the number of the previous
   last tag.

What are such things good for? Well, for demonstrating the stepwise development of a worksheet for example, but mainly they're used to maintain data in an
orderly way. A scientist may want to keep all his observations for the month of
January on one sheet, for February on another etc. A sales manager may like to classify sales by region. All these *could* be kept in one worksheet but it is easier to
maintain on different sheets in one workbook. This makes large amounts of data
easier to keep track of and easier to monitor. It is important to realise that the
separate sheets are not isolated from one another. They can be "taught" to use each
others data without contaminating it. We shall see something of this in the later
stages of our first example. Moving quickly around the worksheet is an important part of spreadsheet work and Excel provides a myriad of methods to navigate

immediately from one part of the screen to another. The tagged worksheet Chp1.XLS on the CD-ROM provides adequate description and material for practice.

We've added an example of another useful feature of the newer versions of Excel on this sheet (**Step1. Raw Data**). Move the mouse cursor to the cell I3 - the one with the small red triangle (comment indicator) in the corner. This signifies that the cell contains a so-called *comment*. Move the cursor over the cell to see the body of the comment. Comments are informative pieces of text hidden, as it were, behind a cell. Actually you can even choose to hide or display both indicators and comments all the time, just as you will. Just click on the right mouse key when the cursor is over a comment to see a popup menu allowing all these niceties. To enter a comment, move the cursor to an empty cell, click the right mouse key and follow the pop-up menu which appears. See the Excel help topic "Hide or display comments and their indicators" for further information. Also visit the *Tools Options* pages to customise these and many other features of Excel. You can use comments as private notes to help you remember features of a worksheet's development or as a note of explanation. Any number of them are allowed on the sheet and you review them in sequence or individually by moving the cursor over one as described above.

At this stage the worksheet looks a little naked. Let us add some descriptive text, this time visible. Spreadsheets refer to nonnumeric data as *labels*. Numeric data are, to them, *values*. You will see much of this nomenclature in this book and elsewhere in the literature. To see our result we move to the tag labelled **Step 2. Adding Text**. Just left-click on the tag named such.

Step 2: Adding Text

We'll start by just adding a couple of column headings to describe our data. A title (T/K) for the single column of temperature data and a cross-column title for the sets of rate constants with their units ($k/M^{-1} s^{-1}$) are the obvious choices. We select the head of the temperatures column and enter T/K into cell C4. Everything seems alright except that the heading T/K looks rather awkward, left-aligned as it is. Excel automatically left-aligns labels unless we instruct it otherwise. We'll change it so that it sits nicely over the numerical data. Excel makes this much easier than older spreadsheet software. At this stage the screen looks like Fig. 2. Notice the three buttons on the formatting bar marked with simulated lines of text:

Move the mouse cursor over them to see a brief explanation (*Tooltip*) of their function. The depressed one (to the right of the *U* button) shows the tip Align Left on pausing the cursor on it. We modify our title to right aligned by simply pressing the right button with the tip Align Right. The label T/K is now sits neatly over the temperature data column. Nothing else to it!

T/K					
700	0.011	0.011	0.012	0.013	0.0114
730	0.035	0.035	0.032	0.0356	0.0349
760	0.105	0.105	0.1055	0.104	0.1045
790	0.343	0.343	0.345	0.342	0.342
810	0.789	0.789	0.787	0.784	0.79
840	2.17	2.17	2.168	2.18	2.19
910	20	20	20.05	20.1	20.09
1000	145	145	146	146	145

Fig. 2. Adding labels to a worksheet

Placing the second title is just a little more complicated as we want it with its superscripts ($^{-1}$) centred over the columns of rate constant data. We enter the titling in the first of the rate data columns (D4). Next we need to select the columns over which the text is to be centred (D4:H4). Now we choose the sequence Format Cells. Then from the Alignment tag we need the Horizontal option of the Text alignment field. Click on the arrow to see its options. The one we want is the Center Across Selection one. Clicking on this and then OK gives us the desired result. Now we'll take care of the superscripts. This is very similar. Again, as before we select the characters (only one –1 at a time since this is all Excel allows) to be superscripted, either in the cell (D4) or on the formula bar. From there we move to the Format Cells menu option. Excel offers us the Font option. From here we choose from the Effects field the Superscript option and click OK. Job done! Repeat this procedure at the second –1 and you'll have both superscripts where they belong. Don't like the text font? Rather have coloured? Or italic? Like almost everything in Excel these too can be altered either just for the current session or, as the so-called default, for all future sessions. The menu items Format and Tools Options are the key things to experiment with to explore what can be done. Use a blank worksheet (File New, click Workbook on the General tag) for practice.

Note: the nomenclature of some menu items has altered from version to version. As we said earlier, we follow here the usage of Excel 97. There may be slight differences in earlier versions.

Step 3: Averaging the Rate Data

The next thing we need to do is to get the data ready for analysis. As a first step we decide to just average the results for each temperature. Later, we'll want to plot the data. For illustration, we opt for the brute-force method of adding each entry and dividing by the number of entries. You'll see why we say "brute-force" in a minute. For this, we need to use what is called in spreadsheet programs, a *formula*. The importance of formulas in the spreadsheet world can not be over-emphasised. In their use lies one of the software's great strengths.

You can follow this discussion on the worksheet **Step 3. Averaging I**. We move the highlight to cell I5 and type, for the 700 K data set:

=D5+E5+F5+G5+H5 ← **Excel calls this a formula**

and press Enter (↵). The = sign is required when entering formulas to differentiate between the formula we want and the text string or label D5+E5+F5+G5+H5 Excel would "think" we meant if we omit it. The value 0.0584 appears in the cell. So far, so good. Five times a value of about 0.012 should be around 0.06. Now to form the average we just divide by five. With G5 selected we double-click on it to enable editing there. Now we can edit the *division operator* (/) into the formula. The final expression in the formula panel is:

=D5+E5+F5+G5+H5/5

Full of expectation we press Enter (↵), only to find the result 0.04925 displayed in cell I5. By hand calculation the average should be about 0.011.012. There must be something wrong! There is! The expression:

=D5+E5+F5+G5+H5/5

means:

=D5+E5+F5+G5+(H5/5)

which is not what we want. We correct the input by editing (double-click on cell I5) the expression to:

=(D5+E5+F5+G5+H5)/5

which obeys Excel's precedence rules and evaluates to 0.01168, which is far more credible. Figure 3 shows the screen at this stage.

We have left room in this worksheet for the reader to continue, writing a new expression for each row of data. However, we show a better way in Step 4. One further thing before we leave this topic. The column titles aligned in Step 2 have been

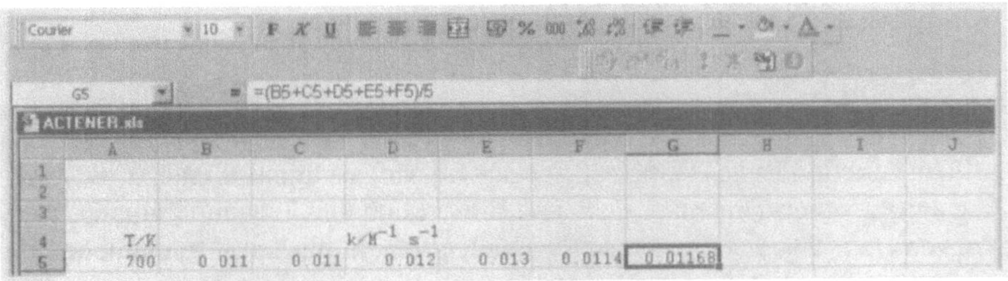

Fig. 3. Averaging the rate data

copied from there to this sheet using the Copy and Paste feature common to all Windows programs. Use the Edit menu or the shortcuts Crtl-C (Cut) and Crtl-V (Paste) to quickly move data around and between worksheets. Remember that, if you unintentionally delete something you need, you can always use the undo button:

to restore the original state of your worksheet. Excel 97 even allows the undoing of several of the most recent operations by way of a pulldown menu (black triangle next to undo button). The worksheet **3.Editing, Searching etc.** in the **Chp1. XLS** workbook provides some material for practice.

Step 4: More on Averaging

At this point it is convenient to introduce another powerful feature of spreadsheet programs, the built-in *functions*. These are a special set of instructions intended to speed up many common operations such as summing, forming averages and much, much more. We shall see a lot of them in the following chapters. In this step we will only indicate the power of this tool.

Before starting this we'll narrow down column I in order to set off the raw data from the averaged values we're going to generate in the next step. For this we just moved the cursor along the column headers (the A, B, C etc. of the so-called A1 cell reference style) to column I. At the point where the cursor changed to a double arrow we dragged this to the desired width (1.75 as displayed by Excel) and let go. That's all. By the way, Excel also allows row **and** column references in the so-called R1C1 format in which both rows and columns are numbered. These can be set via the *Tools Options General Settings R1C1 reference style* sequence.

Note: although previous versions of Excel have also possessed the Function Paste (formerly FormulaWizard) feature the Excel 97 version has been made much smarter. Therefore the description below will only partially apply to version 5.0 and Excel 95 owners. Check Appendix C for more information.

Returning to the actual calculation in **Step 4. Averaging II** we move the highlight to the first cell to the right of the first row of data to be averaged (cell J5), and click on the Paste Function button situated between the AutoSum and Sort Ascending and Sort Descending buttons on the toolbar:

Σ ƒ* A↓ Z↓
        Z↓ A↓

The Function Paste Dialog box appears offering us function categories covering among others Financial, Date and Time, Math and Trig and Statistics. Very convenient is also the Most Recently Used option. In this Excel "remembers" the most recently used functions and displays them in a sorted list. You can save yourself much searching if you look here first. However we click on the Statistics option and see what we're looking for. AVERAGE is offered in the Function name window. We select and click on it. At this stage the screen should look something like Fig. 4. Incidentally, like nearly all Windows boxes this one is also movable and resizable using the mouse. We're only an OK click away from our first average. And how easy it was!

Formula result 116.6764. Fantastic!! Excel (97) has done all the work for us! Hasn't it...? But wait! Isn't this figure a bit too high when all our rate constants are around 0.01? Well, well-intentioned as the Microsoft software designers are, they seem to have overdone it this time. Look at Fig. 4. Why is the Number1 parameter of AVERAGE C5:H5? Why does the number 700 appear in the bracketed list on the right? Answer: Excel has tried to construct the so-called *arguments* of the AVERAGE function itself. Arguments are the values passed to the function. There are no cells downwards, so the cells across are taken as the most sensible candidates. Unfortunately, including the temperature (700)! Well here again the good folks at Microsoft come again to our rescue. Look again at Fig. 4. Find the gridded button on the right with the red arrow pointing to the top left. This allows us to correct false cell addresses manually. Click on it to see how it works. It's called a *collapsible* dialog box. The Paste window closes to a single line and from there you can either edit the cell addresses manually in the Paste window itself or construct a corrected list by dragging the cursor over the cells to be averaged. Don't forget to confirm your choice after using this second method. Maybe at this point you may have noticed something – you have just completed this whole averaging operation using only the mouse! This kind of data entry (anchoring and dragging the highlight) is known as *pointing*. It greatly eases many spreadsheet manipulations by reducing manual keyboard input. Practically all modern spreadsheet software uses it for marking cells and ranges. Most Windows-based spreadsheets, allow whole point operations and more to be carried out using only the mouse.

This is all very well, the attentive reader will be thinking, but after all this effort we still have only one data point calculated. What about the other four? Answer: spreadsheet software provides us with yet another aid for speeding up data pro-

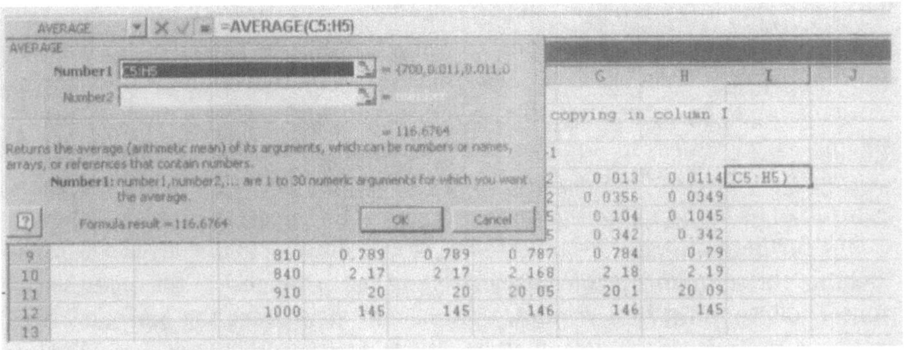

Fig. 4. Using the Function Paste feature

cessing. This is a feature which we shall be using almost ad nauseam in the rest of this book. This is called *copying*. Cells, ranges, labels, values and formulas. They can all be copied in Excel. Some readers may be familiar with copying from word processing so it won't be so new for them. What is new, slightly different and most important is *formula copying*. This is probably one of, if not the most important feature of spreadsheet software. We demonstrate it by walking the reader through such a copying operation. Here goes.

We keep the mouse pointer (hollow cross) on the cell J5. The bottom right corner carries a small, black square know as a *fill handle*. Also you might notice how the pointer changes its form when moving across the selected cell. From the familiar hollow cross, to an arrow and finally to a crosswire when it's over the fill handle. We'll be concerned with the latter. This is what we need for copying. You should try to find out the function of the others by experimenting.

Nothing could be simpler than copying. We just form the crosswire and drag it down to the cell I12 then let it go. A column of numbers appears immediately. Step down the column cell by cell to convince yourself of their contents. Cell J6 contains the formula =AVERAGE(D6:H6), which is just the average of the 730 K data we are looking for. The same applies all the way down the column. Every cell contains the correct formula to calculate the average of the row data to its left. The formulas are correct because spreadsheets use *relative addressing* by default. The term relative addressing refers to the way cell addresses are treated during copying. The spreadsheet tests the cell reference in the original cells formula and replaces it with the reference for the cell in the same relative position in the new cells formula. Other alternatives, *absolute and mixed addressing*, which are discussed in Chapter 2 and elsewhere in this book, allow only part of the address to be updated during copying. The other part, designated by a $ sign, remains fixed.

You can practice the copying process by undoing the copying as described above and redoing it again until you feel comfortable (see the tagged worksheet 3. Editing, Searching etc. for practice material).

The worksheet has now been at this stage, in spreadsheet jargon, *recalculated*. Recalculation is another of the great advantages of spreadsheet software. Data can be entered and results obtained seemingly automatically without rewriting the underlying formulas. By default, recalculation is carried out each time a cell is changed (*Automatic Calculation*). This may however, be altered by the user via the *Tools Options Calculation* tag. Selecting the Calculation *Manual* option forces recalculation to be carried out only on request, Such a request is made by pressing, the F9 key whilst "in" the worksheet or directly from the Calc Now (F9) or Calc Sheet buttons on the *Tools Options Calculation* page. Practical use for this feature can be found in the chapter on Chemical Engineering problem solving.

One further point. If we want it to, Excel will allow us to display the formula "behind" the result in any formula containing cell or range. Use the menu selections *Tools Options View Window* options and then check the Formulas box. Uncheck it to restore the default style. This feature is sometimes useful during the debugging of larger worksheets.

Although we've made some progress we're still a long way from our Arrhenius plot and activation energy.

Step 5: Tidying the Table

Before we turn to this in the next step, let us have a look at the table. We feel it could benefit from a little tidying up. The narrowed down I column could be more attractive. We'll add a little colour and shading. First we drag the hollow cross cursor down the column from I5 to I12 to select the cells to operate on. Then we need to call the *Format Cells Patterns Cell Shading Color* sequence to bring us to Fig. 5.

See if you can work out for yourself how the Average Rate text got aligned the way it is and how each of the headers received its emphasised font. The workbook Chp1.XLS on the CD-ROM contains the sheets **2. Formatting I** and **4. Formatting II** for practice.

Step 6: Generating Data for the Arrhenius Plot

As we saw in the opening section, to determine E_a and A, we must first prepare a plot of ln k vs 1/T. Thus our next step will be to generate this data on the worksheet. Before we go on however, we feel that the table would look better if the 1/T and ln k values were placed to the right of their respective parent columns. For the first we need a column insertion feature. For Excel, no problem. We just place the mouse pointer anywhere in the D column select *Insert Columns* and Excel slides the existing columns one column to the right while at the same time updating their formulas correctly. Check this out by visiting them in the formula bar. While we're in this part of the worksheet we'll generate the 1/T values. We move the highlight to cell D5, type =1/ and move the highlight to the first T value (C5).

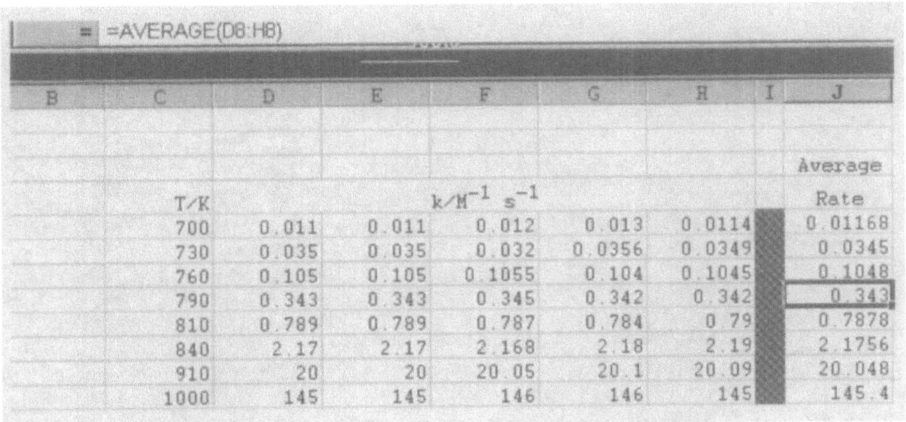

	=AVERAGE(D8:H8)								
B	C	D	E	F	G	H	I	J	
								Average	
	T/K			$k/M^{-1}\,s^{-1}$				Rate	
	700	0.011	0.011	0.012	0.013	0.0114		0.01168	
	730	0.035	0.035	0.032	0.0356	0.0349		0.0345	
	760	0.105	0.105	0.1055	0.104	0.1045		0.1048	
	790	0.343	0.343	0.345	0.342	0.342		0.343	
	810	0.789	0.789	0.787	0.784	0.79		0.7878	
	840	2.17	2.17	2.168	2.18	2.19		2.1756	
	910	20	20	20.05	20.1	20.09		20.048	
	1000	145	145	146	146	145		145.4	

Fig. 5. Setting up the I column

Now we confirm with Enter (↵) and the reciprocal temperature value (0.001428) appears. Excel 97 provides a useful feature here for catching mistakes. If, for instance we had entered =1/C4 (C4 contains T/K) or =1/B5 (B5 is empty) we would have been greeted by the Excel error messages #VALUE! and 'DIV/0! respectively. And rightly so. Excel cannot divide 1 by T/K and obviously division by zero is also not permissible. The new feature shows us where these errors stem from by placing a coloured border round the offending cell or range. Double click on the error message to see it work. Fig. 6 gives an indication of what to expect.

After inserting a new column as for the 1/T column we can begin entering the LN values. We shall use the Paste Function feature again. This time with the Math and Trig Function *category*. We select the LN function and confirm it with OK. The screen will look like Fig. 7. One OK click and the correct result –4.44988 appears in cell L5. We copy the rest of the formulas down their respective columns and arrive at **Step 7. Charting**.

By the way, while we're at this point it is worthwhile browsing through the functions just to see what the most modern Excel version provides by way of library functions. Each one has a brief description which can be seen on selecting it in the Function *name* panel. There will almost certainly be enough there to satisfy even the most discerning of users. You can practice some of this work using the sheet **5. Calculations** in the Chp1.XLS workbook on the CD-ROM.

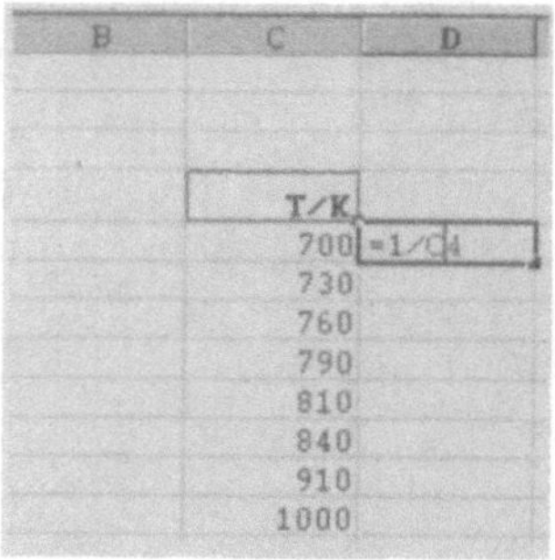

Fig. 6. Associating cells with formulas in Excel 97

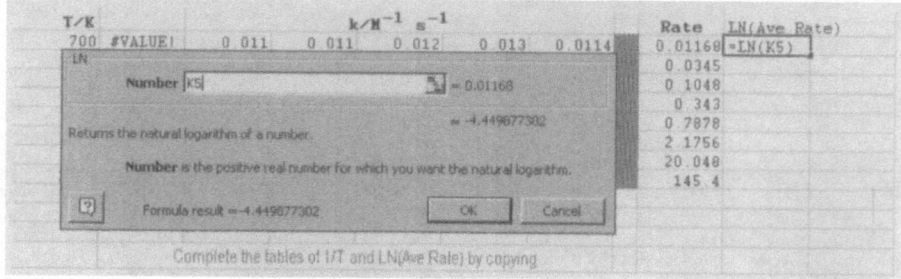

Fig. 7. Generating the LN(Rate) column. Note column L has been widened to avoid overspill into the next column

After all this concentrated effort we decide it is time to take a break and start again later. As most readers will know, it is wise to save one's data before shutting down the computer. Since we have already named the file in an earlier step all we need to do is call *File Save* and leave the file name (ActEner.XLS) unaltered by pressing Enter. However this doesn't have to be. We could change the file name at this stage. By selecting the *File Save As* option, we could enter a new file name. Of course, if we do, the "old" ActEner.XLS still remains safely on the disk. However, any further work we do during the current session will be placed in the new file **NOT** the old one. Wishing to keep life simple we click on the diskette:

and stick with ActEner.XLS.

Step 7: Setting up the Plot

Refreshed after our break we return to our rate data. It's time to see a plot at last! Charting in Excel involves the so-called Chart Wizard. The wizard walks the user through the process and enables a fairly rudimentary plot to be prepared very quickly in only the following four steps.

1. **Chart type:** in this step the best known chart types (Column, Bar, Line) but also some lesser known ones like Radar, Cylinder and Pyramid are all presented in preview form. The user can select one by simply clicking on a preview plot. Standard Types and Custom Types are available, each with a brief description of their usage.
2. **Data to plot:** this step deals with the data to be plotted. If we mark this in advance the Wizard automatically places it in its input fields.
3. **Positioning:** in the third step we determine the position of the legends, titling and whether grids are shown or not
4. **Placement:** in the last step we decide whether to place the chart on the worksheet containing the data or on a separate worksheet.

At the end of the fourth stage the Chart can be further modified. In Excel each component of a plot is an *object*, which means that we have very many possibilities to fine-tune a first rough plot to perfection. We can fine-tune either manually (move the cursor over the Chart to see the identifiers of the various objects like axes, legends etc.). A double click brings us to a detailed editing platform. In the following we'll look at each of the above steps in as far as they are relevant to our ongoing problem.

Calling up Chart Wizard via the *Insert Chart* menu or the main toolbar (blue, yellow, red 3-D histogram icon) we see a screen like Fig. 8.

There you'll see that we've already selected the appropriate chart type for our purposes. This is an XY Scatter chart with the data points joined with smoothed lines. That's Step 1 of chart making over. Pressing the Next> button on the Step

1 dialog and selecting Series Add brings us to the Step 2 :Chart Source Data dialog box below, Fig. 9.

Here it is the collapsible dialog boxes named *X* Values and *Y* Values which interest us. They allow us to enter our X (1/T) and Y (LN Average Rate) by pointing.

Fig. 8. The initial Chart Wizard screen

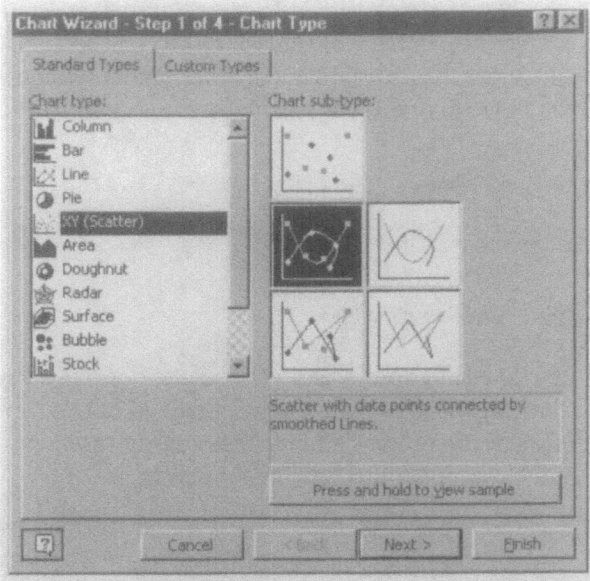

Fig. 9. Step 2 of Chart building

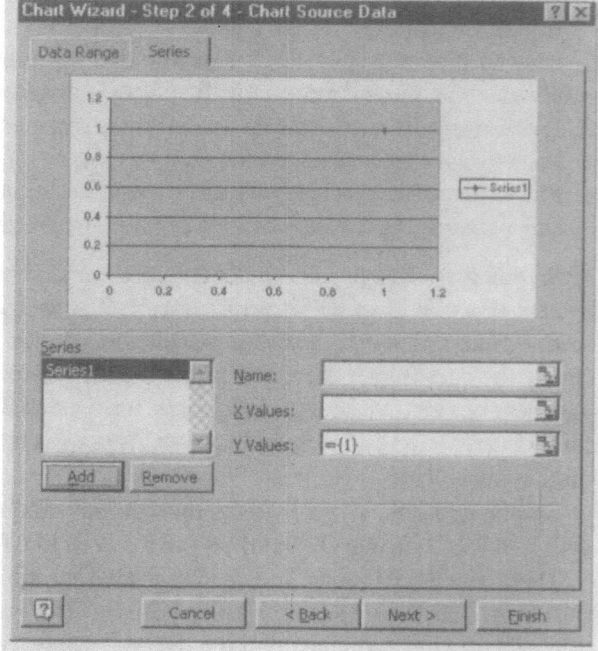

For practice, try entering the coordinates yourself. The completed step should look like Fig. 10.

As you can see from the figure we already have a problem. Excel has not chosen the scale of the X-axis very intelligently. More than $1/2$ of it is empty! Fortunately we have tools to deal with that later. Moving to Step 3 with a click on Next> we see that we deal here with Chart Options, Table 2. We have to decide what to do about Titles, Axes, Gridlines, Legend and Data Labels. Below is a table of the options and their meanings.

Fig. 10. Step 2 completed

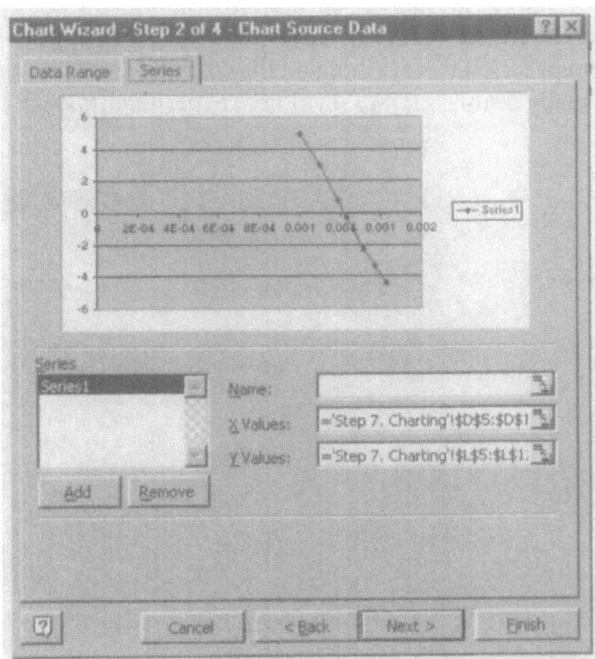

Table 2. Chart options

Tag	Allows	Our Choice
Titles	Placement of chart, X an Y titles	Chart Title: Temperature Depce. of Reaction Rate X Axis: 1/T Y Axis: LN (AveRate)
Axes	Scaling and ticking of X and Y axes	X and Y axes checked
Gridlines	Horizontal and vertical major and minor gridlines	None checked
Legend	If selected, placement of chart legend	None Checked
Data Labels	Either X or Y coordinate values to be placed on Chart	None

Using the choices in the table we arrive at Fig. 11.

In Step 4 following another Next > click we only have to decide where to put our chart. This time our selection is: As object in: Step 7. Charting. Followed by F̲inish and the job is over. At least for now. Our partially completed chart looks like Fig. 12 on the sheet.

The worksheet at this stage looks like **Step 7. Charting**. At this point we leave the reader to complete the remainder of the chart for his or herself. Figure 13 gives a good starting point.

You can get there by clicking anywhere on the Plot or Chart Area or on the object to be edited. This causes the Chart toolbar to appear. From there you have access to practically everything on the chart. The possibilities for modification are virtually limitless. Time spent here experimenting will be well rewarded later.

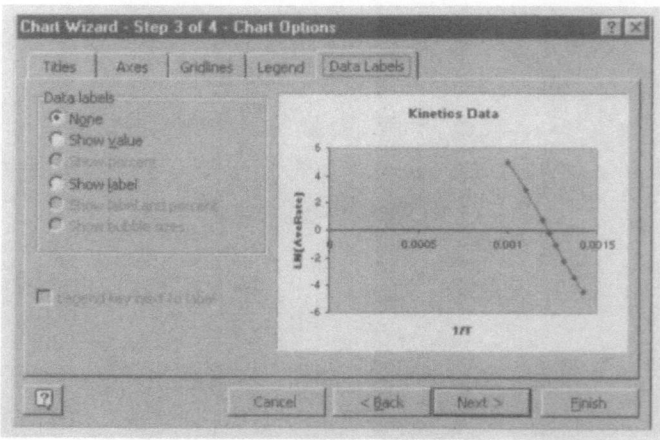

Fig. 11. Step 3 of the Charting process

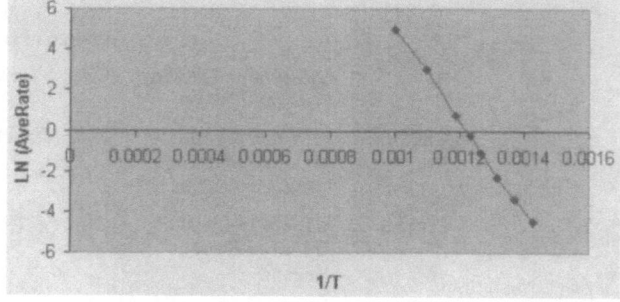

Fig. 12. A first attempt

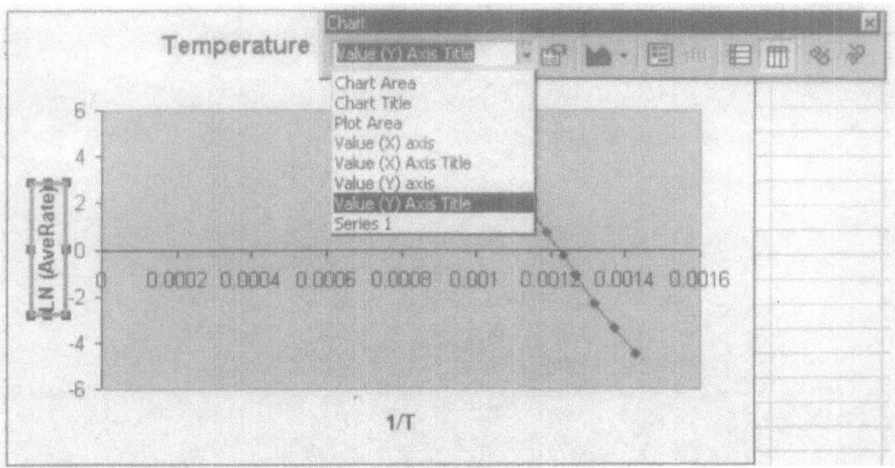

Fig. 13 . Fine-tuning the chart

Fig. 14. The Data Analysis
Toolpak menu

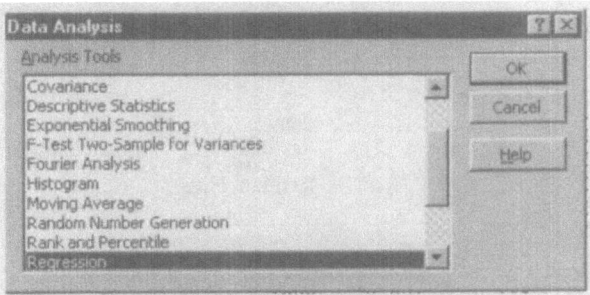

For most readers the prospect of extracting the slope and intercept of a line
from a computer screen will not be pleasant. However, Excel also provides a solution to this and many other data-fitting and statistical problems. This is the
Analysis Toolpak which can be called from the *Tools* menu.

Note: this holds only if you have installed it in the first place. If not, you will
have to install it from your CD-ROM and then call it from the *Tools* Add-*Ins*
...Add Ins Available box. Check the Analysis Toolpak box. If it has worked you
should get a *Data Analysis...* item added to your *Tools* menu. On clicking, this
should produce a screen like Fig. 14.

We'll be using a regression function from the Toolpak.

Step 8: Performing a Data Regression

All the data we need for our A and E_a determinations are now available on the
Step 7 screen. We call up the data regression tool by double-clicking on Regression
or selecting Regression and clicking OK. The screen depicted in Fig. 15 appears.

Fig. 15. The Regression
screen (already filled out)

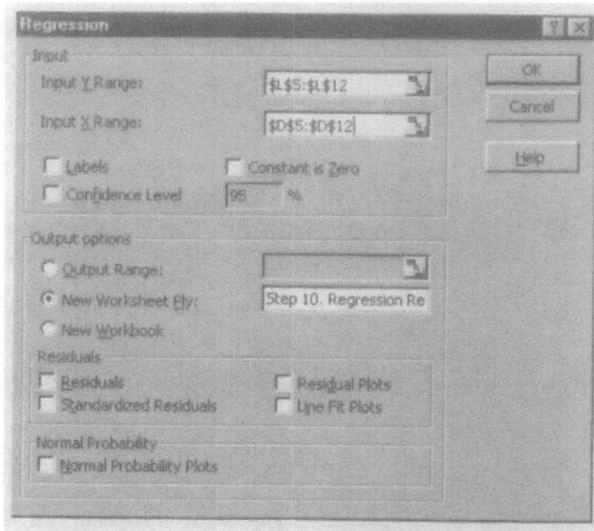

Table 3. Regression options

Constant is Zero	Forces the regression line to pass through the origin
Residuals	Select to include residuals in the residuals output table
Standardized Residuals	Includes standardized residuals in the residuals output table
Residual Plots	Generates a chart for each independent variable versus the residual

The fields seem self-explanatory. The Input X Range: and Input Y Range: entries will be clear– they must be our reciprocal temperature and LN(Ave Rate) values. But what about the Output range? Answer: this is merely the range on the worksheet in which to place the regression results. That is, if we should decide to put them there at all. There is another alternative as we'll see. This is indicated in the box titled New Worksheet *Ply*. It allows us to insert a new worksheet into the current workbook and paste the results there starting at cell A1. Being appropriate for our stepwise approach we've chosen this option and given the new sheet the name Step 9. RegrResults Worksheet. As you'll see Regression has a number of additional options such as those shown in Table 3.

After filling out the Regression input sheet using the point procedure outlined earlier it should look something like Fig. 15. A single click on OK and the results are delivered on our chosen sheet **Step 10. RegrResults**. Figure 16 shows a section of it.

At this point we need to explain some of Excel's nomenclature. What we are looking for is the slope (= E_a/R) and the intercept (= ln A). In Excel's output these are to be found under the title Coefficients as Intercept and X variable 1 respectively. Thus, our best fit results for this data set are:

	A	B	C	D	E	F	G	H	I	J
1	SUMMARY OUTPUT									
2										
3	*Regression Statistics*									
4	Multiple	0.999163								
5	R Square	0.998326								
6	Adjusted	0.998047								
7	Standard	0.140704								
8	Observati	8								
9										
10	ANOVA									
11		*df*	*SS*	*MS*	*F*	*mificance F*				
12	Regressic	1	70.83428	70.83428	3577.899	1.47E-09				
13	Residual	6	0.118786	0.019798						
14	Total	7	70.95306							
15										
16		*Coefficien*	*andard Er*	*t Stat*	*P-value*	*Lower 95%*	*Upper 95%*	*Lower 95*	*Upper 95.0%*	
17	Intercept	27.6395	0.470206	58.78175	1.63E-09	26.48895	28.79006	26.48895	28.79006	
18	X Variabl	-22591.1	377.6793	-59.8155	1.47E-09	-23515.2	-21666.9	-23515.2	-21666.9	

Fig. 16. The RegrResults worksheet

$E_a/R - 22591.1$

$\ln A\ 27.63950$

The correlation coefficient (in Excel-ese R Square) shows the data to be well correlated, a point we noted already from the chart shown earlier.

Step 9: Using Range Names

For the final trivial stage of extracting E_a and A from the above we've set up the worksheet **Step 10. FinalResults**. Here we introduce another very powerful feature of spreadsheets namely, *named ranges*.

Up to now all our references to cells have been made using their addresses. For example, in the worksheet **Step 4. Averaging II** in cell J5 we wrote the formula = AVERAGE(D5:H5). Here AVERAGE's argument was a range, addressed as D5:H5. Most spreadsheet programs provide another way of referring to cells, ranges, values and formulas. Instead of difficult-to-remember cell addresses they give us the option of assigning our cells or ranges descriptive English names. Their use brings a number of advantages:

- We can create them very easily and quickly based on row or column titles or we can enter names ourselves. We can then paste these names directly from the so-called name box directly into formulas
- They can make a worksheet easier to read, and understand (self-documenting)
- We can use the names in formulas, macros and some settings (printing, graphics) instead of cell addresses. In fact, anywhere where we might use a normal reference.
- They make a worksheet robust since formulas containing them are updated when they are moved. Cell addresses are not. This is most important.
- By default, names use absolute cell references

However some simple rules must be obeyed:

- Names cannot be the same as a cell reference, such as Z$100.
- Spaces are not allowed. Underscore characters and periods may be used as word separators for example, First.Quarter or Sales_Tax.
- A name can contain up to 255 characters.
- Microsoft Excel does not distinguish between uppercase and lowercase characters in names. For example, if you create a name Sales and then create another name called SALES in the same workbook, the second name will replace the first one.

Let us now return to our worksheet **Step 10. FinalResults** to see how range names can play a role there. Our first example is just the single cell given the name ScratchPad. Scratch pads are often used in spreadsheeting to provide a place on the sheet for the storage of, for example, intermediate results, named constants, tables of names etc. just as we have done here. They are simply a group of cells usually placed a little off screen so as not to distract from the main purpose of the worksheet. In the case of ScratchPad we chose to name it purely for navigational reasons. We gave the name ScratchPad to the 'Step 10. Final Results'!$T$2 cell quite simply. Note that we chose a cell slightly to the right of the actual named range so that it would be visible after the Go To command. We moved the highlight to T2 on the worksheet Step 10. Final Results and selected Name from the Insert menu. From here we selected the Define option and entered the name Scratchpad into the empty field in the dialog box. Remember you can select the range to be named by reducing the collapsible dialog box to a single line and then pointing to the relevant cells. The screens shown in Figs. 17 and 18 illustrate this point.

Notice how the Refers to box helps when navigating around the worksheet. Navigation around the worksheet is much simpler using range names since we no longer need to remember cell addresses. For example, if we wish to examine the ScratchPad, we only need click on the Name Box on the extreme left of the Formula bar to pull down a list of all names used in the workbook. There we select ScratchPad and double click on it. The tag **Step10. FinalResults** is activated and the name box shows ScratchPad. The cursor jumps automatically to the correct sheet and cell. Even if we had not displayed the formula bar Excel provides another solution. We just press Edit Go To… (or Ctrl+G) and, in the Go to window containing the listing of all named ranges in the workbook, select a destination by double-click. It is important to realise that you can do this from anywhere in the workbook, not just from Step 10.

As a further illustration we have named three other ranges (again only single cells) on **Step 9. RegrResults**, Slope, Intcpt and R Square. As we'll soon see these are convenient for the final calculation. Naming them was slightly different to the previous example in that, after selecting Insert | Name | Define, Excel offers us a default name – the label to the left of the cell to be named. In other words the cells B17 and B18 would, without intervention, receive the names Intercept and

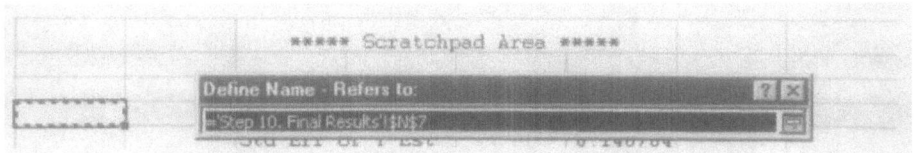

Fig. 17. Defining a named range

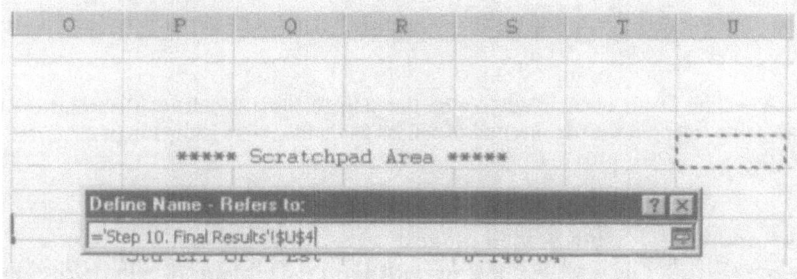

Fig. 18. Collapsing the dialog box to complete a point and click named range definition

X Variable 1. Finding these rather too longwinded we have overwritten them with our own names of Intcpt and Slope. For naming the square of the regression coefficient (R^2) we have accepted the default name of R_Square offered as default name in the *Insert | Name | Define* sequence. This parameter reflects the quality of fit of our regression line. Figure 19 illustrates how they have been used in the worksheet **Step 10. FinalResults.**

There you can see under the heading SCRATCHPAD AREA *Output from Step 9* the three parameters named on the previous sheet. See how the cells R5, R6 and R7 contain the formulas =Intcpt, =Slope and =R_Square. Why this? Well, because we want to have all the working data in one place – in the ScratchPad area where we can work with it. Referring to a name in a = statement allows us to do this. We've named the universal gas constant R_ and placed it under the other parameters for the same reason.

Now we can turn to the question of what all this is good for. The formula in F17:

=Slope*R_/1000)

will make it clear. There is no cell address in this formula – just the named ranges Slope and R_. Compare it to its equivalent:

=R6*Q11/1000)

to see what we meant by self-documentation. Even worse, if we had taken the Slope off the Step 9 sheet it would have looked like this:

=('Step9. RegrResults'!$B$18'*Q11/1000)

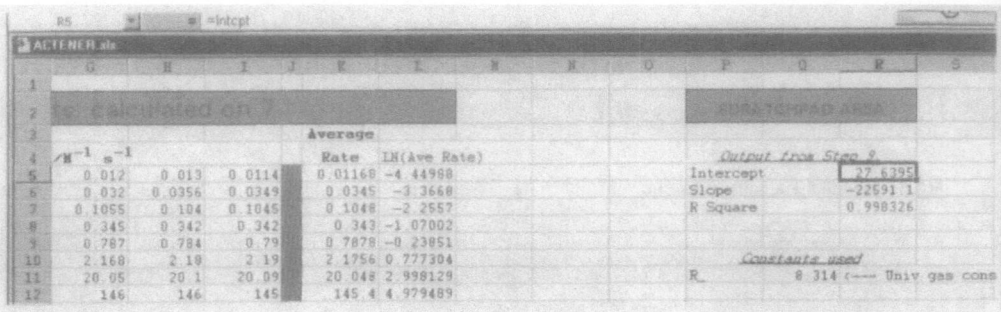

Fig. 19. A section of the ScratchPad area on the Step 10. Final Results worksheet. Note the formula using Intcpt in the background of cell R5 using the range name Intcpt

Reason enough for using named ranges!

Before closing this topic, a few short words on the table, a so-called *range name table,* beginning at P16. Figure 20 shows an extract from it.

It is nothing more than a log of the names we've used on the spreadsheet together with their sheet and cell addresses placed at a position of our choice. Accessed by way of the **Insert** | **Name** | **Paste** | **Paste List** sequence, there are some things to note about it. First, it is not dynamic– range names created after the last | **Paste List** of names will not appear in it. It must be called again to be kept up to date. The second point is that it will overwrite occupied cells without warning. Another reason for the Scratchpad. Obviously, its main use will be in managing range names on larger worksheets, where the numbers and names of ranges might easily lead to confusion.

The cells F19, G17and F18 contain some points of interest. In F19 we've placed the function =IF('Step 9. RegrResults'!B5>0.95;"Well correlated";"Poor Correlation"). This is an example of one of Excel's logical functions. The =IF function, which has the syntax:

IF(LogicalTest,ValueIfTrue,ValueIfFalse)

It returns ValueIfTrue if the LogicalTest is true and ValueIfFalse if it is false. In our case it just serves, very trivially, to display "Well correlated" in the cell if the correlation coefficient turns out to be greater than 0.95. In its turn, the cell G17 contains =IF(R_=8.314;"kJ mol^-1";IF(R_=1.987;"kcal mol^-1";" Check Result-incorrect unit!")). This is a so-called *nested* IF function. The reason is obvious – we have one IF function inside another. Its purpose is to see that the correct units are reported with the final result.

We have used the calculation of the Pre-exp factor (A) in cell F18 to demonstrate another pair of = functions, = FIXED and =EXP and one more operator, &. The use of =EXP, which returns the exponent of its argument, will be clear. It simply yields the A value we're seeking from the intercept (=ln A). FIXED rounds a number to the specified number of decimals, formats it in decimal format using

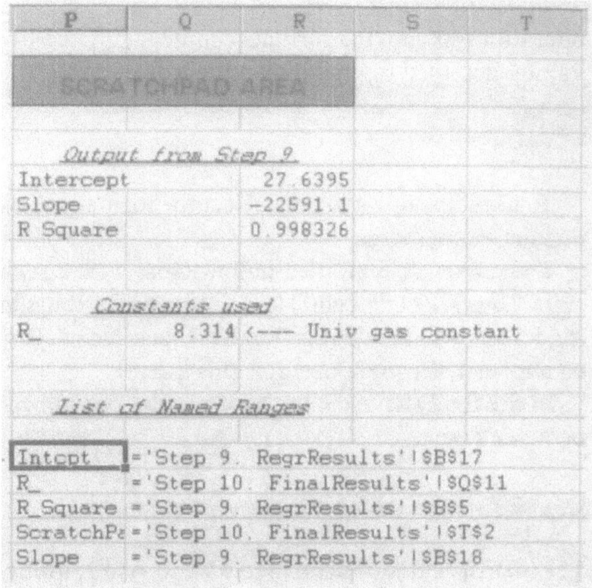

Fig. 20. The range name table on Step 10. Final Results

a period and commas, and returns the result as text. The reason for the = FIXED expression is that, although it's not absolutely necessary, we would like to append the units of A and have the whole appear in one cell. In other words, we wish to combine, Excel says *concatenate*, a numeric value (A) with a label (M^-1s^-1). Excel and most other spreadsheets will not do this without first converting the number to a string (label). For this, it provides the = FIXED function. This has the syntax:

FIXED (Number,NrDecs,NoCommas)

where Number is the number to convert and NrDecs is the number of decimal points to display. NoCommas is a logical value which, if TRUE, prevents FIXED from including commas in the returned text. If NoCommas is FALSE or omitted, then the returned text includes commas as usual. Thus =FIXED(EXP(Intcpt),4, TRUE)&" M^-1 s^-1" returns our A value as a string with four decimal places. The concatenation is accomplished with the special operator &. The result is the string in F18. Note: the cell neighbours E18 and G18 are empty.

The cell F19 contains the function =IF('Step 9. RegrResults'!B5>0,95;"Well correlated";"Poor Correlation"). The explanation for it is similar to that of the nested =IF in G17.

A couple of final words on the formatting in the ScratchPad area and a useful formatting tool. We have formatted the text "SCRATCHPAD AREA" using Format | Cells Alignment Horizontal, Center Across Selection and Vertical, Center. The remainder was done using the Font and Border options. Many everyday operations in copying formats can be facilitated by using the so-called Format Painter accessed by clicking on the icon:

To use it to copy a format select the format you want to copy and click on the Format Painter icon:

The cursor changes to a hollow cross with a replica Format Paint icon to its right. Then select the cell(s) to be formatted by dragging and then let go. The resulting format will be identical to that first selected. Obviously the more complicated a format, the more useful is this feature.

In the final step we'll see how to print out our results and how even this can be made easier.

Step 10: Finishing the Job – Saving and Printing

The main calculations completed we turn to printing our results. Using Excel this is, as usual, a simple task with however, many options. The quickest way to manually print an Excel worksheet like **Step 10. FinalResults** is as follows:

Use the sequence File | Print Area | Set Print Area to select the area to be printed. On the version on the disk we chose the area C3:L20. Note the dashed lines surrounding it on the screen. If you wish, you can alter this by following the sequence File | Print Area | Clear Print Area and then resetting it to your own area using the sequence above.

Next, it is wise to use Excel's Print Preview feature to gain an impression of how the print out will look on paper. It is wise because of the different paper orientations possible on modern printers. Excel uses Portrait as the default. This often leads, when working with wide spreadsheets, to some columns being printed on separate sheets of paper (so-called orphans). In our case Print Preview reveals that we would indeed see the LN column alone on a second printed page. To avoid this Excel allows the print orientation to be switched from Portrait to Landscape and vice versa by using the Setup button. Selecting Landscape and returning to the Page Preview shows that all is well. All columns will be printed on sideways oriented paper (landscape view). We can now elect to either print directly from the Preview page or Close it and return to the main Print option via the File menu. This latter route allows the number of copies and other options to be chosen.

Note the use of the =Today() function in the cell L19 to provide us with a date on the print out. We previously formatted the cell d.mm.yy to arrive at a date in the form 2 Aug 97. Others are of course possible. Today() is a volatile function, that is, it is rewritten each time the sheet is recalculated.

Well, that was the manual way of going about it. Most spreadsheet programs however provide a more automatic way of doing things. For routine operations like this one there is an alternative to typing the File | Print etc. sequence we've used up to here. You will often see it called a *macro* (now rather antiquated) or since the introduction of very comprehensive programming facilities, *code* or *script*. Excel, like most other modern spreadsheets, provides basically two kinds of these. The first kind, which we introduce here, is just a sequence of recorded keystrokes stored under a name. Essentially they are just shortened menu calls. In fact, earlier spreadsheet generations called them just that – *keyboard macros*. Such macros executed "their" sequence of key actions when called up by their names. The macro code itself often looked something like Table 4.

Table 4. Macro code	/fsr~	or	<pprResults~ag

In other words, practically incomprehensible to anybody but their authors and error-prone during development. Newer macro recorders are much more sophisticated and prepare high-quality readable code during recording. We have recorded our printing sequence using Excel's recorder in the following steps.

From the Record Macro Box which appears after the *Tools* | *Macro* | Record New Macro sequence we selected the Macro name (PrintIt). We did *not* select a shortcut key (exercise for reader). At this point a Stop Recording button appeared on the screen.
We then selected the range (C3:L20) to print, then *File* | Prin*t* Area. Excel blacks the background and surrounds the range with dashed lines.
The final steps were just to press Print and OK and to terminate the recording by pressing the Stop recording button.

This simple sequence of keystrokes stores the macro PrintIt for future use. You can call it (and any other macros) from the Tools Macros Macro dialog box (or simpler Alt F8) from the main Excel menu. Click on *R*un to execute it or, if you'd like to see the code generated by the recorder *E*dit. If you choose the latter you'll be deposited in Excel 97's new Visual Basic Editor/Environment (VBE). There you can see that the active section of the code looks like this:

- Range("C3:L20").Select
- ActiveSheet.PageSetup.PrintArea = "$C$3:$L$20"
- ActiveWindow.SelectedSheets.PrintOut Copies:=1, Collate:=True

This, I think you'll agree, looks more appetizing than the <pprResults~ag keyboard macro we saw above.

Good as it is for recording routine tasks, learning VBA and producing "quick and dirty" code you can not use it for actual programming. In addition, the code is rarely optimal, as Excel often adds many more lines than are necessary for a given task. For these and other reasons Excel provides a second feature more allied to conventional programming. Here we use the VBE to enter the vastly more versatile commands of the VBA language.

We illustrate the second kind of macro by simply improving our first attempt. If you've run the macro a couple of times you will have noticed the disconcerting flickering of the screen as it runs. We would like to switch that off. To do this we need to edit the original recorded code. This was done in the Visual Basic Editor. In our CD ROM example the code has already been added at the top of the active section. It is:

ScreenUpdating = False

This simply stops screen updating i.e. the flickering. The lines preceded by ' are comments and are ignored by the computer. They are not active and are used only for informational purposes. For more details on VBA programming consult the Appendix A at the end of this book.

Before we close this topic a final word on one sad aspect of modern computing life. Unfortunately we live in the real world and everything is not sweetness and light. The scourge of computer viruses has not stopped at the door of Microsoft. There are now a large number of so-called macro viruses affecting all VBA-based code. These are variable in their destructiveness. Some will delete all your Windows system files, others your whole hard disk with wide variations in destructive ingenuity in between. Therefore the following warnings: do not open any documents of whose source you are not certain. If you are a Office 97 user your software can carry out a preliminary examination of any documents you are opening with the option of deactivating any viruses it notes. Use this and note however that it does not disinfect. For that you need to get yourself a good antivirus program. It doesn't always happen to the guy next door!

For more information on these and other viruses consult the Symantec Antivirus Research Center web site and links there. The Web address is http://www.symantec.com/avcenter.

2. An Organic Solvent Database

In conducting reactions in organic solvents it is often necessary to choose solvents possessing a certain set of desired properties. For example, for low temperature spectral studies a solvent of low polarity and melting point and high refractive index may be sought. Seeking such a solvent among the around 100 common ones may be a time consuming and error prone job. Using a spreadsheet makes it simple, quick and error-free. This is the background for our demonstration of Excel's database features. Let it be said at the outset that spreadsheet programs do not possess the full functionality of high-end dedicated database programs. Their strength in this context is limited to the simple processing of what are more reasonably called *lists* – series of worksheet rows containing related data – than true databases.

When we perform database tasks, such as finding, sorting, or subtotaling data, Microsoft Excel automatically recognizes the list as a database and uses the following rules to organize the data.

- Each row in the list is a record in the database.
- The columns in the list are the fields in the database.
- The column labels in the list are the field names in the database.

For this reason the following recommendations for working with spreadsheet databases:

- Create column labels in the first row of the list. Microsoft Excel uses the labels to create reports and to find and organize data.
- Use a font, alignment, format, pattern, border, or capitalization style for column labels that is different from the format you assign to the data in the list.
- If you want to separate labels from data, use cell borders not blank rows or dashed lines to insert lines below the labels.

We'll see all of this in action as we look at the example of constructing a database of organic solvent properties. You can follow all of this discussion in the workbook **OrgSolvs.XLS** on the CD-ROM.

Setting up the Database

Setting up a database in Excel is so extraordinarily easy that it is difficult to do anything wrong. It begins with the simple step of entering the headers (field-names) of our list one to a column. As shown in Fig. 21 (see next page).

This done, the solvent data can be entered either directly into the rows or columns or, under the guidance of Excel by means of a data entry form. We chose the latter for safety. For this we only need place the cursor anywhere in the line of headers and select *Data Form*. Excel selects the headers and places them correctly in our form. Figure 22 shows the screen at this stage (see next page).

Now we can start with the actual data entry. Initially the cursor is in the Nr field. We enter 1 and move to the next field (MP) with the TAB, not the Enter (↵) key. We carry on this way until all the data for the solvent (Acetone in our case) is complete and then move to the next one by clicking on New. A new blank form is displayed and we carry on until the whole database has been entered. In the file **OrgSolvs.XLS** on the CD-ROM accompanying this book we have only a small number of solvents (21). It contains more than enough information to demonstrate the main list features. In any case you can add more of your own using the techniques described. Our final database is shown in Fig. 23 (see next page).

So, what can we do with this list of solvents and their properties? Well, one of the simplest things is already provided as default icons on the main panel. These are the icons:

serving descending and ascending sorts respectively. Using these, sorting our list on the basis of any of the headers is as easy as just selecting the header to use as

Fig. 21. Setting up the database fields. Note: we have formatted them bold as recommended

Fig. 22. The data entry mask

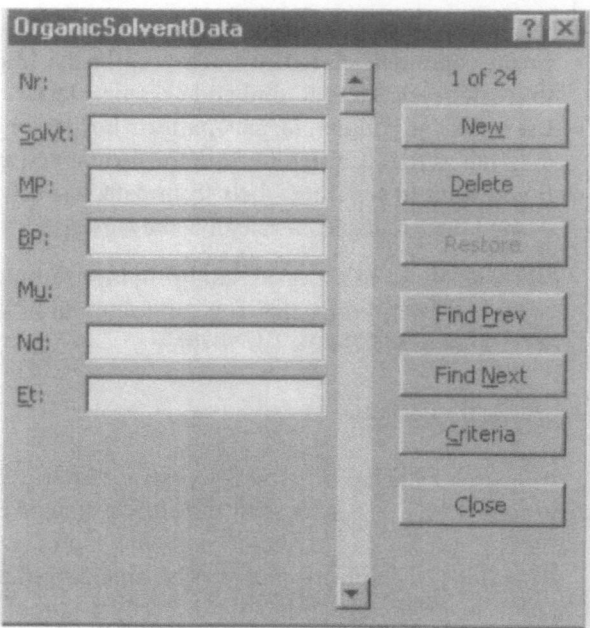

Fig. 23. The OrgSolvts database

	A	B	C	D	E	F	G
1	Nr	Solvt	MP	BP	Mu	Nd	Et
2	1	Acetone	−94.7	56.1	20.56	1.3587	0.355
3	2	Benzene	5.5	80.1	0	1.5011	0.111
4	3	Butanol,1-	−88.6	117.7	17.51	1.3993	0.602
5	4	Carbon Disulfide	−111.6	46.2	0	1.6275	0.065
6	5	Chloroform	−63.5	61.2	3.8	1.4459	0.259
7	6	Cyclohexane	6.7	80.7	0	1.4262	0.006
8	7	Cyclohexanol	25.15	161.1	15	1.4648	0.5
9	8	Diethyl Ether	−116.3	34.4	3.8	1.3524	0.117
10	9	Dioxane,1,4-	11.8	101.3	1.5	1.4224	0.164
11	10	Ethanol	−114.5	78.3	24.55	1.3614	0.654
12	11	Formamide	2.55	210.5	111	1.4475	0.799
13	12	Heptane	−90.6	98.4	0	1.3876	0.015
14	13	Hexane	−95.3	68.7	0	1.3749	0.009
15	14	Methanol	−97.7	64.5	32.66	1.3284	0.762
16	15	Pentane	−129.7	36.1	0	1.3575	0.009
17	16	Piperidine	−10.5	106.2	4	1.4525	0.148
18	17	Pyridine	−41.55	115.25	12.91	1.5102	0.302
19	18	Quinoline	−14.85	237.1	7.3	1.6273	0.269
20	19	Tetrahydrofuran	−108.4	66	7.58	1.4072	0.207
21	20	Toluene	−95	110.6	2.38	1.4969	0.099
22	21	Water	0	100	78.3	1.333	1

sort criterion and clicking one or the other of the sort icons. For example, to sort our database in order of ascending melting point (MP). We:

- select the cell labelled MP
- click on the ascending icon

The database now looks as in Fig. 24:

Nothing else to it! In fact, if you choose the route Data Sort *menu* option you can conduct more complicated searches based on multiple field headers. There are two ways to convert our sorted database back to the original form in Fig. 23. What are they?

We can also use the input form to search for database elements with certain properties. For example, we might want to find the properties of the solvent piperidine. Easy. We:

- Select *D*ata *F*orm from the main menu and on the entry form click *C*riteria
- In the field *S*olvt: we type piperidine and click Find *P*rev or Find *N*ext until the element sought appears

Note you can also use so-called wild cards. For example, try, instead of typing piperidine just pip*. For further practice you might like to see what happens with d* and t*. Remember to *C*lear the criteria between runs.

	A B	C	D	E	F	G
1	Nr Solvt	MP	BP	Mu	Nd	Et
2	15 Pentane	−129.7	36.1	0	1.3575	0.009
3	8 Diethyl Ether	−116.3	34.4	3.8	1.3524	0.117
4	10 Ethanol	−114.5	78.3	24.55	1.3614	0.654
5	4 Carbon Disulfide	−111.6	46.2	0	1.6275	0.065
6	19 Tetrahydrofuran	−108.4	66	7.58	1.4072	0.207
7	14 Methanol	−97.7	64.5	32.66	1.3284	0.762
8	13 Hexane	−95.3	68.7	0	1.3749	0.009
9	20 Toluene	−95	110.6	2.38	1.4969	0.099
10	1 Acetone	−94.7	56.1	20.56	1.3587	0.355
11	12 Heptane	−90.6	98.4	0	1.3876	0.015
12	3 Butanol,1−	−88.6	117.7	17.51	1.3993	0.602
13	5 Chloroform	−63.5	61.2	3.8	1.4459	0.259
14	17 Pyridine	−41.55	115.25	12.91	1.5102	0.302
15	18 Quinoline	−14.85	237.1	7.3	1.6273	0.269
16	16 Piperidine	−10.5	106.2	4	1.4525	0.148
17	21 Water	0	100	78.3	1.333	1
18	11 Formamide	2.55	210.5	111	1.4475	0.799
19	2 Benzene	5.5	80.1	0	1.5011	0.111
20	6 Cyclohexane	6.7	80.7	0	1.4262	0.006
21	9 Dioxane,1,4−	11.8	101.3	1.5	1.4224	0.164
22	7 Cyclohexanol	25.15	161.1	15	1.4648	0.5

Fig. 24. The organic solvents database sorted in order of increasing melting point

Filtering Without the Form

At this point we'll turn to searching without the entry form. Using the so-called AutoFilter we can force Excel to display only those members in the database range possessing the properties we wish to see. To do this we:

- Place the cursor anywhere in the list (a header or list member will do)
- Select the menu option *Data Filter* AutoFilter

Now our screen looks like Fig. 25.

Each one of the pull-down arrows represents a separate criterion. For example, assume we want to see the five solvents with the lowest melting points. We click on the MP arrow and from the ensuing menus (Fig. 26) select Top 10...

Note the periods ... after a menu item imply further submenus. After negotiating these we arrive at our search results (Fig. 27).

To restore the full list to the screen click on the pull-down arrow and click on (All).

With this we have still not exhausted all of Excel 97's database (list) features. There are still a couple more hidden under the Data AutoFilter menu option. We'll first have a brief look at custom searches and then at the Advanced Filter... option.

Custom searches are accessed, as we said, via the AutoFilter pull-down menu shown in Fig. 26 (shown as (Custom...)). This allows us to display database items which fulfil complex selection criteria including AND and OR conditions. Let us say we wish to find those solvents having a dielectric constant greater than twenty but less than a hundred. The completed Custom AutoFilter dialog box for this search looks like Fig. 28 and the results as in Fig. 29.

Note that, on your computer screen, the filtered cells have been given a blue row marker.

If you need to apply three or more conditions to a column, use calculated values as your criteria, or copy records to another location, you'll have to use the next option provided by Excel, the so-called Advanced Filter. Using this we can structure our searches with much more flexibility. Setting it up however is a little more strenuous. We need the following:

1. A *List range* with which to work. In this example this will be our list of organic solvents with some of their common properties. It could be an address book, a set of literature references or pretty well any large set of data requiring rapid scanning, searching or sorting.
2. A *criteria range* into which we place our search conditions. The criteria range must contain one row of criteria labels and at least one row which contains the comparison criteria defining the search conditions. Comparison criteria can be a series of characters you want to match, such as "Formamide" or "formamide" or an expression, such as ">300." Note: When evaluating data, Microsoft Excel does not distinguish between uppercase and lowercase characters.

Fig. 25. The AutoFilter feature

Fig. 26. Using the Auto Filter
feature

Fig. 27. The lowest melting solvents in the list

In this example these are the properties we expect from our solvent. Such conditions specify exactly what we are looking for. For example, paraphrased into simple English they would typically look like this: "Find me the properties of acetone, toluene and formamide", "Show me all solvents with boiling points lower than 50" or "Copy all solvents with boiling points lower than 75 and dielectric constant greater than 15". Some users like to place this and the output range on a separate worksheet.

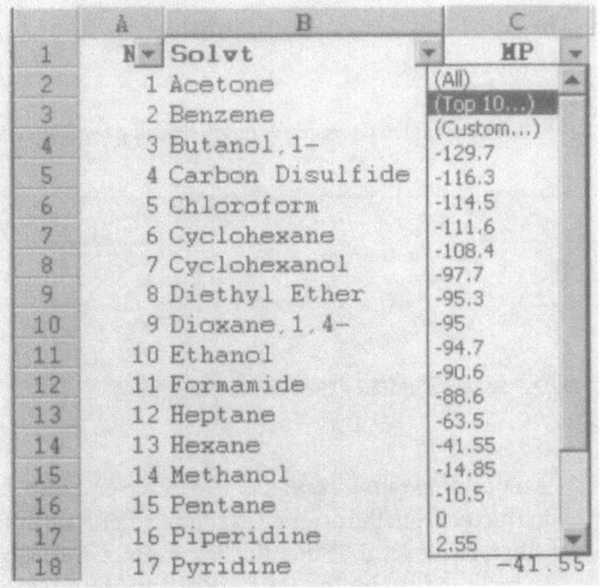

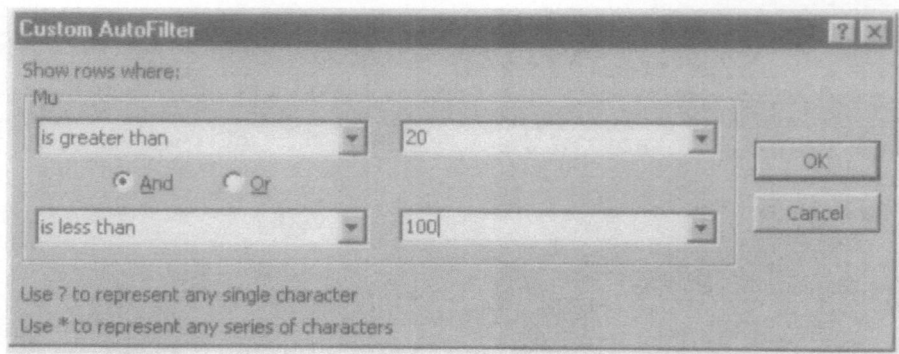

Fig. 28. The completed Custom AutoFilter dialog box (Mu > 20 and < 100)

	A	B	C	D	E	F	G
1	N	Solvt	MP	BP	Mu	Nd	Et
2	1	Acetone	−94.7	56.1	20.56	1.3587	0.355
11	10	Ethanol	−114.5	78.3	24.55	1.3614	0.654
15	14	Methanol	−97.7	64.5	32.66	1.3284	0.762
22	21	Water	0	100	78.3	1.333	1

Fig. 29. Solvents fulfilling the criteria in Fig. 28

3. An optional *output range* to which to send our "hits" if we wish to have them extracted from the original list. Excel offers two options when using the Advance Filter. These are the **Filter the list, in-place** and **Copy to another location** actions to be found under the *Data Filter* Advanced Filter option. Earlier spreadsheet programs called these more descriptively *Find* and *Extract*.

Firstly, a word of definition on the difference between the two options. The difference is simple. Copy to another location copies hits into the predefined output range where they remain for the current session or until overwritten by a subsequent search using the same output range. They are saved with the worksheet when it is saved. Filter the list, in-place displays the hits within the list range while at the same time hiding those not fitting the criteria. It allows upward and downward scrolling of the results.

Let us now see how these features can be used on our OrgSolvs.XLS database. We want the find the properties of formamide, water and methanol and display them in the original list range. In other words, we want an in-place filter. OrgSolvs.XLS contains all necessary ranges predefined and ready for use. Let us examine their contents and positions and identify their roles.

- List range (A1:G22): begins with the first cell in the field labels row and ends in the last field of the last record. As you'll see it contains, as it must, the field labels.

- Criteria range (B28:B31): at installation this should be an empty worksheet area below or to the right of the Input range. We have placed it below the database. It need contain only those field names which are to be searched.
- An Output range is not necessary for this first example. The one you see on the worksheet starting at cell A51 is used in a later example. It does not disturb searches not using it. As mentioned earlier the output range should be below or to the right of the database and criteria ranges or on another worksheet to avoid inadvertent overwriting of existing data. Only field names for which output is expected, that is actively searched, need be listed. For simplicity one often copies the complete row of database labels to the Criteria and/or Output ranges and ignores or deletes unused fields. This point will be clearer after we look at some more examples.

All three ranges were each set up by dragging and dropping using the Data Filter Advanced Filter dialog boxes as shown in Fig. 30. Almost all our work on searching will be done in this dialog box. Take some time to become familiar with it.

Spadework complete we can now carry out our Find and/or Extract operations. First however, we must tell Excel for what to search. The program needs

	Nr	Solvt	MP	BP	Mu	Nd	Et
1	Nr	Solvt	MP	BP	Mu	Nd	Et
2	1	Acetone	-94.7	56.1	20.56	1.3587	0.355
3	2	Benzene	5.5	80.1	0	1.5011	0.111
4	3	Butanol,1-	-88.6	117.7	17.51	1.3993	0.602
5	4	Carbon Disulfide	-111.6	46.2	0	1.6275	0.065
6	5	Chloroform	-63.5	61.2	3.8	1.4459	0.259
7	6	Cyclohexa				1.4262	0.006
8	7	Cyclohexa				1.4648	0.5
9	8	Diethyl E				1.3524	0.117
10	9	Dioxane.1				1.4224	0.164
11	10	Ethanol				1.3614	0.654
12	11	Formamide				1.4475	0.799
13	12	Heptane				1.3876	0.015
14	13	Hexane				1.3749	0.009
15	14	Methanol				1.3284	0.762
16	15	Pentane				1.3575	0.009
17	16	Piperidin				1.4525	0.148
18	17	Pyridine				1.5102	0.302
19	18	Quinoline	-14.85	237.1	7.3	1.6273	0.269
20	19	Tetrahydrofuran	-108.4	66	7.58	1.4072	0.207
21	20	Toluene	-95	110.6	2.38	1.4969	0.099
22	21	Water	0	100	78.3	1.333	1
26	Criteria Range for use with Advanced Filter						
28	Nr	Solvt	MP	BP	Mu	Nd	Et
29		formamide					
30		water					
31		methanol					

Advanced Filter dialog box:

Action:
- Filter the list, in-place
- Copy to another location

List range: Data!$A$1:$G$22
Criteria range: $B$28:$B$31
$A$54:$G$54
☐ Unique records only

OK   Cancel

Fig. 30. The search criteria range in OrgSolvsXLS

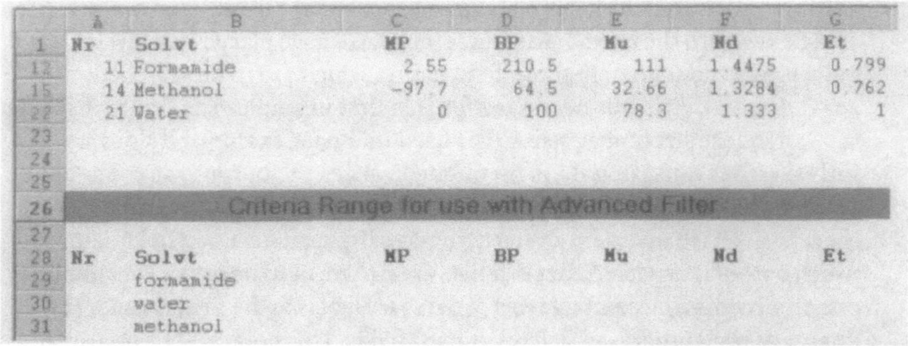

Fig. 31. The Search results in OrgSolvs.XLS

Table 5. Problem Solving

Problem	Excel Solution
Find properties of ethanol	Set Criteria Range as B28:B29.Type ethanol in Criteria Range under Solvt. Use the in-place option.
Filter all solvents prefixed with cyclo	Type cyclo* in Criteria Range under Solvt. Use the in-place option.
Filter all solvents not prefeixed with cyclo	Type<>cyclo* in Criteria Range under Solvt. <> means "not equal to". * is a wild-card and means "it doesn't matter what follows".
Filter all solvents with BP>200	Set new Criteria Range D28:D29. Enter formula >200 in D29. Use the in-place option. Answer: Formide, Quinoline.
Extract all solvents with dielectric constant >15	Enter formula >15 in E29. Set Criteria Range to E28:E29. Use the Copy to another location option. Output range at A51 contains hits.
Extract all solvents with dielectric constant >15 *AND* refractive index >1.5	Set Criteria Range to E28:F29. Enter formulas <15 and >1.5 into E29 and F29.Use the Copy to another location option. Output range at A51 contains hits.
Extract all solvents with dielectric constant >15 *OR* refractive index >1.5	Set Criteria Range to E28:F29. Enter formulas <15 and >1.5 into E29 and F29.Use the Copy to another location option. Output range at A51 contains hits.
Extract all solvents whose boiling point >160 *OR* whose dielectric constant is >20	Set Criteria Range to D40…E42. Enter formulas <160 into D29 and >20 into E30.Use the Copy to another location option. Output range at A51 contains hits.

criteria. It gets them from typed information in the appropriate field label in the Criteria range. The screen snapshot shown in Fig. 30 shows the relevant worksheet sections:

Clicking OK produces our search results as shown in Fig. 31

To restore the list to its original state after completing a search select *Data Filter Show All.*

Fortunately, Excel can handle a variety of search criteria making it quite a powerful tool for operations on small to medium size databases. We illustrate the main operations by solving the following problems. Table 5.

The last three examples show two powerful features of the Excel database facility:

- The ability to place logical operators into a single cell to sharpen the search criteria
- The possibility of executing AND or OR conditions by placing criteria for different fields on the same or different rows in the criteria range.

We recommend the reader to work through these examples of Excel's sometimes rather clunky database feature.

Readers wishing to follow these examples in a DOS environment can consult the authors previous book *Spreadsheets for Chemists* (VCH, 1995). There he will see that, in the Lotus 1-2-3 environment, the basic processes were not so much different to today's as described above.

Epilogue

Well, dear Reader that concludes our whistle-stop look at the barebones of Excel 97 spreadsheeting. It may seem a lot for the first course though, as I warned at the outset, in fact we haven't been able to do more than scratch the surface. I hope it has whetted your appetite to dig deeper. Spreadsheet software has come of age and has achieved a complexity and versatility hardly dreamed of only a few short years ago. In the remainder of the book you will find more than enough examples to support this statement. So -

Once more unto the breach, dear friends, once more. And don't forget he

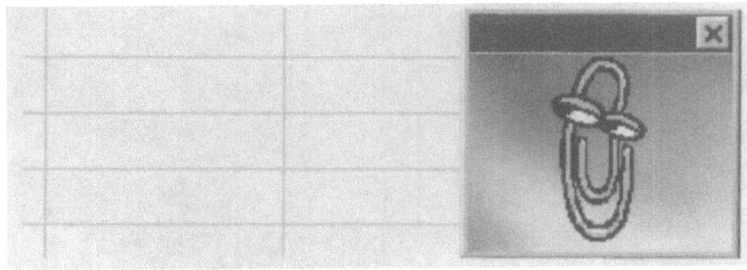

is always there to help you!

Applying Spreadsheets in Physics and Electronic Engineering

W. J. Orvis

Introduction

The use of spreadsheets in Physics and Electronics Engineering does not really form two separate topics in mathematical analysis. These two disciplines are closely related in both the types of problems solved and in the methods used for finding solutions to those problems. While calculations in electronics engineering tend to overlap the branches of physics that deal with electrostatics, fields, and waves, the equations being solved overlap many other fields. For example, the physical equations of heat flow and temperature in a solid are identical to those for electrostatic potential and electric field in a solid. In one place a variable is named temperature and in the other it is potential, but in both cases the equations have the same form.

Since the equations are the same, you would rightly expect that the worksheet methods used to solve those equations are also the same. They differ only in the problem being solved, not in the methods used to solve them.

Spreadsheet programs have got to rank as one of the most useful tools for scientists and engineers to come along since the hand calculator. While VisiCalc, the first true PC spreadsheet program, did not have a great impact on Physics and Electronics Engineering, Lotus 1-2-3 quickly entered our toolboxes as an indispensable part of our analysis capabilities. It had good accuracy and a reasonable set of functions to work with. Most everything you needed was either available or could be calculated without too much effort. Plus, data management was handled by the worksheets and graphing was built-in. Today's modern spreadsheet programs contain everything that the older programs had and much, much more. Many of the advanced analytical procedures and engineering functions are built in. Special functions like Bessel and gamma functions are also built in or are available as an add-in.

In this chapter, you will learn how:

- spreadsheets are used in physics.
- spreadsheets are used in electronics engineering.
- to perform different aspects of numerical analysis with a spreadsheet.

All the examples discussed in this section are included in the directory /PhysEE on the CD-Rom accompanying this book. This chapter will give you only a taste of how to perform many types of numerical analysis with a spreadsheet program. Hopefully, it will point you in the right direction to solve other similar problems using the same methods shown here. For a more complete description of numerical methods with a spreadsheet and for many more examples, see the references at the end of this section.

Spreadsheets in Electronics Engineering

The work of Electronics Engineers generally deals with two types of quantitative information; experimental data, and equations. With this information in hand, the engineer must analyze it, to try and gain a better understanding of what the data is trying to tell him, or put it in a form that makes it useful in the design process.

A spreadsheet program is an ideal vehicle for analyzing both of these types of information. The built-in graphing capabilities of modern spreadsheet programs makes visualizing data almost trivially easy to do. Changing the type of graph, the limits, the style of lines and markers takes only seconds. With the addition of a compatible graphing package, such as DeltaGraph Pro, the resulting graphs are also suitable for publication.

Analyzing experimental data and calculating the values of equations is also a straightforward process. Using the worksheets calculational capabilities, there are few numerical methods that can not be performed. The inclusion of Visual Basic for Applications as a macro language in the Excel Spreadsheet makes even those methods that need to be implemented with a traditional programming language available for use on a worksheet. We'll see a couple of examples of this later in the chapter.

Spreadsheets in Physics

Physicists also deal with experimental data and simple functions, but tend to also need solutions to much more complex functions including sets of coupled, nonlinear equations and differential equations. As most complex equations are extremely difficult or impossible to solve analytically, numerical solutions are often the only reasonable alternative for getting results. While a scientist can and often does write special purpose computer programs to calculate numerical solutions to these problems, many of them can be solved with a spreadsheet taking much less effort. Even differential equations are amenable to solution on a spreadsheet, with results that are often much more understandable than those calculated with a special purpose computer program. This increase in understanding is attained because the results of all the intermediate calculations are plainly visible on the worksheet.

Promoting an Understanding of Numerical Methods

Most numerical methods used by Physicists and Electronic Engineers can be implemented on a spreadsheet, or are already available as built-in or add-in functions. Excel version 7 includes an add-in that contains a complete set of science and engineering functions. By implementing numerical methods on a worksheet, all of the methods inner workings are made visible to the student or researcher. Linkages between formulas become linkages between cells as the formulas in each of the cells build on those calculated before them. You can then see those inner workings and get a much better feel for how the method works, what are its advantages, and what are its shortcomings. Much of this understanding is missing when the same functions are implemented in a high level computer language like FORTRAN or C. With a spreadsheet you actually see the calculation take place, with its inner workings laid out before you.

References

The following references contain more detailed descriptions of how to perform many more types of numerical methods on a worksheet than can be presented in this chapter. Note that the book for 1-2-3 users is out of print, but is occasionally available in used book stores. Another useful source of examples are the Worksheets columns of the *Computers in Physics* journal. Every month for the first three or four years of its publication, that column contained a different physics calculation implemented on a worksheet. Currently, the column is not published every month. For general numerical methods, the books by Gerald and Press are very good. Press' book is especially useful as the FORTRAN listings given in the book are easily translated to the Visual Basic for Applications language built-into Excel.

W. J. Orvis, 1–2–3 for Scientists and Engineers, 2nd. Ed., Sybex, Alameda, CA, 1991, ISBN: 0–89588–733–9 (Note: While this book is out of print, it is occasionally available in used book stores.)

W. J. Orvis, Excel for Scientists and Engineers, 2nd. Ed., Sybex, Alameda, CA, 1996, ISBN: 0–7821–1761–9

R. Dory ed, "Worksheets" column, Computers In Physics Journal, American Institute of Physics, most issues

C. Gerald, Applied Numerical Analysis, 2nd. Ed., Addison Wesley, Reading, Mass, 1978

W. H. Press, et al., Numerical Recipes; The Art of Scientific Computing, Cambridge University Press, Cambridge, UK, 1986

S. C. Bloch, Spreadsheet Analysis for Engineers and Scientists, John Wiley, New York (1995)

Modeling

Probably the most used capability of worksheets by scientists and engineers is modeling devices and processes. Most modeling consists of finding simple equations that describe a device or process and then calculating and plotting their values. The values calculated from the models are then used to either validate the models, or to guide the development of new devices or processes. Most model equations are not terribly complex, and can be evaluated with a single table of values. Others can be extremely complex, containing multiple inputs and spanning multiple equations.

Calculating Simple Functions

As you might expect, simple functions are also simple to implement on a worksheet. The independent variable goes in one column and formulas that calculate the dependent variable go in an adjacent one. If the function has adjustable coefficients, they can be pulled out of the function and placed in cells by themselves so that they can be adjusted to make the results of the function fit some known values. The example in this section shows how to model simple functions using a worksheet.

Drift Velocity and Mobility of GaAs

When an electric field (E) is applied across a semiconductor, the free electrons within the material begin to drift along the field lines. The velocity of the drifting electrons is limited by their collisions with the crystal lattice, and quickly reaches a collision dominated drift velocity. Because it is collision dominated, this velocity is a constant at any applied field. The proportionality constant between the drift velocity and the applied electric field is the electron's mobility. In GaAs, the drift velocity (v) and mobility (μ) are modeled with the following set of equations:

$$v = \frac{\mu_1 E\left(1 + BF^k\right)}{1 + F^k}$$

$$\mu = \frac{v}{E}$$

$$F = \frac{E}{E_0}$$

where: $E_0 = 4{,}000$ V/cm

$\mu_0 = 8{,}000$ cm^2/V-s

$k = 4$

$B = 0.05$

Figure 1 shows a worksheet created to calculate the mobility and drift velocity for electric fields between 0 and 10,000 V/cm. Figure 2 shows both of those functions on a single plot with two vertical axes. Note how the velocity curve flattens at high applied fields indicating that the velocity has reached saturation. The velocity overshoot at about 3000 V/cm is caused by carriers moving from a high mobility to a lower mobility conduction band. This example also shows how to pull the coefficients of the calculation out of the formula and into a separate table. When done in this way, the coefficients can be adjusted as needed in the table and all of the formulas will automatically recalculate. If the coefficients had been left in the formulas, you would have to redo all the formulas in order to change a single coefficient.

Create the worksheet by following these steps:

1. On a new worksheet type the following values in the indicated cells.
   - A1 **Drift velocity and mobility in GaAs**
   - G2 **Fitting Parameters**
   - B3 **E field**
   - B4 **(V/cm)**
   - C3 **F**
   - C5 **=B5/E0**

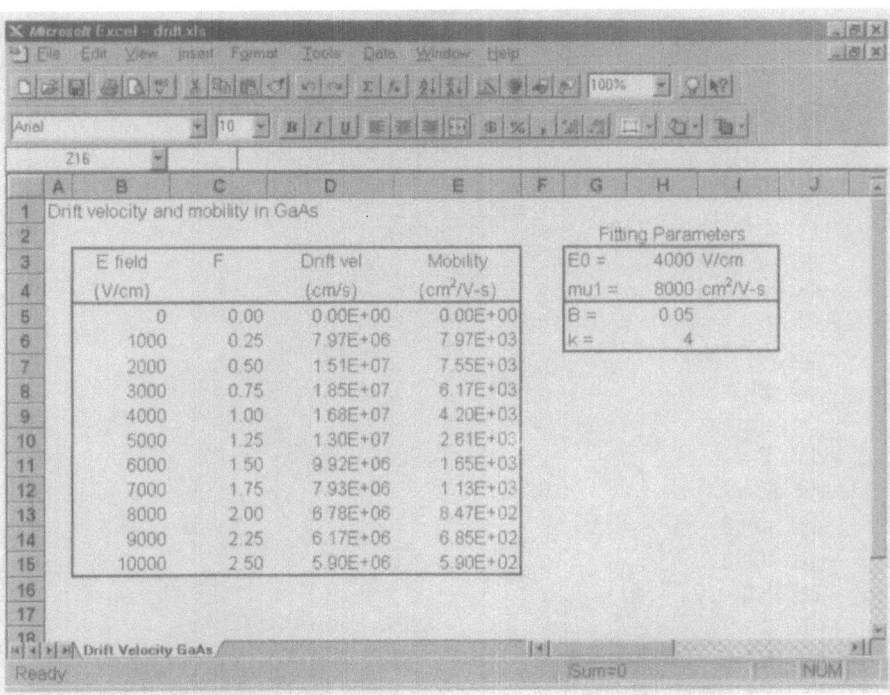

Fig. 1. A worksheet to calculate the mobility and drift velocity for electrons in GaAs under applied fields between 0 and 10,000 V/cm

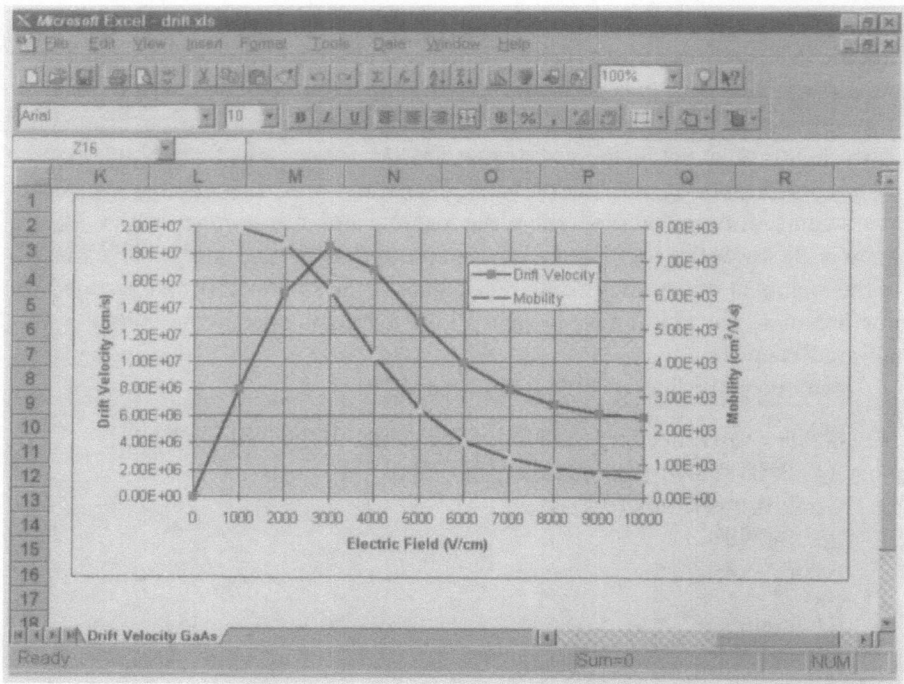

Fig. 2. A plot of the functions calculated in Fig. 2

- D3 **Drift vel**
- D4 **(cm/s)**
- D5 **=MU1*B5*(1+B*C5^K)/(1+C5^K)**
- E3 **Mobility**
- E4 **(cm2/V-s)**
- E5 **0**
- E6 **=D6/B6**
- G3 **E0 =**
- G4 **mu1 =**
- G5 **B =**
- G6 **k =**
- H3 **4000**
- H4 **8000**
- H5 **0.05**
- H6 **4**
- I3 **V/cm**
- I4 **cm2/V-s**

2. Name cells H3:H6 as E0, MU1, B, and K.
3. In cells B5:B15 place the numbers 0 through 10,000 in steps of 1000. Do so

automatically by placing **0** in B5, **1000** in B6, selecting B5:B6 and dragging the fill handle down to cell B15.
4. Copy the formulas in cells C5:D5 down the columns to cells C15:D15 by selecting cells C5:D5 and dragging the fill handle down to cell D15.
5. Copy the contents of cell E6 down into cells E7:E15 by selecting E6 and dragging the fill handle down to E15.
6. Draw boxes around the table as shown in Fig. 1.
7. Format cells B5:B15 as Number with 0 decimals.
8. Format cells C5:C15 as Number with 2 decimal places.
9. Format cells D5:E15 as Scientific with 2 decimal places.
10. Select cells B5:E15 and use the ChartWizard to create the chart shown in Fig. 2. Open the chart and delete the curve for column C.

The example file on the disk is called Drift.XLS.

Calculating More Complex Functions

The example in the last section creates a worksheet for a function with a single independent variable and a single dependent variable. However, many functions have two or more independent variables that have to be considered and included in the calculation. For two independent variables, the most obvious layout is a square with one variable along one edge and the other across the top. The calculated functions go in the body of the table and use the values of the independent variables that are on the same row and column as the function.

For more than two variables, one variable can be placed across the top of the table but the rest must go down the side in adjacent columns, or all the variables can go in adjacent columns down the side. The unfortunate result of this type of layout is that you must repeat values in order to get every combination of every independent variable.

Planck Distribution for Blackbody Radiation (as a Worksheet)

The spectral emission of a heated black object or blackbody eluded most physicists for many years. In 1879, Josef Stefan deduced that the total spectral emission from a blackbody (M^b) was proportional to the fourth power of the temperature (T).

$$M^b(T) = \sigma T^4$$

It took until 1900 for Max Planck to apply the new quantum theory to the problem and derive the formula for the spectral emittance (M^b_λ) at any wavelength (λ). This equation is known as the Planck distribution for blackbody radiation.

$$M^b_\lambda(T) = \frac{2\pi hc^2}{\lambda^5}\left(e^{\frac{hc}{\lambda kT}} - 1\right)^{-1}$$

Here, h is Planck's constant and k is Boltzmann's constant. Integrating this equation over all possible wavelengths, results in the Stefan-Boltzmann Law, which, indeed shows the expected relationship to the fourth power of the temperature.

$$M^b(T) = \frac{2\pi^5 k^4}{15 h^3 c^2} T^4$$

The Planck distribution contains two independent variables, the temperature (T) and the wavelength (λ), which are combined to calculate the emittance. Figures 3 and 4 show a worksheet and chart that calculate the Planck distribution every 500 degrees from 500 K to 2000 K. The worksheet places the temperature across the top of the table and the wavelength down the left side. The body of the table contains the Planck distribution calculated using the coefficients in the table at the top of the page.

The chart type shown in Fig. 4 is what you would commonly expect to find in a text book on the subject. However, a more intuitive chart is shown in Fig. 5, where the distribution is plotted in two dimensions instead of just one. This two dimensional chart is calculated from the same data set as those in Fig. 3, just a different chart type was chosen in the ChartWizard.

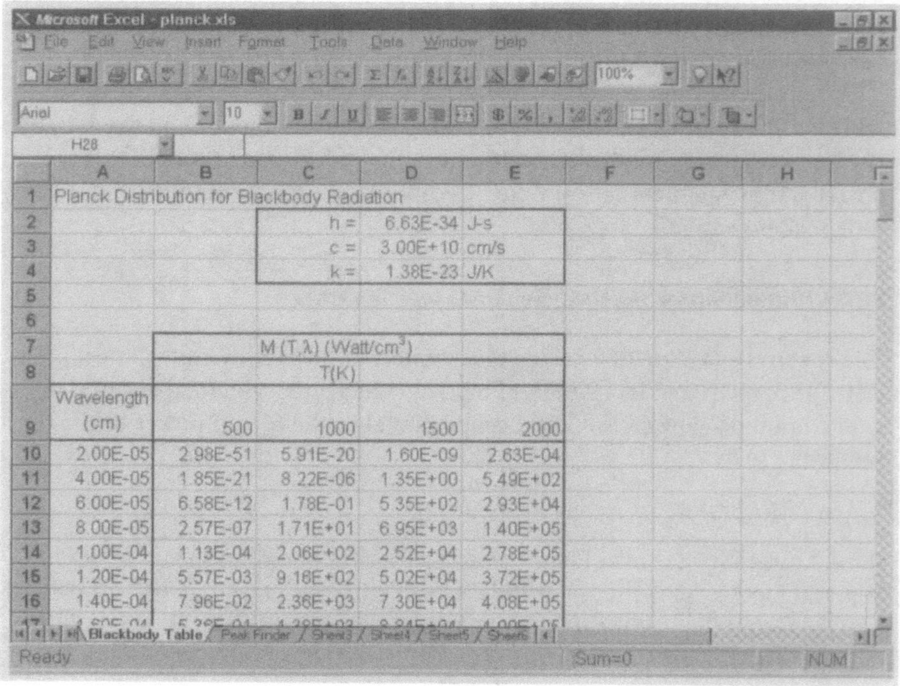
Microsoft Excel - planck.xls

Planck Distribution for Blackbody Radiation

h =	6.63E-34	J-s
c =	3.00E+10	cm/s
k =	1.38E-23	J/K

Wavelength (cm)	M (T,λ) (Watt/cm³) T(K) 500	1000	1500	2000
2.00E-05	2.98E-51	5.91E-20	1.60E-09	2.63E-04
4.00E-05	1.85E-21	8.22E-06	1.35E+00	5.49E+02
6.00E-05	6.58E-12	1.78E-01	5.35E+02	2.93E+04
8.00E-05	2.57E-07	1.71E+01	6.95E+03	1.40E+05
1.00E-04	1.13E-04	2.06E+02	2.52E+04	2.78E+05
1.20E-04	5.57E-03	9.18E+02	5.02E+04	3.72E+05
1.40E-04	7.96E-02	2.36E+03	7.30E+04	4.08E+05

Blackbody Table / Peak Finder / Sheet3 / Sheet4 / Sheet5 / Sheet6
Ready Sum=0 NUM

Fig. 3. A worksheet for calculating the Planck distribution for spectral emittance from a blackbody

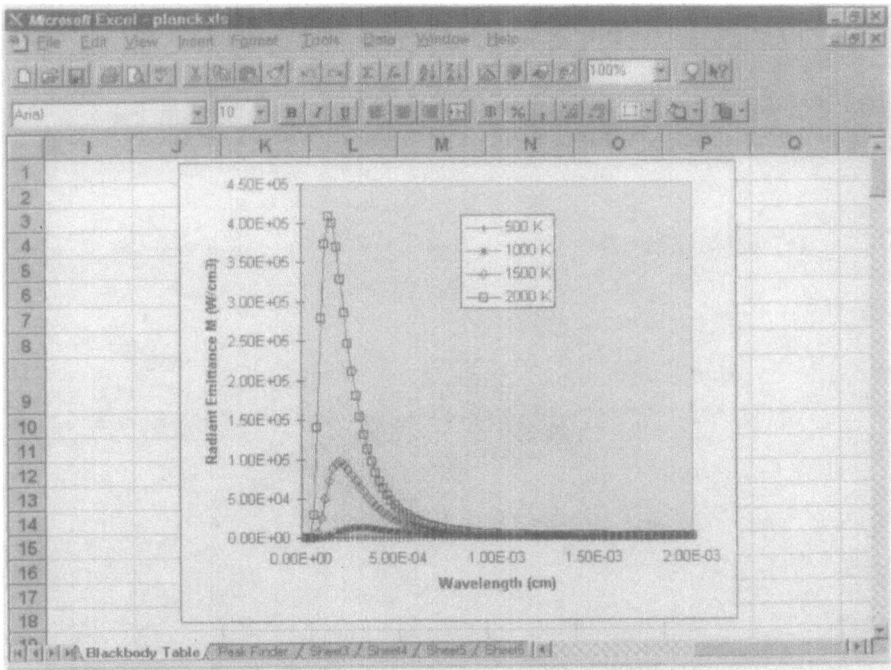

Fig. 4. The spectral emission of a blackbody versus wavelength for four different temperatures

Details of this and the following treatments of the problem can be found in the disk file Planck.XLS. To create the worksheet, perform the following steps.

1. Start with a new worksheet and insert the following values in the indicated cells.
   - D2 **6.63E-34**
   - D3 **3.00E+10**
   - D4 **1.38E-23**
   - B9 **500**
   - C9 **1000**
   - D9 **1500**
   - E9 **2000**
2. Insert labels as shown in Fig. 3. Cells C2:C4, and C8 are right justified and cell A9 is centered.
3. Name cells D2:D4 as **h, c_,** and **k.** The name c_ is used for cell D3 because the name c is reserved in Excel.
4. In cells A10:A109 place the numbers 2×10^{-5} through 2×10^{-3} in steps of 2×10^{-5}. Do so automatically by placing **2E-5** in A10, **4E-5** in A11, selecting A10:A11 and dragging the fill handle down to cell A109.
5. In cell B10, type **=(2*PI()*h*c_^2/($A10^5))*1/(EXP(h*c\_/($A10*k*B$9))-1)**

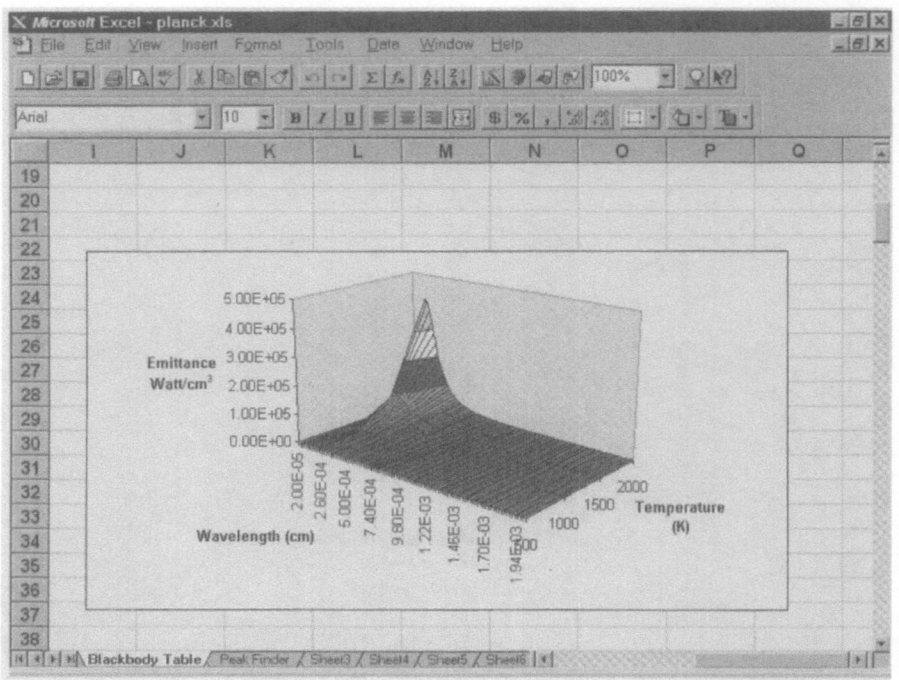

Fig. 5. A three dimensional graph of the spectral emission of blackbody radiation

6. Copy the contents of cell B10 right into cells C10:E10 by selecting cell B10 and dragging the fill handle right to cell E9.
7. Copy the contents of B10:E10 down into cells B11:E109 by selecting cells B10:E10 and dragging the fill handle down to cell E109.
8. Outline the cells as shown in Fig. 3.
9. Select cells A9 through E109, click the ChartWizard button and follow the instructions to create a line plot of the contents of those cells like Fig. 4. Use column A for the *x* labels and row nine for the legend text.
10. Select the same cells and again run the ChartWizard but this time create a 3D surface plot like that in Fig. 5. Use column A for the x labels and Row 9 for the y labels.

The following section shows a more elegant way of approaching this problem.

Creating Custom Functions With Visual Basic for Applications

A unique capability of Microsoft Excel is the inclusion of Visual Basic for Applications (VBA) as the macro language for spreadsheet automation. Visual Basic for Applications is a subset of the Visual Basic language that has been optimized for editing worksheets. Being related to the BASIC language, it is easy to learn and program with. In addition to automating the worksheet, you can create custom functions for use on the worksheet.

If you already know the Visual Basic language, you will have no trouble programming with Visual Basic for Applications. Check the on-line help included with Excel for any differences in the language elements and functions. In addition to the standard Visual Basic language elements Visual Basic for Applications can control OLE Automation Objects which include worksheets and the contents of cells on a worksheet. However, these special access capabilities are not needed for creating custom worksheet functions. You need only name your special function for it to be available for use on a worksheet. If you know a modern form of BASIC, most of this will be familiar. Again, you should check the User's Guide and on-line help to see the differences between Visual Basic for Applications and the version of BASIC you know. In the event you do not know any version of BASIC, you will need to learn that language first before attempting to create functions using it. A complete description of the Visual Basic language is beyond the scope of this chapter, but there are several good books available such as, Excel for Scientists and Engineers, Orvis, (Sybex, 1996), or Visual Basic for Applications by Example, Orvis, (Que, 1994). The Appendix at the end of this book provides more information on this topic.

Planck Distribution for Blackbody Radiation (as a Custom Function)

As an example, consider the Planck distribution function developed in the last section. Instead of copying the complete function into each cell of a worksheet, you can create a custom function and use it in each cell to calculate the distribution function.

To create the worksheet, perform the following steps.

1. Start with a copy of the Planck distribution worksheet from the last section.
2. Create a custom function by choosing the Insert, Macro, Module command to insert a new module sheet, then type the following basic code into it.

> **Option Explicit 'Force all variables to be declared.**
>
> **Function PlanckBB(dblWavelength As Double, dblTemp As Double) As Double**
> **'Calculate the Planck blackbody distribution function**
> **'Inputs:**
> **' dblWavelength wavelength (cm)**
> **' dblTmp temperature (K)**
> **'Outputs:**
> **' PlanckBB Planck black body distribution (Watts/cm^3)**
>
> **'Define some physical constants.**
> **Const h = 6.63E-34 'J-s Planck's constant.**
> **Const k = 1.38E-23 'J/K Boltzmann's constant.**

Const c = 30000000000# 'cm/s Speed of light.
Const pi = 3.1415926 'Pi

'Calculate and return the value of the blackbody distribution.
PlanckBB = (2 * pi * h * c ^ 2 / (dblWavelength ^ 5)) * 1 / (Exp(h * c/
   (dblWavelength * k * dblTemp)) - 1)

End Function

3. Name the macro sheet Planck Macro using the Edit, Sheet, Rename command.
4. Select the copy of the Planck Distribution worksheet and clear all the contents
   of cells C2:E4.
5. Change cell B10 to =**PlanckBB($A10,B$9)**
6. Copy cell B10.
7. Select cells B10:E109 and choose Edit, Paste Special.
8. In the Paste Special dialog box, choose the Formulas option button and click
   OK. The formula is copied into all the cells without changing the formatting.

The resulting worksheet is identical to Fig. 3, minus the table of Physical constants
in cells C2:E4, showing that the function gives results that are identical to the worksheet formulas. An additional benefit, is that the custom function, once defined,
is available to use on any other worksheet. Simply have the worksheet that contains it open at the same time or copy the function's macro sheet into the workbook where you need it. You could also copy it to Excel's global macro sheet, making it available for any open worksheet though this should only be done for functions you use often.

   Taking a closer look at the VBA code, note that several of the lines start with
a single quotation mark (') or have one after the executable part of a statement.
Anything following a single quotation mark is a comment and has no effect on
the executing code. The first executable line of code contains the Option Explicit command. This command forces you to define every variable before using
them, which helps prevent errors caused by misspelling a variable name. All variables used in the program must be either defined in the Function statement at
the beginning of a function or in Dim statements.

   At the beginning of the function is the following Function statement

   Function PlanckBB(dblWavelength As Double, dblTemp As Double) As
   Double

This statement names the function (PlanckBB) with the name you will use to call
it in a worksheet, and declares the type of the input and output values. In this
case, there are two double precision floating point input values (dblWavelength
and dblTemp), and the value returned by the function is also double precision
floating point. Double precision allows numbers in the astronomic range of
approximately E-324 to E+308 to be represented in tha computer. The dbl prefix
on the variable names has no effect on the type of value stored in the variable. It

is there only to help you keep track of the type of value you have specified to be stored in a variable.

Following the function statement are four Const statements that define four constants to be used within the function. Following the constants is a statement that calculates the Planck distribution and equates it to the function name. Whatever value is last equated to the function name is the value that is returned by the function when it ends. By convention all VBA functions must end with an End Function statement.

Complex Custom Functions

Visual Basic for Applications is not limited to calculating simple functions such as that shown in the previous section. Visual Basic for Applications is a fully functional programming language that is capable of being used to perform complex calculations. When you find you need a special function that is not available as a built-in function, find a book on FORTRAN algorithms that contains the function you are interested in. A function written in FORTRAN is easily converted to Visual Basic for Applications. An extremely useful book in this respect is: Press, et al., Numerical Recipes, Cambridge Univ. Press, Cambridge, (1986).

Using Compiled Functions

In addition to using built-in functions created with Visual Basic for Applications, fully compiled functions created with other applications can also be used with an Excel worksheet. For example, if you have an integration that takes a long time to calculate with the interpreted Visual Basic for Applications language, you can achieve a tremendous speed increase by recreating it in C or FORTRAN and producing a fully compiled function stored in a dynamic link library (.DLL). To use that function on the worksheet, you have only to declare it using the Declare statement in a module. Once declared, you can call it by name like any other function. See the User Manual and the on-line Help for the syntax of the Declare statement and for more information about using compiled functions.

Analyzing Experimental Data

The second most common data analysis task of the scientist or engineer is analysis of experimental data. In most cases, the first step is to convert from the measurements (voltages, displacements, and so forth) to engineering units (pressures, temperatures, velocities, and so forth). This task is accomplished in exactly the same way as the modeling tasks described in the last few sections. The experimental data is the independent variable and the equation is the conversion function that turns the raw data into engineering units.

The next analysis task is to analyze the data to help you better understand that data and what it is telling you about the system you are experimenting with.

Analysis of the data generally consists of graphing, statistics, and curve fitting or modeling.

Graphing the Data

In most cases, the first step in data analysis is to simply graph the data to see the shape of the experimental curve. Very often this experimental curve will show features that are lost when you are just looking at the numbers. The curve may also show you where your experiment is going wrong, such as peaks in the data when your measuring instrument changes ranges are indicative of either an out of range measuring instrument or a measuring affecting the measurements.

With a spreadsheet program, graphing the data is almost a trivial undertaking. When the data has been typed into the worksheet, simply select it and paste it into a chart sheet. With Excel, the easiest way to create a new chart is to use the ChartWizard. The ChartWizard takes the currently selected information on the worksheet and steps you through the options for creating a chart. Chart options include both two and three dimensional chart types, labeling of the axes, linear or logarithmic scales and scale limits. After the ChartWizard completes, other options can be set such as line and marker colors, grids and axes label formats.

Residential Electric Power Usage – Graphing Data

Before you can graph some data, you need some data to graph. I just happen to have monthly data on electric power usage at a residence (my house) in Livermore, California. That data shown in the following table, covers 6 years, Table 1. The file covering this topic is named Euse.XLS on the disk.

This data is typed directly into a worksheet as shown in Fig. 6. To plot it, select the contents of cells A4:G16, click the ChartWizard button, and follow the instructions to make the chart in Fig. 7. Note that all the chart types except for XY (scatter) assume that the x data is equally spaced and that any x data is only used

Table 1. Daily average electric power usage by month

Month/Year	Daily Average Usage (kWh/day)					
	1990	1991	1992	1993	1994	1995
Sep	20.9	19.2	23.5	21.4	27.1	28.1
Oct	19.4	21.6	22.1	25.3	26.1	29.9
Nov	18.9	23.7	24.3	28.9	28.1	30.2
Dec	20.4	24.5	25.3	28.7	33.1	34.0
Jan	26.3	29.2	27.8	35.4	38.6	42.4
Feb	26.4	32.2	29.9	32.5	36.5	36.9
Mar	23.0	27.2	25.2	31.9	34.4	34.0
Apr	21.5	23.2	23.4	28.5	33.7	32.5
May	20.2	22.5	21.6	25.8	27.7	31.6
Jun	19.9	20.4	22.8	24.2	28.1	30.8
Jul	18.7	20.5	23.3	23.5	28.0	30.1
Aug	20.3	19.9	23.4	24.9	29.5	33.6

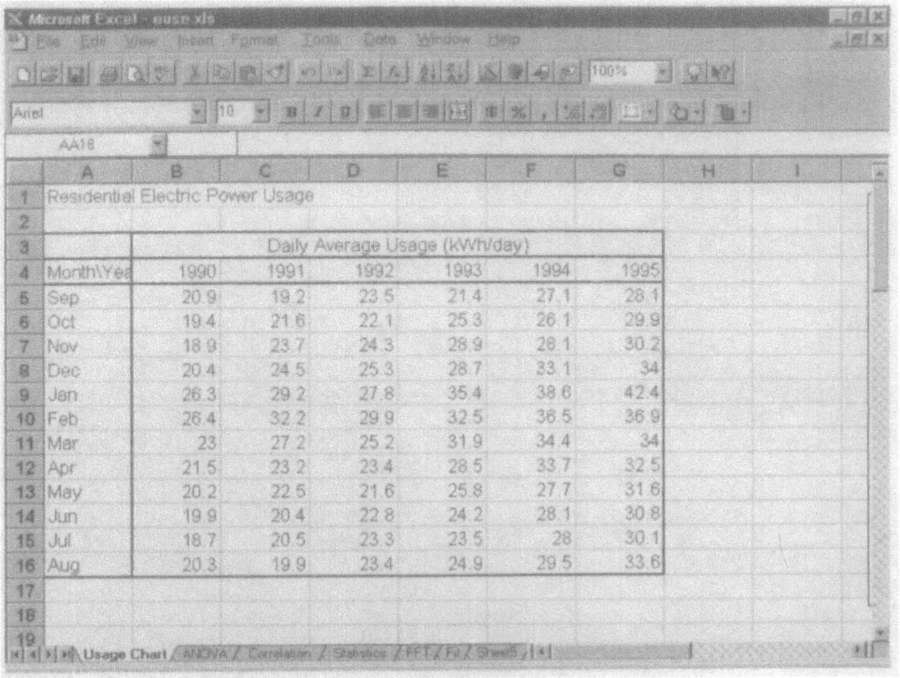

Fig. 6. Experimental data typed into a worksheet

as labels. Interesting chart: it clearly shows the annual changes between summer and winter; it also shows a steady increase in total energy use (must be the kids!).

Calculating Statistics

The next part of analysis of experimental data is to calculate statistics on the data, to answer questions about how the data changes from year to year and how it changes over the course of a year. You expect that the data is increasing, but you should check to see if the apparent increase is statistically significant.

To do statistics on the data, use the built-in statistical package in the Data Analysis add-in. The package includes a descriptive statistics package, analysis of variance (ANOVA), correlation, fast Fourier transforms (FFT), curve fitting and several others.

Residential Electric Power Usage – Statistics

To calculate some basic statistics on the residential electric power usage data, use the Descriptive Statistics package in the Data Analysis Toolpack. Descriptive statistics include the largest and smallest values, Mean, Mode, range and other descriptive values. Descriptive statistics are not meant to prove anything, the indi-

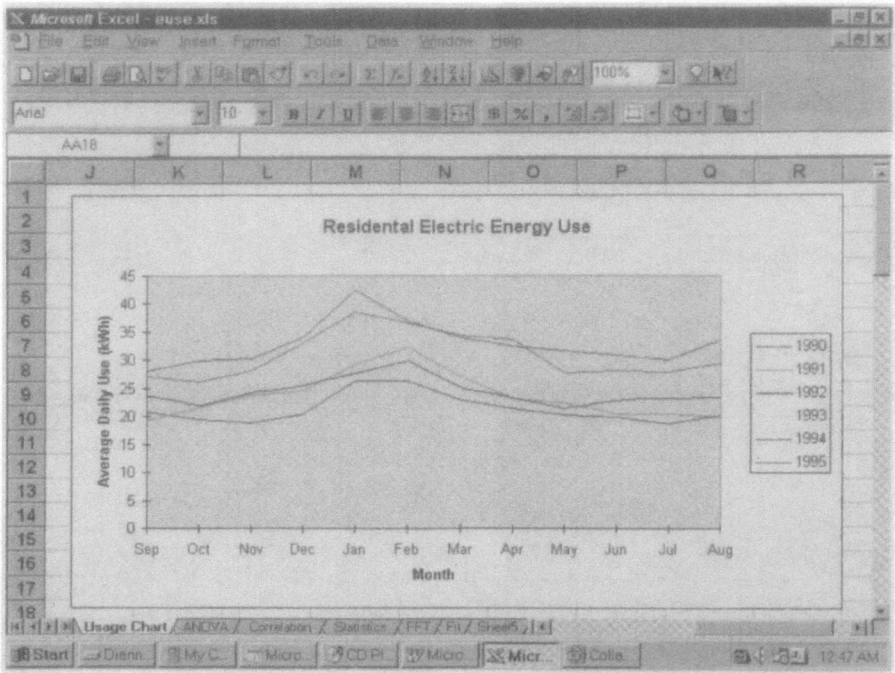

Fig. 7. The completed chart of the residential electric power usage

cated statistical calculations are simply applied to the data and the result print-ed out, no matter if they are appropriate or not.

To calculate the descriptive statistics (assuming the data analysis toolpack is already installed), perform the following steps.

1. Open the data sheet with the data shown in Fig. 6.
2. Choose the Tools, Data Analysis command to display the Data Analysis dia-log box.
3. Choose the Descriptive Statistics option and click OK.
4. In the Descriptive Statistics dialog box that appears, click in the Input Range box, select the data in the table plus the top row containing the column labels (B4:G16), check the Labels In First Row check box, click all the statistics options to include, click the Output Range option button, click in the Output Range box and choose cell B4 on the next worksheet (named Statistics here). (Fig. 8 shows the filled out dialog box.)
5. Click OK to calculate the correlation coefficients.

When the calculation is done, the table shown in Fig. 9 appears on the worksheet. Note that a separate table is calculated for each year. If a calculation can not be performed, NA appears in the cell, such as for the Mode calculation, where the data must be ordered first before the mode can be calculated.

Fig. 8. Fill in the Descriptive Statistics dialog box with the input and output ranges

	B	C	D	E	F	G	H	I
4	1990		1991		1992		1993	
5								
6	Mean	21.325	Mean	23.675	Mean	24.38333	Mean	27.5833
7	Standard Error	0.754795	Standard E	1.158965	Standard E	0.690721	Standard E	1.19872
8	Median	20.35	Median	22.85	Median	23.45	Median	27.1
9	Mode	#N/A	Mode	#N/A	Mode	23.4	Mode	#N/A
10	Standard Deviation	2.614688	Standard D	4.014774	Standard D	2.392729	Standard D	4.15250
11	Sample Variance	6.836591	Sample Va	16.11841	Sample Va	5.725152	Sample Va	17.2433
12	Kurtosis	0.643492	Kurtosis	0.339481	Kurtosis	1.547964	Kurtosis	-0.5196
13	Skewness	1.298704	Skewness	1.030644	Skewness	1.336231	Skewness	0.45891
14	Range	7.7	Range	13	Range	8.3	Range	1
15	Minimum	18.7	Minimum	19.2	Minimum	21.6	Minimum	21
16	Maximum	26.4	Maximum	32.2	Maximum	29.9	Maximum	35
17	Sum	255.9	Sum	284.1	Sum	292.6	Sum	33
18	Count	12	Count	12	Count	12	Count	1
19	Largest(1)	26.4	Largest(1)	32.2	Largest(1)	29.9	Largest(1)	35
20	Smallest(1)	18.7	Smallest(1	19.2	Smallest(1	21.6	Smallest(1	21
21	Confidence Level(95.0%)	1.661294	Confidenc	2.550867	Confidenc	1.520268	Confidenc	2.6383
22								

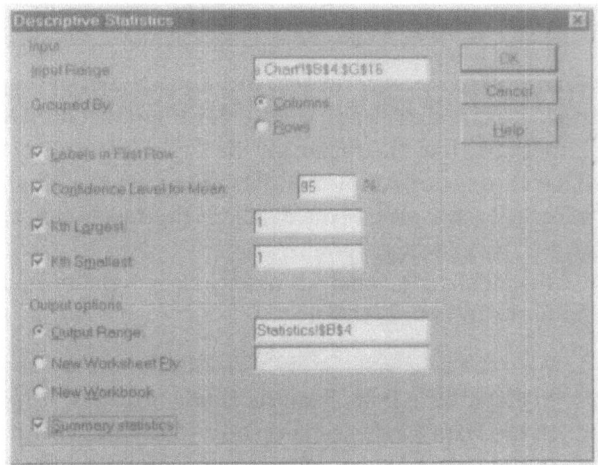

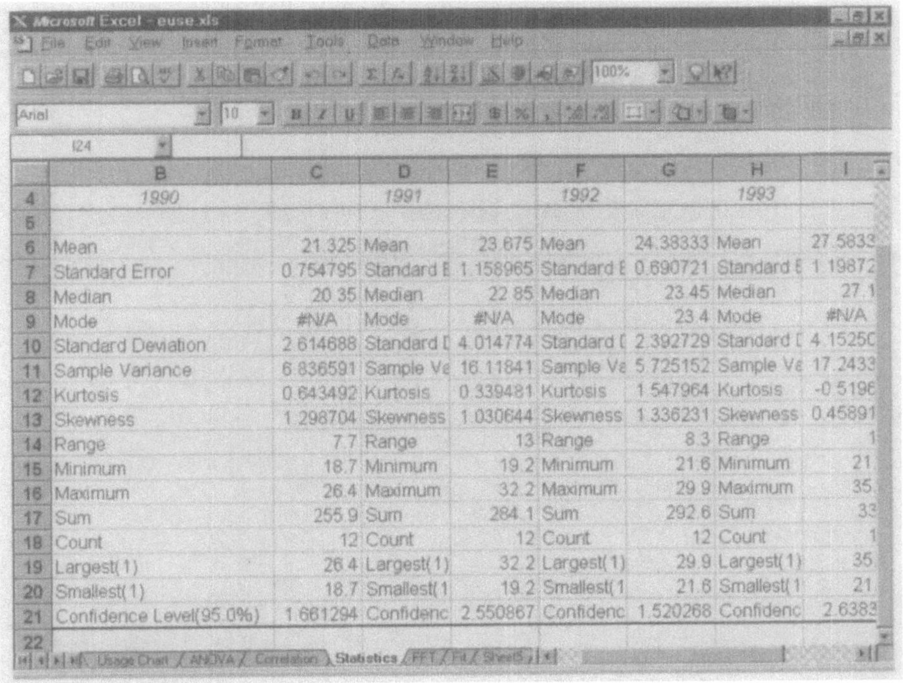

Fig. 9. The Descriptive Statistics calculated for the residential electric power usage data

Residential Electric Power Usage – ANOVA

Figure 10 shows a worksheet with the ANOVA package applied to the residential power usage data. Here, we use the Single-Factor ANOVA calculation to test the hypothesis that the means of the five yearly sets of data are statistically equal. That is they are samples of data drawn from a single pool of data with a single mean. If, as we expect, the data fails this test the means are not equal and are not drawn from a single pool.

To calculate the analysis of variance on the residential power usage data, perform the following steps.

1. With the data sheet in Fig. 6 visible, choose the Tools, Data Analysis command to display the Data Analysis dialog box.
2. In the Data Analysis dialog box, select Single Factor ANOVA and click OK.
3. In the Single Factor ANOVA dialog box that appears, click in the Input Range box, select the data in the table plus the top row containing the column labels (B4:G16), check the Labels In First Row check box, click the Output Range option button, click in the Output Range box and choose cell B4 on the next worksheet (named ANOVA). Fig. 11 shows the filled out dialog box.
4. Click OK to complete the calculation.

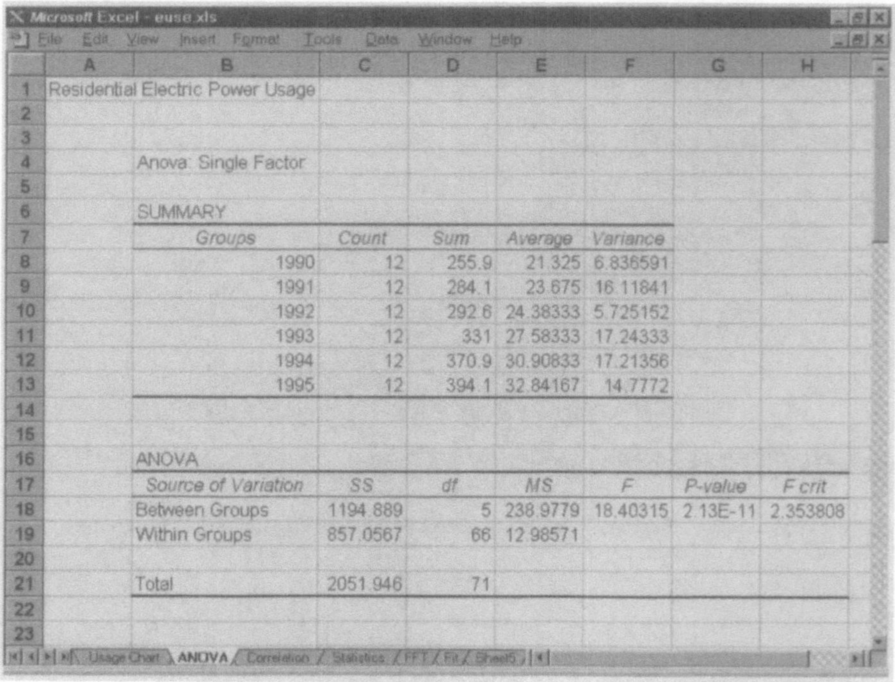

Residential Electric Power Usage

Anova: Single Factor

SUMMARY

Groups	Count	Sum	Average	Variance
1990	12	255.9	21.325	6.836591
1991	12	284.1	23.675	16.11841
1992	12	292.6	24.38333	5.725152
1993	12	331	27.58333	17.24333
1994	12	370.9	30.90833	17.21356
1995	12	394.1	32.84167	14.7772

ANOVA

Source of Variation	SS	df	MS	F	P-value	F crit
Between Groups	1194.889	5	238.9779	18.40315	2.13E-11	2.353808
Within Groups	857.0567	66	12.98571			
Total	2051.946	71				

Fig. 10. ANOVA statistical analysis of the residential electric power usage data

Fig. 11. Fill in the Single Factor ANOVA dialog box with the input and output ranges

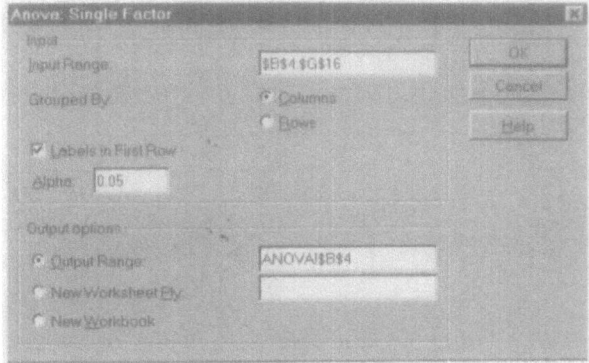

When you click the OK button, the ANOVA shown in Fig. 10 is inserted in the output range. If you examine the summary table, you see that the group averages are definitely increasing, but the group variances are changing wildly. The results of the ANOVA test are in the ANOVA table. The important values to examine are F and F_{crit}. F is the cumulative F statistic calculated from the data and F_{crit} is the critical F value from a table of F statistics. For the hypothesis to be true (all the group averages come from the same population) the value of F must be less than F_{crit}. As this is definitely not true, the test fails and the change in the group means is statistically significant.

To learn more about the mathematics behind the ANOVA and the other statistics given in the table, see, for example Statistical Methods for Scientists and Engineers by Bethea Duran, and Boullion (Marcel Dekker, New York, 1975).

Residential Electric Power Usage – Correlation

Another useful statistical calculation to be performed on the residential electric power usage data is to do a correlation analysis from one year to the next. You would expect the correlation to be high as the relative power usage tends to be seasonal.

To calculate the correlation between the years, perform the following steps.

1. Open the data sheet with the data shown in Fig. 6.
2. Choose the **Tools, Data Analysis** command to display the Data Analysis dialog box again and choose the Correlation option. Click OK.
3. In the Correlation dialog box that appears, click in the Input Range box, select the data in the table plus the top row containing the column labels (B4:G16), check the Labels In First Row check box, click the Output Range option button, click in the Output Range box and choose cell B4 on the next worksheet (named Correlation here). Figure 12 shows the filled out dialog box.
4. Click OK to calculate the correlation coefficients.

Fig. 12. Fill in the Correlation dialog box with the input and output ranges

Residential Electric Power Usage
Correlation

	1990	1991	1992	1993	1994	1995
1990	1					
1991	0.86461	1				
1992	0.865391	0.892457	1			
1993	0.789776	0.924366	0.781622	1		
1994	0.882661	0.859167	0.840036	0.89388	1	
1995	0.848094	0.788455	0.765275	0.86274	0.901091	1

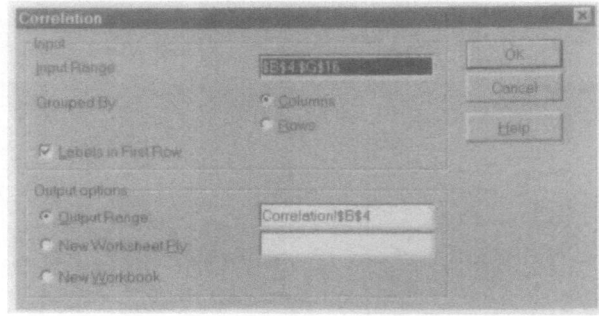

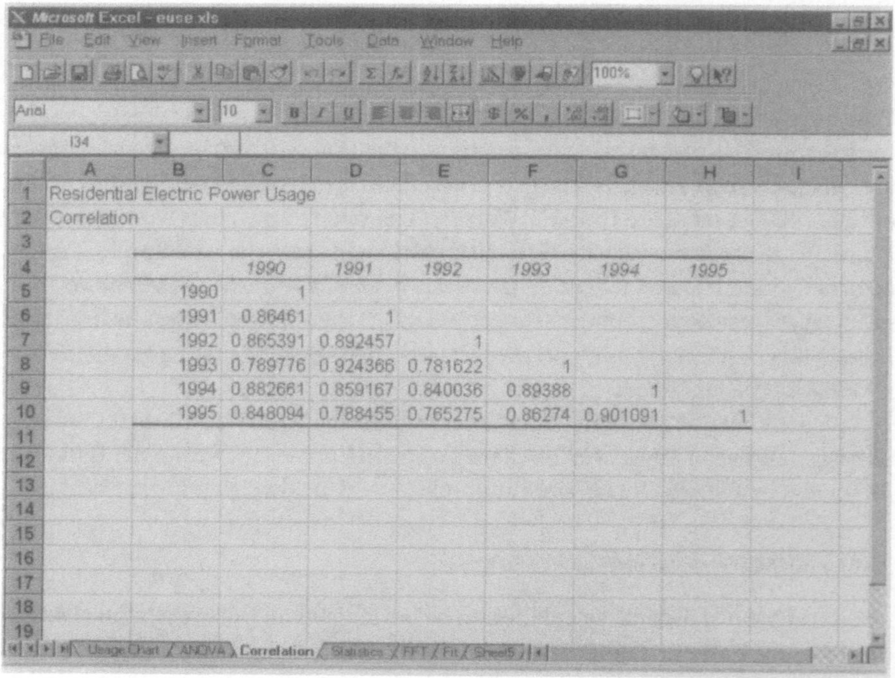

Fig. 13. Table of correlations between the data sets

When the calculation is complete, the table shown in Fig. 13 appears on the worksheet. The table lists all the datasets across the top and down the first column. The body of the table contains the correlation coefficient for each dataset with each other dataset. Only half the table is filled in as the correlation coefficients are symmetric. That is, the correlation of group 1 on group 2 is the same as group 2 on group 1. The correlation coefficients are calculated such that a coefficient value of 1.0 indicates that two curves have exactly the same shape. The values in the table are all between 0.75 and 1.0, indicating a high degree of correlation between years.

Residential Electric Power Usage – FFT

The Fast Fourier Transform (FFT) is an analysis technique that identifies cyclical patterns in experimental data by fitting the data with a series of sine functions. When the transform is applied, it calculates the coefficients and phase of the sine functions. The size of these coefficients indicates the strength of the frequency associated with that term.

To apply the FFT to the data, it must be combined into a single data set instead of six linked sets. This is done by simply copying the data from adjacent columns to the bottom of the second column (the first column contains the month). The FFT package from the Data Analysis Toolpack can then be applied to the data.

To calculate the FFT of the residential power usage data, perform these steps.

1. On a blank worksheet, type the following values in the indicated cells.
   - A1 **Residential Electric Power Usage**
   - B4 **Month**
   - C4 **Daily Average Usage (kWh/day)**
   - E4 **Fourier Coefficients**
2. Open the worksheet shown in Fig. 6, select B5:B16, copy the selection, switch to the FFT worksheet select C5 and click paste.
3. Do the same for all the other columns in the table, copying each to the bottom of the current list in column C.
4. In cell B5, type **Sep-90**.
5. In cell B6, type **Oct-90**.
6. Select cells B5:B6 and drag the fill handle down to cell B76, filling the column with the dates.
7. In cell D5, type **0**.
8. In cell D6, type **1**.
9. Select D5:D6 and drag the fill handle down to cell D68 to fill the column with the integers from 0 to 63.
10. Choose the Tools, Data Analysis command, select Fourier Analysis in the dialog box and click OK.
11. In the Fourier Transform dialog box that appears, click in the Input Range box, select the data in column C (C5:C68). Note that we have not chosen all of the data, but only the first 64 elements. The FFT algorithm can only be applied to a list that contains a power of 2 elements. Click the Output Range option button, click in the Output Range box and choose cell E5. Fig. 14 shows the filled out dialog box.
12. Click OK to calculate the FFT.
13. Add outlines to the table as shown in Fig. 15.

The results of the calculation are shown in Fig. 15. The output of the FFT is a series of complex numbers that contain the magnitude and phase of the particular term. The location of the term in the list indicates its frequency. To make these more useful, we need to calculate the actual magnitudes of the terms and

Fig. 14. Fill in the Fourier
Transform dialog box with
the input and output ranges

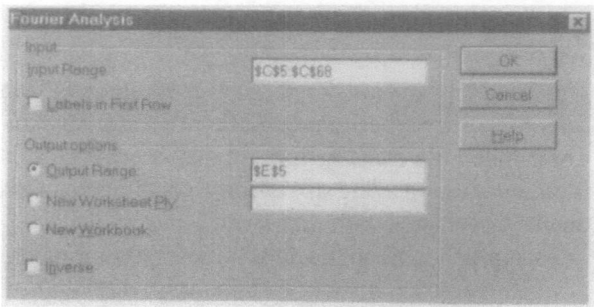

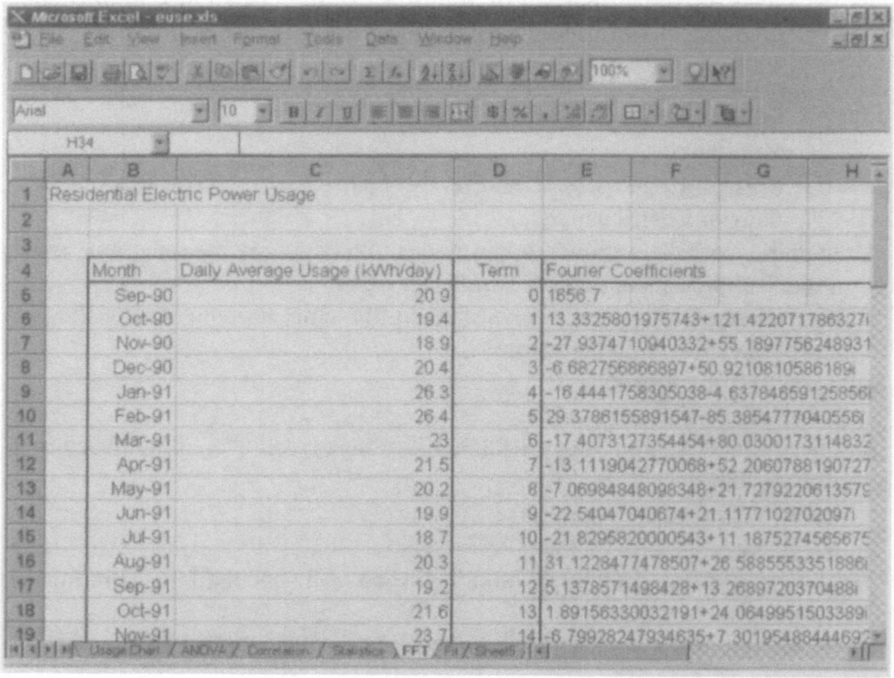

Fig. 15. The Fast Fourier Transform of the residential electric power usage data

their frequency. The magnitude of a complex number is normally calculated by
taking its square and then taking the square root. As the complex number is stored
in a string, this could be a problem, but the appropriate calculation is available
as a function, making the calculation simple to set-up.

To convert the FFT results to magnitude and frequency, perform these steps.

1. Type the following values in the indicated cells.
   - I2 **n=**
   - J2 **64**
   - I4 **Magnitude**

- J4 **Frequency**
- K4 **Period**

2. In cell I5, type **=IMABS(E5)**
3. Select copy the formula in cell I5 down into cells I6:I68 by selecting cell I5 and dragging the fill handle down to cell I68.
4. In cell J5, type **0.**
5. In cell J6, type **=D6/$J$2**
6. Copy the formula in cell J6 down into cells J7:J68.
7. In cell K5, type **Infinite**
8. In cell K6, type **=1/J6**
9. Choose the Format, Cells command, Number tab and set the format to Number with 1 decimal place.
10. Copy the formula in cell K6 down into cells K7:K68.
11. Outline the cells as shown in Fig. 16

The IMABS() function used in cell I5 automatically calculates the absolute value (real magnitude) of an imaginary number stored as text. The frequency of the sine wave is found by dividing the term number by the number of terms, which is from the definition of the FFT. The period is then just the inverse of the frequency. Figure 17 shows a graph of the magnitudes of the terms versus the period. As you can

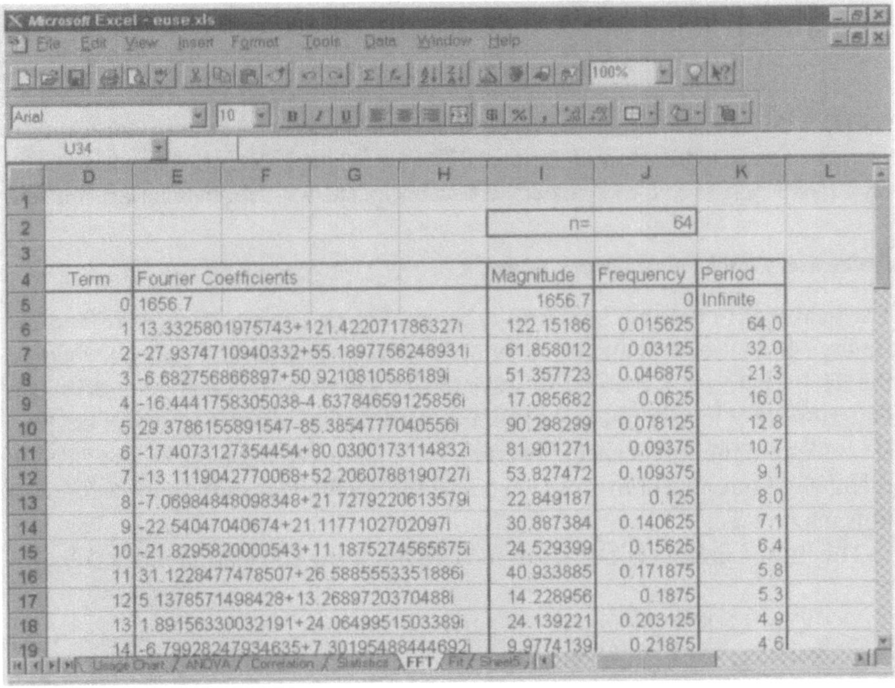

Term	Fourier Coefficients	Magnitude	Frequency	Period
0	1656.7	1656.7	0	Infinite
1	13.3325801975743+121.422071786327i	122.15186	0.015625	64.0
2	-27.9374710940332+55.1897756248931i	61.858012	0.03125	32.0
3	-6.682756866897+50.9210810586189i	51.357723	0.046875	21.3
4	-16.4441758305038-4.63784659125856i	17.085682	0.0625	16.0
5	29.3786155891547-85.3854777040556i	90.298299	0.078125	12.8
6	-17.4073127354454+80.0300173114832i	81.901271	0.09375	10.7
7	-13.1119042770068+52.2060788190727i	53.827472	0.109375	9.1
8	-7.06984848098348+21.7279220613579i	22.849187	0.125	8.0
9	-22.54047040674+21.1177102702097i	30.887384	0.140625	7.1
10	-21.8295820000543+11.1875274565675i	24.529399	0.15625	6.4
11	31.122847747850 7+26.5885553351886i	40.933885	0.171875	5.8
12	5.137857149842 8+13.2689720370488i	14.228956	0.1875	5.3
13	1.89156330032191+24.0649951503389i	24.139221	0.203125	4.9
14	-6.7992824793463 5+7.30195488444692i	9.9774139	0.21875	4.6

Fig. 16. The FFT data converted to magnitudes and frequencies

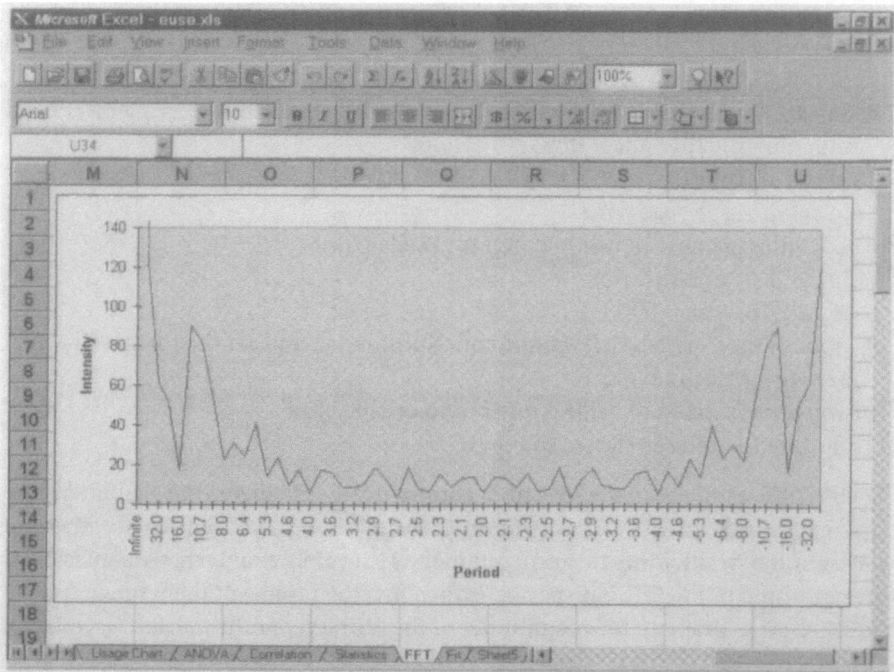

Fig. 17. The magnitude of the FFT terms versus the period

see from the data, the largest terms occur around a year (12.8 months) with another peak around 6 months (5.8 months). The actual periods do not appear exactly on a 1 year period or 6 month period because of the way frequencies are chosen.

Residential Electric Power Usage – Curve Fit

The next step in data analysis it to fit the data to an analytical curve to find the coefficients of the curve and to gain more understanding about the Physics behind the data. There are a number of ways to fit a curve to data using a worksheet. Functions to calculate linear curves, and power series using regression analysis are built-into the spreadsheet program. To fit a more general curve, you need to calculate a goodness of fit and then adjust the coefficients until that goodness of fit is maximized.

The most common goodness of fit criterion is the correlation index (r^2),

$$r^2 = 1 - \frac{\sum_{i=1}^{n}\left(y_i - y(x_i)\right)^2}{\sum_{i=1}^{n}\left(y_i - \langle y_i \rangle\right)^2}$$

here, x_i and y_i are the data pairs, $y(x_i)$ is the model being fit to the data, and $<y_i>$ is the average of the y_i data values. The better a curve fits the data, the closer r^2 is to 1. Values of r^2 of 0.9 or larger indicate very good fits to the data.

The built-in curve fitting functions automatically calculate r^2 when a curve is fit, but you must calculate it yourself in a general curve fitting worksheet. We expect the residential electric power usage data will have a sine wave like shape with a one year period, so a first attempt at a fit is to use a formula like the following.

$$\text{Energy Usage} = A*\text{SIN}(B*t + C) + D$$

here, t is the date, A, B, C, and D are coefficients to be determined, such that A is the magnitude of the sine wave, B is the frequency, C is the phase and D is the magnitude of the constant part of the data.

To fit the residential electric power usage data with a sine wave, perform the following steps.

1. On a new worksheet, type the labels shown in Fig. 18.
2. Name cells B3:B6 as **A, B, C_**, and **D**. B5 must be named C_ instead of C because C is reserved in Excel.
3. Switch to the FFT worksheet and copy the data in cells B5:C76.
4. Switch back to the curve fit worksheet, select cell A10 and paste the residential electric power usage data.
5. In cell C10 type **1** and in C11 type **2**. Select C10:C11 and drag the fill handle down to cell C81 to insert the integers from 1 to 72.
6. Name cells C10:C81 as **t**.
7. In cell D10, type the model equation **=A*SIN(B*t+C_)+D**
8. In cell E10, type the equation for the square of the residual (the difference between the data and the model) **=(D10-B10)^2**
9. In cell F10, type the equation for the square of the deviation of the data about the average (the average will be in cell E4) **=(B10-$E$4)^2**
10. Select cells D10:E10, and drag the fill handle down to E76 to apply the equations to all the data.
11. In cell E82, type a formula to calculate the sum of the squared residual **=SUM(E10:E81)**
12. In cell F82, type a formula to calculate the sum of the squared deviation of the data about the average **=SUM(F10:F81)**
13. In cell E3, type a formula to bring the sum of the squared residual to the top of the page **=E82**
14. In cell E4, type a formula to calculate the average of the data **=AVERAGE (B10:B81)**
15. In cell E5, type a formula to copy the sum of the squared deviation of the data about the average to the top of the page **=F82**
16. In cell E6, type a formula to calculate the correlation index **=1-E3/E5**
17. Add boxes about the data and adjust the column widths and heights as shown in Fig. 18

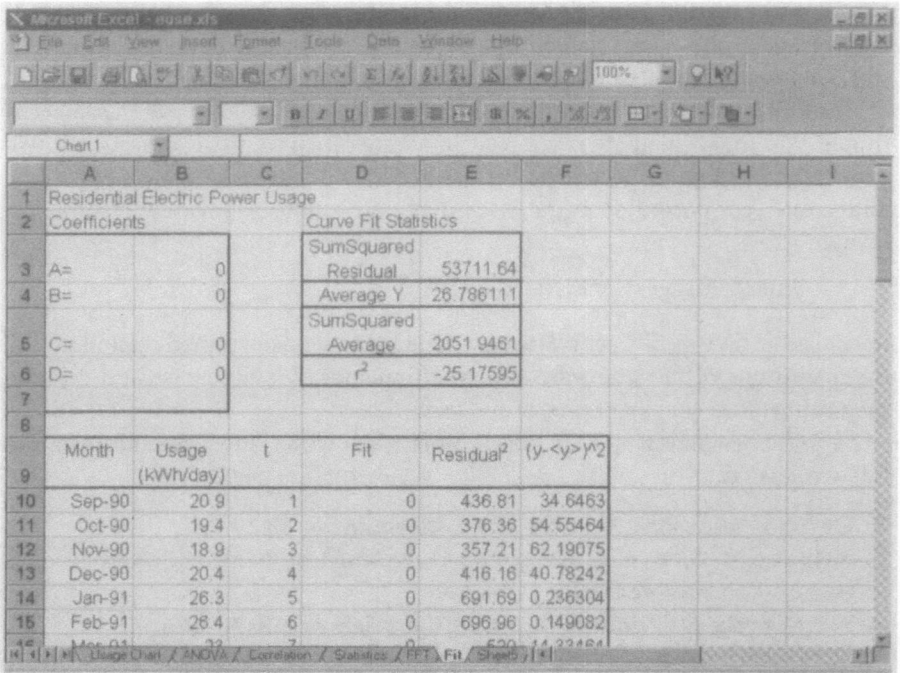

Fig. 18. The worksheet to fit a sine curve to the residential electric power usage data

The worksheet now looks like Fig. 18. All that is needed are the values for A, B, C, and D that make the curve best fit the data. You could adjust the values by hand until you find the right values, but Excel has another add-in program known as the Solver that is designed to do this kind of adjustment.

To find the best fitting values, perform the following steps.

1. It turns out that there are several possible solutions to this problem that depend on the starting point of the Solver. These other solutions are local maxima of r^2, that make the Solver think it has found the true solution. To get the solution you want, you need to set the initial solution somewhere close to the final solution. We believe the period to be one year, so set the initial value of B (in cell B4) to 0.5. Set the other initial values in B3, B5 and B6 to 1. If you don't know where the true solution is, you need to try different starting points to see if they all result in the same solution. If not, you need to search around to find the true solution.

2. Choose the Tools, Solver command to display the Solver dialog box. In the dialog box, set the target cell to $E$6, the Equal To option to Max, and the By Changing Cells box to $B$3:$B$6. This tells the Solver to adjust cells B3:B6 until E6 is maximized. The filled out Solver is shown in Fig. 19.

Fig. 19. The Solver dialog box filled out to fit the sine curve to the residential electric power usage data

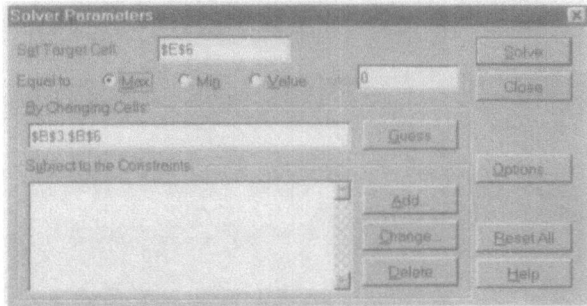

Fig. 20. The worksheet after the Solver has found a solution

3. Click OK to calculate the solution. In a moment, the Solver displays a dialog box indicating that it has found a solution and asks you if you want to keep it. Click OK and the solution is stored in Cells B3:B6 as shown in Fig. 20.

If you examine the value of r^2, it appears that the sine curve does not fit very well. Plotting the data and the fit on the same chart (Fig. 21) shows that the Solver did find a good solution for that model, but that the model does not fit the data. In addition to the Sine wave, a growth function must be added to show the growth of the average from year to year. To handle this, change the model formula to the following, Fig. 21.

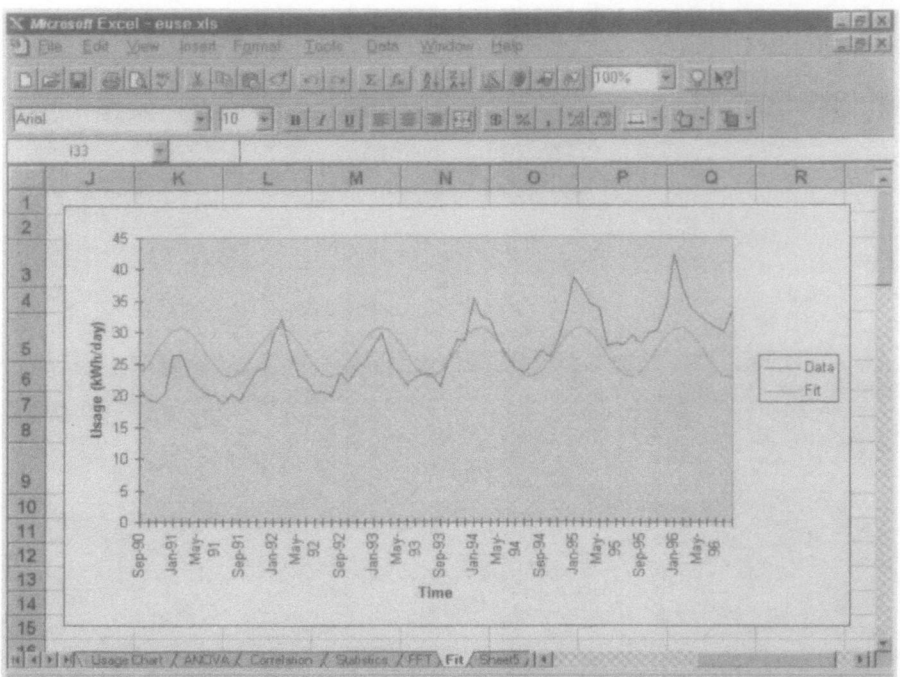

Fig. 21. A graph of the curve fit and the data it was fit to

The model used is lacking a growth term such as:

$$\text{Energy Usage} = A*SIN(B*t + C) + D + E*t$$

To apply the new formula, perform the following steps.

1. In cell A7, type **E=**
2. Name cell B7 as **E**
3. Edit the formula in cell D10 to **=A*SIN(B*t+C_)+D+E*t**
4. Copy the formula in cell D10 into cells D11:D81.

The worksheet is now ready to be used. Run the Solver again, initializing the values as before to find the solution. Add B7 to the list of cells to adjust and the worksheet in Fig. 22 appears, with a chart of the data shown in Fig. 23 (s. p. 66). Note that the value of r^2 is now very close to 1 and that the model does a relatively good job of fitting the general trend of the data.

Solving Equations

After working with experimental data, probably one of the most often needed numerical capability of a Physicist or Electronics Engineer is that of solving one or more equations to find specific roots. Excel provides three main methods for

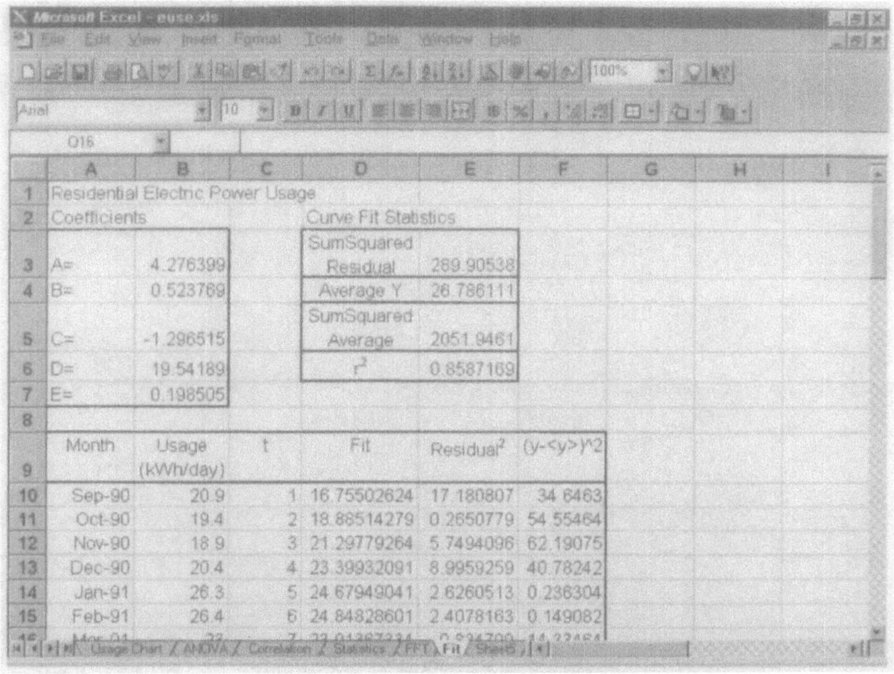

Fig. 22. The worksheet after adding a growth term to the model shows a reasonably good fit to the data

equation solving: the Solver, described in the previous section and in addition, matrix and iteration operations. Matrix solutions are well suited for solving systems of linear equations while nonlinear equations can usually be solved with iteration techniques.

Using the Solver to find Solutions

The Solver is an equation solver that uses the method of steepest descents to find the solution to a system of equations. As you have seen, to use the Solver you must have a cell containing a goal. In addition, that goal must be sensitive to the solution so that its value can be used to guide the Solver to the correct solution. The Solver works by changing the input cells and examining the change in the goal. It then chooses those changes in the input cells that cause the goal cell to move most quickly towards the goal value.

A difficulty with using the Solver to solve systems of equations is that there must be a single goal. The problem with this arises in systems where you have several equations that must all simultaneously be equal to zero. Now you can add them up and use the sum as the goal, but this allows solutions where one equation has a large positive value and another has a large negative one. The sum may

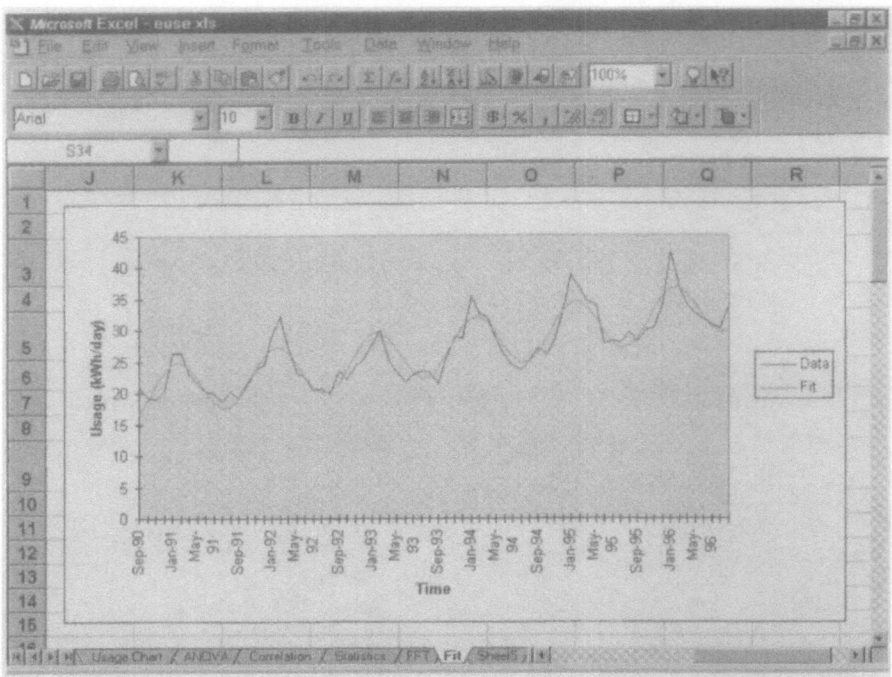

Fig. 23. A graph of the curve fit using the model with the sine wave and a growth term

go to zero but the individual equations do not. You could fix this by adding the absolute values, but then the goal cell is no longer completely responsive to changes in the inputs, which may make it impossible for the Solver to find a solution. For systems of equations, it is usually better to use an iterative technique.

Planck Distribution for Blackbody Radiation – Peak Finding

The Planck distribution function calculated previously has peaks whose position depends on the temperature of the blackbody. A useful calculation is to locate the top of a peak. The worksheet in Fig. 24 does just that. The Solver is used to find the peaks for each function by setting it to find the arguments to the function that result in the largest value of the function.

To create the worksheet, perform the following steps.

1. On a new worksheet, type the following values in the indicated cells.
   - E2 **6.63E-34**
   - E3 **3.00E+10**
   - E4 **1.38E-23**
   - B6 **500**
   - C6 **1000**

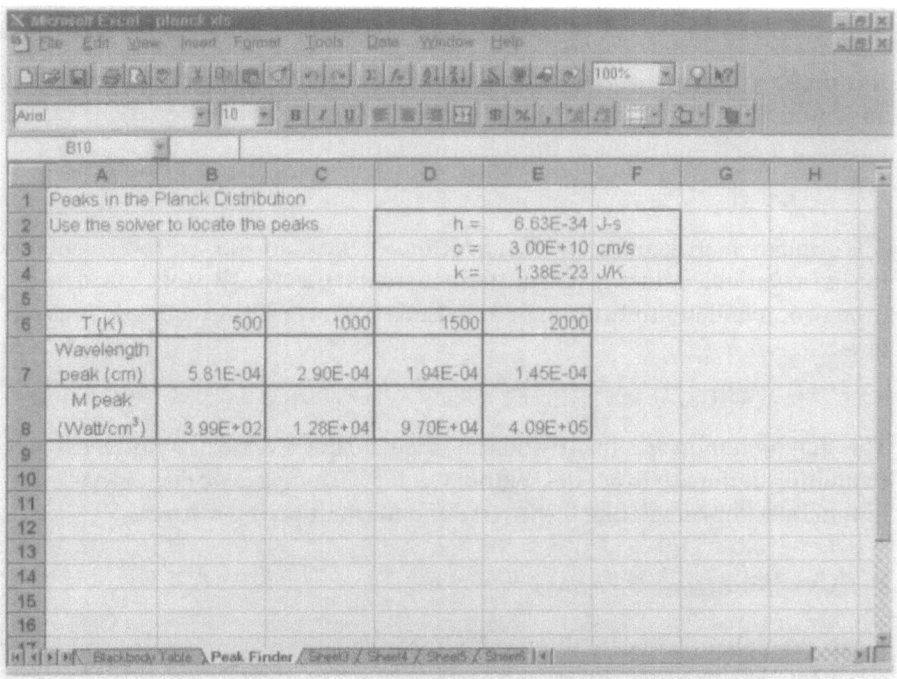

Fig. 24. Finding the peaks in the Planck distribution using the Solver

  - D6 **1500**
  - E6 **2000**
2. Type labels shown in Fig. 24.
3. Name cells E2, E3, and E4 as **h**, **c_**, and **k**.
4. Outline the cells and set the cell heights and widths according to Fig. 26.
5. In cell B8, type the formula for the Planck distribution function
   =(2*PI()*h*c_^2/(B7^5))*1/(EXP(h*c_/(B7*k*B6))-1)
6. Copy this formula into cells C8:E8.
7. Place **1E-4** in cells B7:E7.
8. Open the Solver and set **B8** as the target cell, **B7** in the By Changing Cell field, click the Max option button and click OK.

This setup sets the Solver to adjust the value in cell B7 until the value in cell B8 is maximized, which is the peak of the distribution function for the particular temperature. The Solver must be run again, once for each of the other columns in the table to find the maximum of the distribution function for each of the temperatures. To run it again for the other columns, change the cell reference in the Target and By Changing Cell fields of the Solver to the references appropriate to the column of the table you want to calculate.

Solving Linear Equations with a Matrix

Any problem that can be posed as a matrix can be solved by inverting the matrix and multiplying through by that inversion. Most linear sets of equations can be posed as matrix equations in the following form.

$$Ax = b$$

A is a square matrix and x and b are vectors. A and b are known and x contains the list of variable values that make the equations true. To solve this, calculate the inverse of A and multiply through from the left. The resulting equation is as follows.

$$x = A^{-1}b$$

This equation indicates that if you can calculate the inverse of A you have only to multiply it times b to get the solution vector x. The Excel worksheet has built-in functions for calculating the inverse and for multiplying matrices.

Matrix Solution of Linear Equations

As an example of a matrix solution of linear equations, consider the following set of coupled, linear equations:

$$2x + 5y + 14.6z = 353.65$$

$$17x + 6.7y + 3z = 764.55$$

$$9.5x + 14y + 22.8z = 1038.85$$

These equations can be posed in the following matrix form:

$$\begin{vmatrix} 2 & 5 & 14.6 \\ 17 & 6.7 & 3 \\ 9.5 & 14 & 22.8 \end{vmatrix} \begin{vmatrix} x \\ y \\ z \end{vmatrix} = \begin{vmatrix} 353.65 \\ 764.55 \\ 1038.85 \end{vmatrix}$$

To solve this system of equations on the worksheet, perform the following steps.

1. On a new worksheet, type the labels and values of the matrix (B8:D10) and result vector (F8:F10) as shown in Fig. 25.
2. In cell D11, type =**MDETERM(B8:D10)**
   This cell contains the determinant for the matrix. If the determinant is 0, the matrix is singular and does not have a solution.
3. Select cells B13:D15 and type the following into the first cell =**MINVERSE (B8:D10)**
4. Hold down Control and Shift when clicking on the check box to create a matrix equation inserted in the whole selected range. The formula should appear in the cells surrounded with curly brackets.

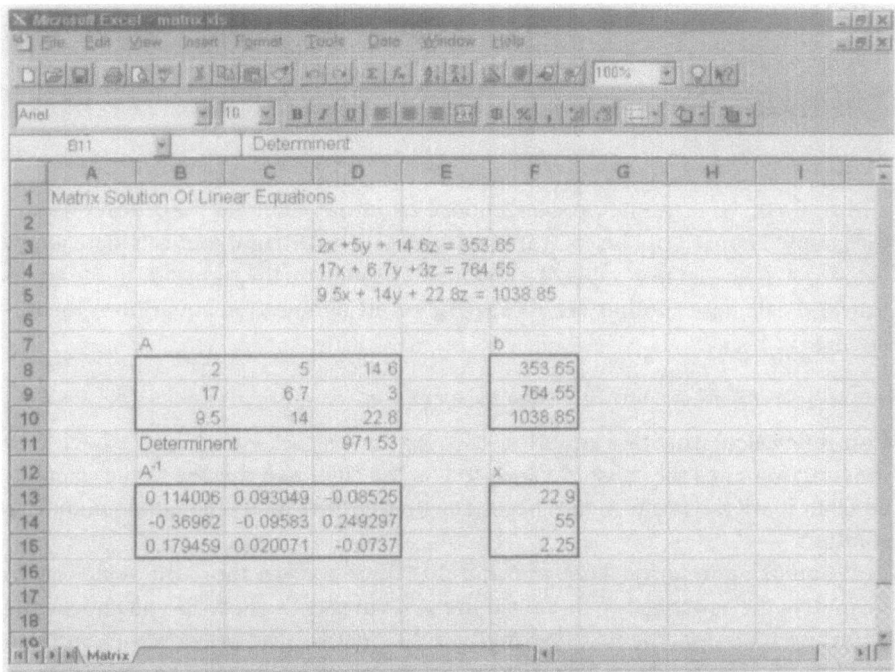

Fig. 25. Matrix solution of three coupled linear equations

5. Select cells F13:F15, and type the following into the first cell =MMULT (B13:D15,F8:F10)
6. Hold down Control and Shift when clicking the check box to insert it as a matrix equation into the whole selection.

This completes the worksheet. As soon as the last equation is inserted, the worksheet recalculates and gives the results as shown in Fig. 25. This technique can easily be expanded to many equations. Experience has shown that the built-in matrix solvers are good up to about ninety equations. After that, the accuracy degrades and a different solver must be created using Visual Basic.

You can practice using other constants or equations with the disk file Matrix.XLS.

Solving Linear and Nonlinear Equations with Iteration and Relaxation

When the Solver and matrix methods fail, you are left with the iterative methods. Luckily, the iterative methods can solve both linear and nonlinear equations, including transcendental equations such as $x = \sin(x)$. To solve equations with iteration, the equations need to be posed in the form,

$$x = f(x)$$

or for coupled equations,

$$x = f(x,y,z)$$
$$y = f(x,y,z)$$
$$z = f(x,y,z)$$

and so forth. To start the process, choose an initial value for x, (or for $x, y, z, \ldots$ for coupled equations) insert it in the function on the right and calculate a new value of x. Take that new value of x and insert it again on the right side of the equation and calculate another value of x. The resulting iteration equation is the following.

$$x_{i+1} = f(x_i)$$

Continue calculating this equation, copying the calculated value of x back into the function until the value of x inserted on the right and the new value calculated from it are the same. At this point, the method has converged and you have a solution.

When using iteration, keep in mind that there is often more than one way to formulate the equations to be solved in the form shown above. If you choose the wrong formulation, the iteration method diverges instead of converges and you will not find the solution. The rule for convergence is that the absolute value of the derivative of the equation is less than 1.

$$|f(x)'| < 1$$

You could calculate this for every potential solution method, but it is generally easier to simply try a solution and if it diverges, reformulate the problem to use a different one.

In addition to iterating to find the solution, you often add over or under relaxation to the solution method to control the amount of change in the solution from one iteration to the next. Relaxation is needed when the equations being iterated converge either too fast or too slow. If the iterations converge too fast, they tend to overshoot the solution and start oscillating about it. Relaxation in a problem of this type reduces the amount of change each iteration applies to the problem and stops the oscillations. If the iterations converge too slow, it just takes a long time to reach the solution and relaxation can speed it up.

To apply relaxation to an iteration problem, calculate the amount of change an iterate applies to the solution and reduce (or increase) it by some fraction. The resulting iteration equation is as follows.

$$x_{i+1} = c(f(x_i) - x_i) + x_i$$

Here, c is the relaxation factor being applied to the method. If c is one, this reduces to a standard iteration method. If c is less than one, this is an under relaxation method. If c is greater than one, this is an over relaxation method. The actual value of c used in a problem is rarely less than 0.5 or greater than 2.

Iterative Solution of Coupled Equations

To demonstrate the iterative solution method, solve the same set of three coupled equations as were solved with the matrix method.

$$2x + 5y + 14.6z = 353.65$$

$$17x + 6.7y + 3z = 764.55$$

$$9.5x + 14y + 22.8z = 1038.85$$

The first step of the solution is to reformulate the equations one for each of the three variables. To maintain the accuracy, you generally try to solve for the term with the largest coefficient.

$$z = (353.65 - 2x - 5y)/14.6$$

$$x = (764.55 - 6.7y - 3z)/17$$

$$y = (1038.85 - 9.5x - 22.8z)/14$$

I originally solved the third equation for z instead of the first, but the problem diverged, so I reformulated the problem to solve the first equation for z, which worked.

To solve these three equations on the worksheet using simple iteration, place the three formulas above in three worksheet cells, with each formula referencing the other two. This creates a circular reference which the worksheet program complains about. Choose the Tools, Options command, Calculation tab and click the manual calculation option and check the iteration check box. Set the number of iterations to 1. Whenever you press F9 or Control - =, the worksheet is recalculated once. When the worksheet converges, it looks like that in Fig. 26.

To create this worksheet, perform the following steps.

1. Choose the Tools, Options command, Calculation tab and set the problem to Manual calculation with an iteration value of 1.
2. Type the following values on a new worksheet in the indicated cells and type the labels shown in Fig. 26.
   - B4 **FALSE**
   - B12 **1.5**
3. Name cell B4 as **reset** and B12 as **relax**.
4. In cell B7, type =IF(**reset**,1,(353.65–2*B8–5*B9)/14.6)
5. In cell B8, type =IF(**reset**,1,(764.55–6.7*B9–3*B7)/17)
6. In cell B9, type =IF(**reset**,1,(1038.85–9.5*B8–22.8*B7)/14)
7. In cell B14, type =C14+**relax***((353.65–2*C15–5*C16)/14.6–C14)
8. In cell B15, type =C15+**relax***((764.55–6.7*C16–3*C14)/17–C15)
9. In cell B16, type =C16+**relax***((1038.85–9.5*C15–22.8*C14)/14–C16)
10. In cell C14, type =IF(**reset**,1,B14)
11. In cell C15, type =IF(**reset**,1,B15)
12. In cell C16, type =IF(**reset**,1,B16)

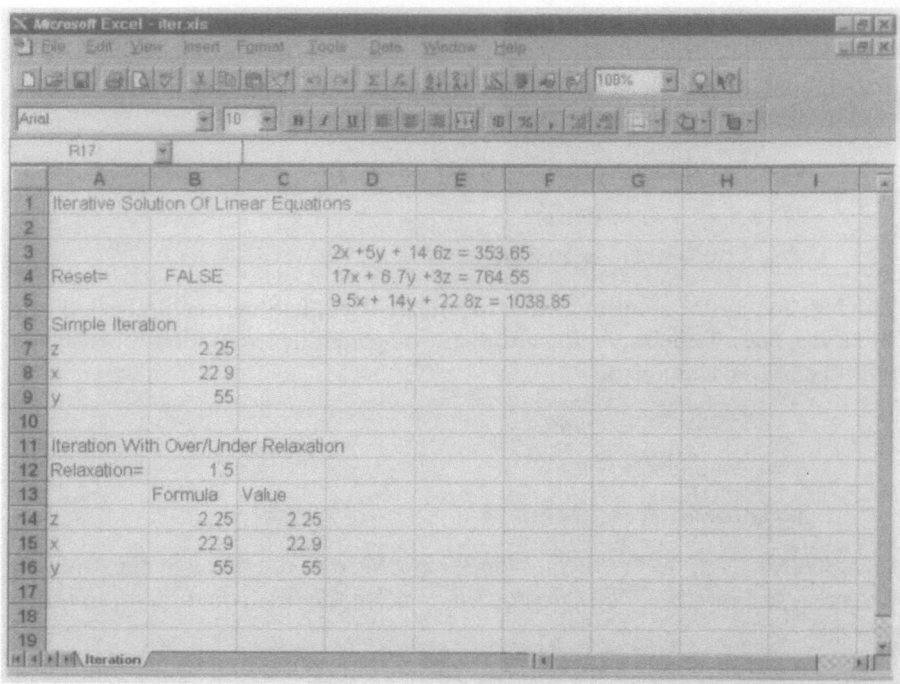

Fig. 26. Solving three coupled equations using iteration and iteration with relaxation

To use this worksheet, set the value of B12 (reset) to True and press F9 once to reset the worksheet, setting all the values to 1. Change B12 to False and press F9 to calculate the first iteration of the worksheet. Continue pressing F9 until the values converge.

The worksheet solves the problem two ways, with simple iteration at the top and with iteration and relaxation at the bottom. The iteration method places the three iteration formulas in IF statements. The IF statements are needed to reset the problem to a known starting point. They are also needed to clear any error values from the problem. If any equation develops an error the circular references in the problem cause all the equations to change to an error value. A reset capability is the only way to clear an error.

In this case, only three cells are needed to set-up iteration of the three equations. This is because the iteration formula for any variable only involves the other two. The more general problem where all the iteration formulas involve all the variables would require two cells for each formula. One to hold the formula and one to store the previous value. This second cell is needed because a formula in a cell may not refer to the current value of the cell that contains it.

At the bottom of the page is the iteration and relaxation setup (B14:C16). Here the relaxation factor (relax) is 1.5, so this is an over relaxation problem, but an under relaxation problem would be setup in exactly the same way. Two cells are needed for each equation being solved; one to hold the relaxation formula and one to store the value of a calculation for the next iteration. The IF statement that resets the problem is in the storage cells, but could just as easily have been on the equation cells as in simple iteration. You can glean more details by working through the example on the disk. The file name is Iter.XLS.

Finding Numerical Solutions to Differential Equations

As few differential equations that describe real systems can be solved analytically, numerical solutions are the only practical way to evaluate them. Spreadsheet programs are a reasonable platform for calculating numerical solutions to several ordinary and partial differential equations. Especially two-dimensional field problems where the field equations can be inserted in a single cell and the arrangement of the cells represents the physical arrangement of the actual problem. This physical arrangement is not required for a solution, but it makes the results much more intuitive. Most field problems can be solved using the iterative methods of the last section, as the field equations, written as finite differences, simply constitute a coupled set of equations. Other solution methods such as the Euler method are also amenable to a worksheet solution.

Poisson Distribution Between Two Slits

As a first example, consider an electron accelerator consisting of two charged metal plates with slits in them. Electrons passing through the slits are accelerated by the fields between the plates as shown in Fig. 27. If you need to know the fields within this device, you must solve the Poisson equation within the interior of the accelerator.

$$\nabla^2 \varphi = -\frac{q\rho}{\varepsilon}$$

Where φ is the potential, q is the electron charge, ρ is the charge density and ε is the dielectric constant. For problems with no internal charges, the Poisson equation reduces to the Laplace equation:

$$\Delta^2 \varphi = 0$$

Fig. 27. An electron accelerator consisting of two charged places with slits for the electrons

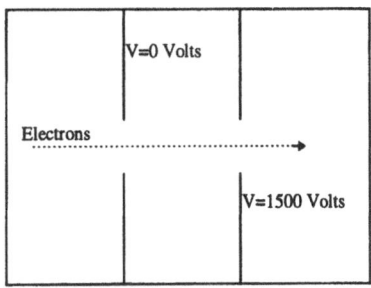

V=0 Volts

Electrons

V=1500 Volts

If the problem is an infinitely deep slit, we can describe this system with a two-dimensional Cartesian coordinate system. In this system, the Laplace equation becomes:

$$\frac{\partial^2 \varphi}{\partial x^2} + \frac{\partial^2 \varphi}{\partial y^2} = 0$$

We then discretized the problem by dividing the problem into a grid of locations. The equations are discretized by replacing the derivatives with central difference formulas.

$$\frac{\varphi_{1,j+1} - 2\varphi_{i,j} + \varphi_{i,j-1}}{h_x^2} + \frac{\varphi_{i+1,j} - 2\varphi_{i,j} + \varphi_{i-1,j}}{h_y^2} = 0$$

where i and j are grid indices in the x and y direction, and h_x and h_y are the grid spacings in the x and y directions. If we make the grid spacings the same in both directions and solve this equation for the potential at any grid point, the following equation results.

$$\varphi_{i,j} = \frac{1}{4}\left(\varphi_{i+1,j} + \varphi_{i-1,j} + \varphi_{i,j+1} + \varphi_{i,j-1}\right)$$

which shows that the potential at any grid point is just equal to the average of the potential at the surrounding four points. This equation is ideally suited for spreadsheet analysis. Figure 28 shows a worksheet created to solve this problem geometry.

To create this worksheet, perform the following steps. Or, if you prefer, try following its development on the disk file Poisson.XLS.

1. Choose the Tools, Options command, calculation tab and set the problem to Manual calculation with an iteration value of 1.
2. Type the following values in the indicated cells.
   - A1 **Laplace solution around two charged slits**
   - I4 **V1=**
   - J4 **0**
   - I5 **V2=**
   - J5 **1500**
   - A6 **y\x**
3. Name cells J4 and J5 as **V1_** and **V2_**.
4. In cells C6:O6, type the integers 1 through **13**.
5. In cells A8:A19, type the integers 1 through **12**.
6. In cell B8, type **=D8** and copy the cell down into cells B9:B19. This creates a derivative equals zero boundary condition at that side.
7. In cell C7, type **=C9** and copy the cell into cells D7:O7 creating another derivative equals zero boundary condition.

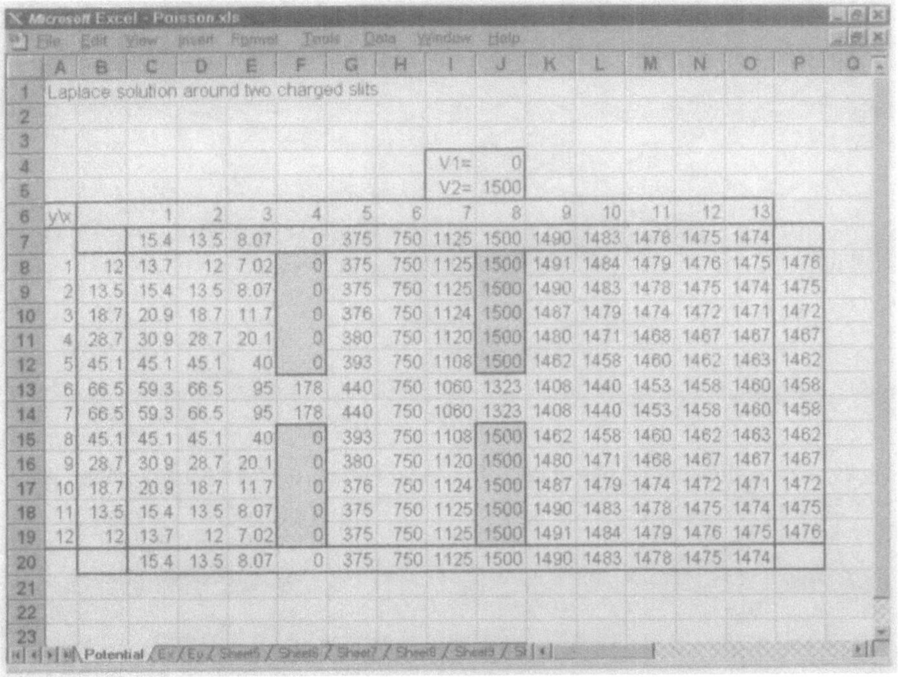

	A	B	C	D	E	F	G	H	I	J	K	L	M	N	O	P
1	Laplace solution around two charged slits															
2																
3																
4									V1=	0						
5									V2=	1500						
6	y\x		1	2	3	4	5	6	7	8	9	10	11	12	13	
7			15.4	13.5	8.07	0	375	750	1125	1500	1490	1483	1478	1475	1474	
8	1	12	13.7	12	7.02	0	375	750	1125	1500	1491	1484	1479	1476	1475	1476
9	2	13.5	15.4	13.5	8.07	0	375	750	1125	1500	1490	1483	1478	1475	1474	1475
10	3	18.7	20.9	18.7	11.7	0	376	750	1124	1500	1487	1479	1474	1472	1471	1472
11	4	28.7	30.9	28.7	20.1	0	380	750	1120	1500	1480	1471	1468	1467	1467	1467
12	5	45.1	45.1	45.1	40	0	393	750	1108	1500	1462	1458	1460	1462	1463	1462
13	6	66.5	59.3	66.5	95	178	440	750	1060	1323	1408	1440	1453	1458	1460	1458
14	7	66.5	59.3	66.5	95	178	440	750	1060	1323	1408	1440	1453	1458	1460	1458
15	8	45.1	45.1	45.1	40	0	393	750	1108	1500	1462	1458	1460	1462	1463	1462
16	9	28.7	30.9	28.7	20.1	0	380	750	1120	1500	1480	1471	1468	1467	1467	1467
17	10	18.7	20.9	18.7	11.7	0	376	750	1124	1500	1487	1479	1474	1472	1471	1472
18	11	13.5	15.4	13.5	8.07	0	375	750	1125	1500	1490	1483	1478	1475	1474	1475
19	12	12	13.7	12	7.02	0	375	750	1125	1500	1491	1484	1479	1476	1475	1476
20			15.4	13.5	8.07	0	375	750	1125	1500	1490	1483	1478	1475	1474	

Fig. 28. Solving the two-dimensional Laplace equation around an electron accelerator

8. In cell C20, type **=C18** and copy the cell into cells D18:O18 creating another derivative equals zero boundary condition at that side.
9. In cell P8, type **=N8** and copy the cell into cells P9:P19 creating another derivative equals zero boundary condition at that side.
10. In cell F8, type **=V1_** and copy the cell into cells F9:F12, and F15:F19 creating the fixed boundary condition at the first slit.
11. In cell J8, type **=V2_** and copy the cell into cells J9:J12, and J15:J19 creating the fixed boundary condition at the second slit.
12. In cell C8, type **=(C7+D8+C9+B8)/4** and copy it into all of the other interior points.
13. Draw lines around the problem and shade cells as shown in Fig. 30

To solve the problem, press F9 until the values of the formulas stop changing. The worksheet applies the finite difference equation at all the interior points. At the metal slits, fixed boundary conditions equal to the applied voltage are inserted. At all the other boundaries set the normal derivative of the potential to zero. Setting the derivative equal to zero is accomplished by inserting another grid of cells just outside of the existing grid. The formulas in these cells are then set to equal the value at the cell two cells in from the edge. After the iteration has reached convergence, the worksheet looks like Fig. 28. Figure 29 shows a three dimen-

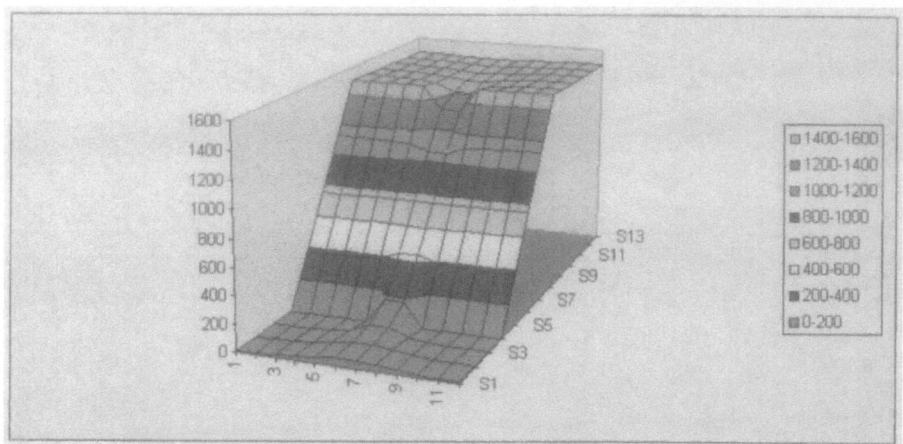

Fig. 29. A three-dimensional plot of the field solutions calculated for the interior of the electron accelerator

	A	B	C	D	E	F	G	H	I	J	K	L	M	N	O	P
1	X component of the electric field.															
2																
3																
4																
5					h=	1										
6		y\x	1	2	3	4	5	6	7	8	9	10	11	12	13	
7																
8		1	0	3.33	5.98		-375	-375	-375		8.14	6.2	3.97	1.9	0	
9		2	0	3.68	6.77		-375	-375	-375		8.7	6.34	3.9	1.81	0	
10		3	0	4.6	9.36		-375	-374	-375		10.6	6.55	3.47	1.46	0	
11		4	0	5.37	14.3		-375	-370	-375		14.3	5.85	1.96	0.54	0	
12		5	0	2.52	22.5		-375	-358	-375		20.9	0.57	-2	-1.2	0	
13		6	-0	-18	-56	-173	-286	-310	-286	-174	-58	-22	-9.3	-3.5	0	
14		7	-0	-18	-56	-173	-286	-310	-286	-174	-58	-22	-9.3	-3.5	0	
15		8	-0	2.52	22.5		-375	-358	-375		20.9	0.57	-2	-1.2	0	
16		9	-0	5.36	14.3		-375	-370	-375		14.3	5.85	1.96	0.54	0	
17		10	-0	4.6	9.36		-375	-374	-375		10.6	6.55	3.47	1.46	0	
18		11	-0	3.67	6.77		-375	-375	-375		8.7	6.34	3.9	1.81	0	
19		12	-0	3.33	5.97		-375	-375	-375		8.14	6.2	3.97	1.9	0	
20																
21																
22																

Fig. 30. The x component of the electric field for a two slit electron accelerator

sional plot of the potential within the device. Note how the accelerating field exists only between the two plates.

Calculating Electric Field Components from the Potential

When you know the potential everywhere within the volume of interest, it is a relatively simple matter to take the derivative of the potential to get the two electric field components. The electric field is defined as the negative gradient of the potential.

$$\begin{aligned} \mathrm{E} &= -\nabla\varphi \\ &= -\frac{\partial\varphi}{\partial x}\hat{\mathrm{i}} - \frac{\partial\varphi}{\partial y}\hat{\mathrm{j}} \end{aligned}$$

Each of these two components can be calculated using central differences.

$$E_x = \frac{\varphi_{i+1,j} - \varphi_{i-1,j}}{2h}$$

$$E_y = \frac{\varphi_{i,j+1} - \varphi_{i,j-1}}{2h}$$

Y component of the electric field

h= 1

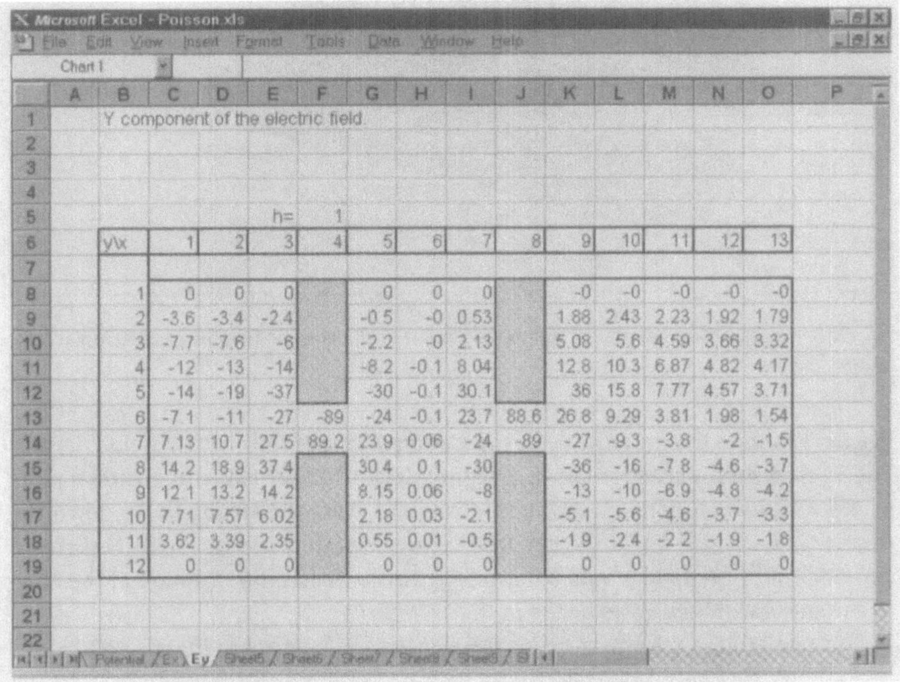

y\x	1	2	3	4	5	6	7	8	9	10	11	12	13
1	0	0	0		0	0	0		-0	-0	-0	-0	-0
2	-3.6	-3.4	-2.4		-0.5	-0	0.53		1.88	2.43	2.23	1.92	1.79
3	-7.7	-7.6	-6		-2.2	-0	2.13		5.08	5.6	4.59	3.66	3.32
4	-12	-13	-14		-8.2	-0.1	8.04		12.8	10.3	6.87	4.82	4.17
5	-14	-19	-37		-30	-0.1	30.1		36	15.8	7.77	4.57	3.71
6	-7.1	-11	-27	-89	-24	-0.1	23.7	88.6	26.8	9.29	3.81	1.98	1.54
7	7.13	10.7	27.5	89.2	23.9	0.06	-24	-89	-27	-9.3	-3.8	-2	-1.5
8	14.2	18.9	37.4		30.4	0.1	-30		-36	-16	-7.8	-4.6	-3.7
9	12.1	13.2	14.2		8.15	0.06	-8		-13	-10	-6.9	-4.8	-4.2
10	7.71	7.57	6.02		2.18	0.03	-2.1		-5.1	-5.6	-4.6	-3.7	-3.3
11	3.62	3.39	2.35		0.55	0.01	-0.5		-1.9	-2.4	-2.2	-1.9	-1.8
12	0	0	0		0	0	0		0	0	0	0	0

Fig. 31. The y component of the electric field for a two slit electron accelerator

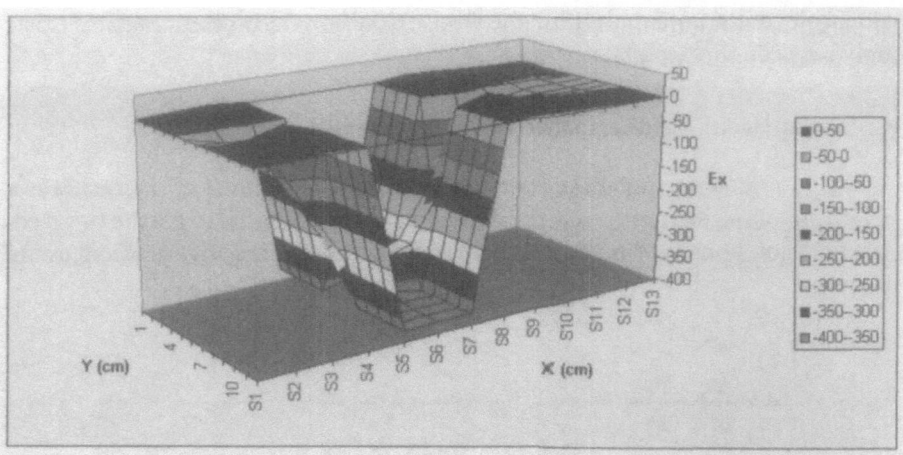

Fig. 32. A two-dimensional plot of the x component of the electric field in an electron accelerator

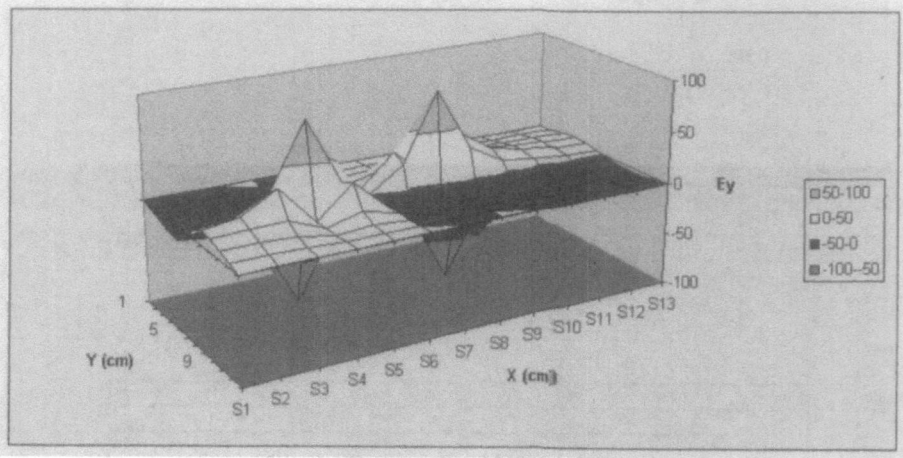

Fig. 33. A two-dimensional plot of the y component of the electric field in an electron accelerator

Figure 30 shows the magnitude of the x component of the electric field (E_x) and Fig. 31 shows the y component (E_y). Figures 32 and 33 are two-dimensional plots of the x and y components of the electric field. Note the strong fields down the center of the device where the electrons are accelerated. In Fig. 33, the transverse component of the field would be 0, except for the field enhancement at the edge of the slit that shows up as peaks in the field plots.

To calculate the x component of the electric field, perform these steps.

1. On a clean sheet in the same workbook as the solution of the Poisson equation, type the following values in the indicated cells.

- A1: **X component of the electric field.**
- E5 h=
- F5 1
- B6 y\x

2. Name cell F5 as h
3. In cells C6:O6, type the integers between 1 and 13.
4. In cells B8:B19 type the integers between 1 and 12.
5. In cell C8, type =-(**Potential!D8-Potential!B8)/(2*h)**
6. Copy the contents of cell C8 to all the cells in the interior of the problem.
7. Outline, shade and set the column widths as shown in Fig. 30.

To calculate the *y* component of the electric field, perform these steps.

1. On a clean sheet in the same workbook as the solution of the Poisson equation, type the following values in the indicated cells.
   - A1: **Y component of the electric field.**
   - E5 h=
   - F5 1
   - B6 y\x

2. Name cell F5 as **h_**
3. In cells C6:O6, type the integers between 1 and 13.
4. In cells B8:B19 type the integers between 1 and 12.
5. In cell C8, type =-(**Potential!D8-Potential!B8)/(2*h_)**
6. Copy the contents of cell C8 to all the cells in the interior of the problem.
7. Outline, shade and set the column widths as shown in Fig. 31.

Recalculate the worksheets by pressing F9 to complete the solution.

Oscillations of a Cantilever Beam: Euler and Modified Euler Methods

In this section we will numerically solve the following differential equation giving the actual position of the tip of a cantilever at any point in time:

$$m\frac{d^2 y}{dt^2} + \left(\frac{3EI}{l^3}\right) y = 0$$

This equation is an initial value problem, that is we know the position and velocity of the cantilever at the initial time and want to integrate the equation some time into the future to find the final position. One of the most straightforward methods for solving a differential equation of this type is the Euler method. The Euler method uses the differential equation to give the acceleration which, along with the initial position and velocity, can be used to project the position a short time into the future. The amount of time that you can project a solution into the future is limited both by the mathematics of the problem, but its Physics as well. To maintain accuracy, no step may be larger than the time constant for any process of interest. Mathematically, the steps must be small enough to keep the solution stable, but not too small so as to cause round off error to destroy the accuracy. Round-off error is caused by the fact that a computer uses numbers with a discrete number of digits with which to calculate. If the difference between two numbers divided by either of those numbers is smaller than the number of digits in the computer's numbers than the difference is meaningless.

To solve the equation above using the Euler method, the differential equation is solved for the acceleration.

$$\frac{d^2 y}{dt^2} = -\left(\frac{3EI}{ml^3}\right) y$$

The acceleration is combined with the initial velocity (v_i) to calculate a final velocity (v_{i+1}) at some time (dt) in the future.

$$v_{i+1} = v_i + \frac{d^2 y}{dt^2} dt$$

Knowing the velocity and position (x_i) at the beginning of the step, we can calculate the position at the end of the step (x_{i+1}).

$$x_{i+1} = x_i + v_i dt$$

We then use the new position and velocity to take another step, and so on until we have reached the time of interest. Figure 34 shows a worksheet that calculates the position of the tip of the cantilever versus time using the Euler method. Figure 35 shows a plot of that solution with a line drawn at the location of the end of a period calculated from the analytical solution for the frequency.

To create this worksheet, perform the following steps.

1. On a new worksheet, type the following values in the indicated cells, and the labels shown in Fig. 34.
   - G3   2.90E+07
   - G4   426
   - G5   144
   - G6   20000
   - G12 0.1
2. Right justify cells F3:F6, C8, F8, and F12.
3. In cell D8, type the analytic frequency =(1/(2*PI()))*SQRT(3*E*I/(m*L^3))
4. In cell G8, type the analytic period =1/D8
5. Name cells G3:G6 as E, I, L, and m.
6. Name cell G12 as delt.
7. In cell F15, type the initial time 0
8. In cell G15, type the initial position 1
9. In cell H15, type the initial velocity 0

The initial position and velocity assume that we depress the cantilever one inch and release it.

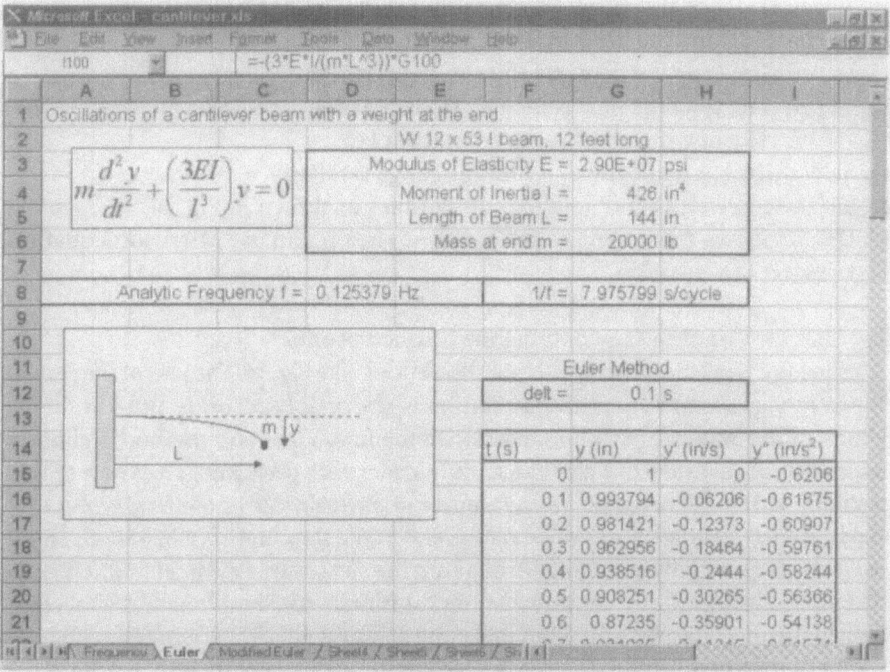

Fig. 34. Solving the cantilever problem using the Euler method

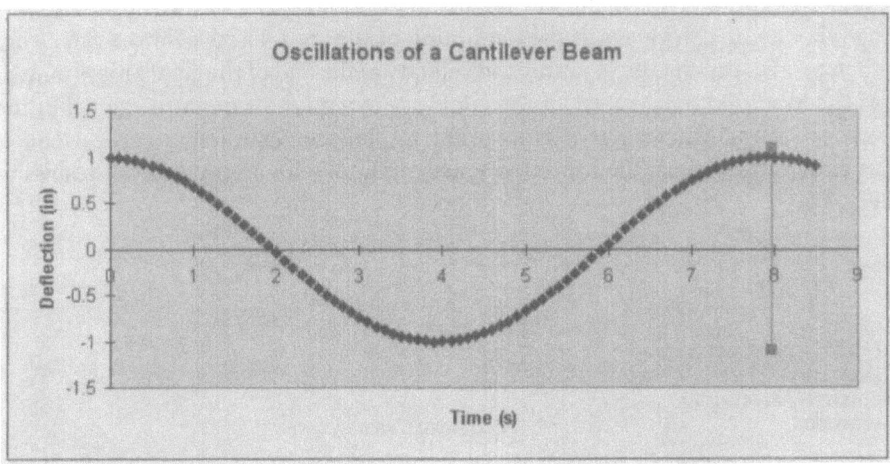

Fig. 35. A graph of the position calculated with the Euler method. A line crosses the curve at the analytical position of the end of a period

10. In cell I15, type the initial acceleration =-(3*E*I/(m*L^3))*G15
11. In cell F16, type the new time =F15+delt
12. In cell G16, type the new position =G15+H16*delt
13. In cell H16, type the new velocity =H15+I15*delt
14. In cell I16, copy the new acceleration from I15.
15. Select G16:I16 and copy that into the range G17:I100
16. Set the cell widths and heights, and outlines as shown in Fig. 34.
17. Use Windows Draw to draw the cantilever beam and use Microsoft equation
    to insert the equation.

The disk file Cantilev.XLS contains the finished product.

    When the worksheet recalculates, it should look like Fig. 34. The plot of the results shown in Fig. 35 shows that the calculation is reasonably accurate, but has slightly missed the analytic period marked with the line. The Euler method is slightly inaccurate at each step and that inaccuracy grows with each step. The cause of this problem is that the slope of the acceleration and velocity at the beginning of a step is used to estimate those values at the end of the step as shown in Fig. 36. These two slopes are actually changing throughout the step, making the calculation miss the end each time. If the curvature of the solution is constantly in one direction, then all the inaccuracies add up to make the solution get slowly worse and worse.

    A better estimate of the slope of the acceleration and velocity is to use the average slope across the step to predict the values at the end of the step. The problem with this is that we are trying to predict the value at the end of the step, but would need it to calculate the average. The way out of this problem is to use the Euler method to predict the values of the slopes at the end of the step and to then average them with the values at the beginning of the step. These average values are then used to predict the position and velocity at the end of the step. This is known as the Modified Euler method. You could try to apply this method again, but the error in the solution starts causing problems. Figure 37 shows the same problem as above solved with the Modified Euler method, with a graph of the results in Fig. 38.

Fig. 36. Inaccuracy in the Euler method leads to progressively inaccurate solutions

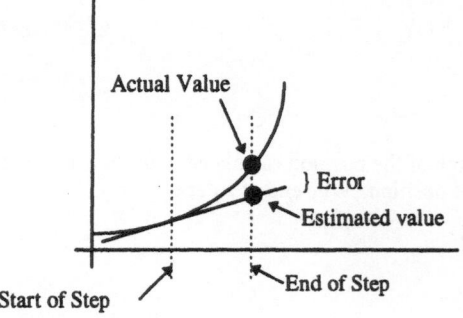

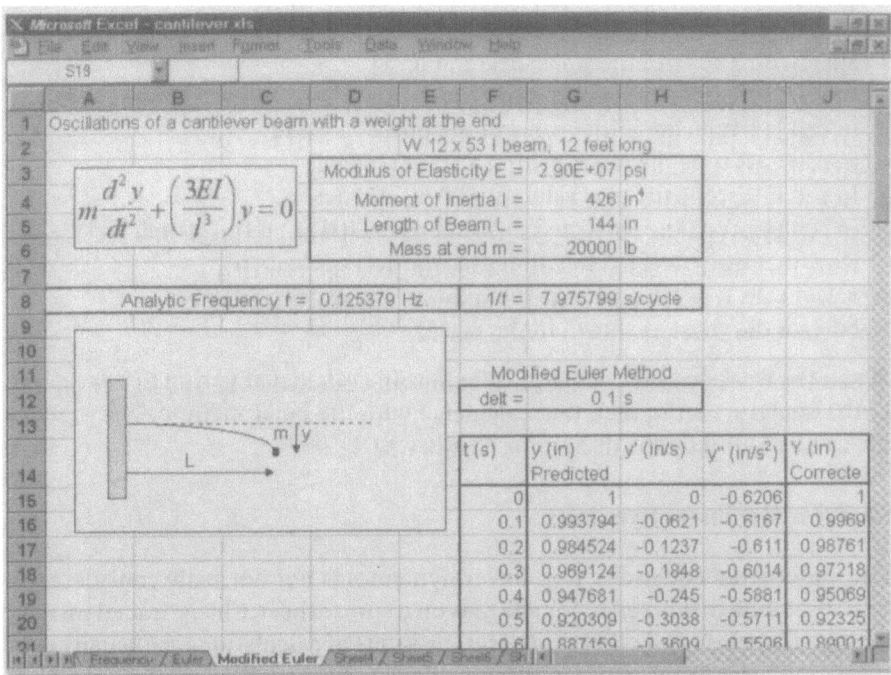

Fig. 37. Solving the cantilever problem using the Modified Euler method

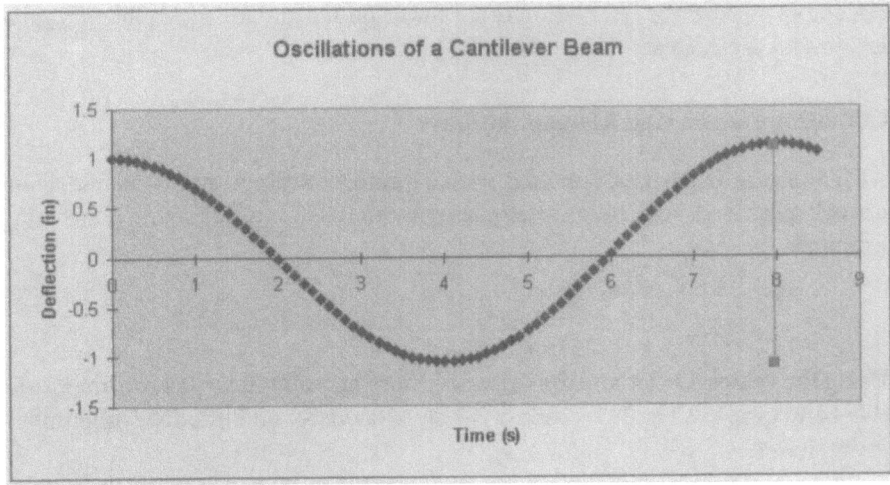

Fig. 38. A graph of the position calculated with the Modified Euler method. A line crosses the curve at the analytical position of the end of a period

To calculate the Modified Euler method, perform these steps.

1. Start with a copy of the worksheet in Fig. 34.
2. In cell J14, type **Y (in) Corrected**
3. In cell J15, type the initial value of the position **=G15**
4. In cell G16, type the new position **=J15+H16*delt**
5. In cell H16, type the new velocity. **=H15+I15*delt**
6. In cell J16, type the corrected position **=J15+((H15+H16)/2)*delt**
   Note that the corrected position uses the average velocity.
7. Select cells G16:J16 and copy them down into cells G17:J100
8. Format the sheet as shown in the Fig. 37.

When the worksheet recalculates, note how the calculated period better match-
es the analytic period than that calculated with the Euler method. The practice
file can be found on the diskette as Cantilev.XLS.

Studying Nonlinear Dynamics

Like differential equations, non-linear dynamics is not normally considered to
be a discipline that is suited for analysis on a spreadsheet. Chaos, fractal images,
strange attractors and other more esoteric kinds of calculations are usually considered to be in the realm of mainframe calculations. Actually, calculation of fractal images are not well suited for direct calculation on a worksheet, but are better
performed by a Visual Basic function or a compiled function owing to the large
number of calculations required to calculate a complete image. However, strange
attractors require the calculation of far fewer points and are amenable to calculation on a worksheet.

Calculating a Henon Map: A Strange Attractor

A Henon map is a fractal function with a strange attractor that is calculable on
a worksheet. A Henon map is calculated by iterating the following generating
function:

$$x_{n+1} = 1 + y_n - Ax_n^2$$
$$y_{n+1} = Bx_n$$

Using the values $A = 1.4$ and $B = 0.3$, and a starting point of $x = 0, y = 0$, the worksheet and graph in Fig. 39 is created. Other values of A and B produce other interesting maps.

The actual values calculated by the generating functions are chaotic in nature,
but form the characteristic shape shown in the figure. The Henon map shape
shown in the figure is called a strange attractor. It is an attractor because no matter what the starting point, the points either all fall on the curve shown in the figure or they tend to infinity. It is strange, because the figure is not simple such as
a point or a circle.

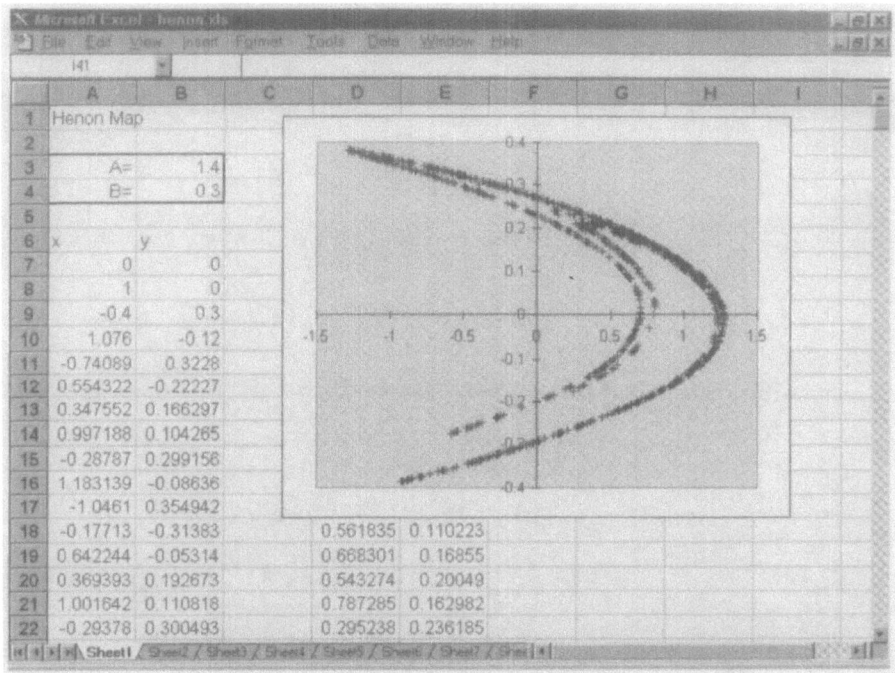

	A	B	C	D	E	F	G	H	I
1	Henon Map								
2									
3	A=	1.4							
4	B=	0.3							
5									
6	x	y							
7	0	0							
8	1	0							
9	-0.4	0.3							
10	1.076	-0.12							
11	-0.74089	0.3228							
12	0.554322	-0.22227							
13	0.347552	0.166297							
14	0.997188	0.104265							
15	-0.28787	0.299156							
16	1.183139	-0.08636							
17	-1.0461	0.354942							
18	-0.17713	-0.31383		0.561835	0.110223				
19	0.642244	-0.05314		0.668301	0.16855				
20	0.369393	0.192673		0.543274	0.20049				
21	1.001642	0.110818		0.787285	0.162982				
22	-0.29378	0.300493		0.295238	0.236185				

Fig. 39. Calculating a Henon map on a worksheet

The worksheet is a simple calculation of two simple functions and is left as an exercise for the reader. It is also included on the disk as Henon.XLS.

Summary

In this chapter, you have seen the spreadsheet in action as a tool for the Physicist or Electronics Engineer. Hopefully, I have given you an idea about how you approach science and engineering problems using a spreadsheet program. For simple problems of calculating and displaying a function, a spreadsheet is ideal. More complex problems can also be solved on a worksheet by judicious arrangement of the formulas that are being calculated. In addition to calculating and displaying functions, a spreadsheet can be used to solve non-linear equations, differential equations, and even fractal equations.

Spreadsheets as Tools in Mathematical Modeling and Numerical Mathematics

E. Neuwirth

Among scientists and mathematicians spreadsheets are still not fully appreciated. Too many scientists think that spreadsheets are essentially an accountant's tool. Therefore, in this chapter we will try to show how spreadsheets can be used as powerful tools for mathematical modeling. We will also show certain techniques for using graphical user interface elements which are relatively unknown but extremely helpful when dealing with mathematical models.

In the following examples we will demonstrate a few basic techniques of how to use spreadsheets for small to intermediate scale numerical methods. Since we are concentrating more on how spreadsheet programs can be used for modeling rather than on the modeling itself, the models will be a rather simple. Also, since we have to show examples of how this can be done with real programs we have to select a program as our base for the detailed instructions. Since Microsoft Excel is at the moment the most widely used spreadsheet program all our instructions will be given for this platform. Since the basic techniques for all spreadsheet programs are, however, very similar, all these techniques should work (with very small modifications) for almost any spreadsheet program available today.

All example material can be found in the directory \Chap3 on the CD-Rom accompanying this book. The file MathMod.XLS contains all of the early models in tabbed worksheets. The later examples, Brown, Coinflip, Logistic and Pense have their own XLS files. Additional information on the last named file can be found in the files Pense.DOC, in WinWord format and Pense.HTM, in HTML format.

A Simple Dynamical System

Let us introduce some of the basic concepts with a very simple model which we will later refine. We will set up a simple growth model defined by a difference equation. Assuming a state variable $x(t)$ and a constant growth rate c we can write our assumptions as:

$$\Delta x(t) = x(t + 1) - x(t) = c.x(t)$$

Now let us set up a spreadsheet implementing this equation with a constant

$$c = 0.05.$$

We set up the following table

t	x(t)	$\Delta x(t)$

Since we want to solve our system numerically we have to give initial values for *t* and *x(t)*. This is done easily, we just enter numerical values into the appropriate cells:

t	x(t)	$\Delta x(t)$
0	100	

Now in a next step we calculate $\Delta x(0) = c.x(0)$. It seems to be useful to represent this formula by the following diagram:

t	x(t)	$\Delta x(t)$
0	● 100	0.05* ▲

Of course, this is not what you will see on your screen. It is just a pictorial representation of the fact that the increment for the variable *x* is calculated by multiplying the current value with 0.05. The "real" content of the cell is a formula. How this formula is displayed on screen depends on the spreadsheet program in use (that is, whether it's Excel, Lotus 1–2–3 etc) and on the configuration of this program, but in most cases the formula will look something like =0.05*B2.

The reference to the cell is written as B2 because the standard way of giving cell addresses is assigning a letter to each column and a number to each row. So B2 is the cell in column B, i.e. the second column, and in row 2. There are alternative ways of writing cell addresses, but the A1 notation (that is how it is called colloquially) is the most widely used form.

It is possible to enter formulas this way directly from the keyboard, but this implies that the user does "mental bookkeeping" for cell addresses. One of the most important user interface elements of spreadsheet program is that formulas can be created by a "point and click" procedure. Incidentally, the method we are about to describe has been available since the very first generation of spreadsheet programs, so it is a precursor to the graphical and mouse based user interfaces. To enter our formula using this method we proceed as follows:

1. Enter =0.05* from the keyboard
2. Use either the mouse or the cursor keys to move the location of the "selected cell" to cell B2. The formula "adjusts itself" and displays =0.05*B2.
3. Press the Return-key

Using this "point and click" method to create formulas has the advantage that one is operating with the abstract formula and concrete values at the same time. Usually it is much easier to avoid mistakes because one is pointing at the objects to be referred to in the formula. It takes some time to get used to the point and click method when coming from a program language oriented background, but in many cases after a short time creating "webs of formulas" it becomes much more efficient than typing cell addresses manually. Additionally, we might say that by using the point and click method we represent relations between quantities not by names, but by spatial arrangement.

But now, let us continue constructing our worksheet. When we constructed our model, the computer screen of course did not display the arrows we used in the previous diagram to represent the formula, the screen immediately displayed the value:

t	x(t)	Δx(t)
0	100	5

Now we have to extend the model for $t = 1$. So the value in the t-column is best characterized by the expression "add 1 to the value above". The next value in the $x(t)$ columns is calculated by $x(t+1) = x(t) + \Delta x(t)$. We can depict both of these formulas in the following graphical way:

t	x(t)	Δx(t)
0	100	5
+ 1	+	

Now we have to calculate $\Delta x(1)$. The formula is $\Delta x(1) = 0.05 * x(1)$. This formula is very similar to the corresponding formula for $\Delta x(0)$, which was $\Delta x(0) = 0.05 * x(0)$. In both cases the number we have to calculate is produced by multiplying the number to the left with the constant 0.05. "Visually speaking", here are our two formulas:

t	x(t)	Δx(t)
0	100	*0.05
1	105	*0.05

So the lower formula is just a copy of the upper formula with the references shifted down together with the formula. And this is one of the most important concepts in spreadsheets: *relative references*. When a standard formula is copied from one cell to another cell, the references will be shifted horizontally and vertically by the same amount as the formula. So to create the formula for $\Delta x(1)$ in our example we just have to copy the formula down. We will indicate this graphically by shading the cell containing the original formula in dark gray and the cell containing the copy of the formula in light gray:

t	x(t)	Δx(t)
0	100	*0.05
1	105	*0.05

To extend the model for $t > 1$ we observe that our colloquial way of expressing the formula yielding the values for the t-columns applies in all the rows for this column: "just add 1 to the value above". Therefore, we can simply copy the formula producing $t = 1$ from $t = 0$ down:

t	x(t)	Δx(t)
0	100	*0.05
+ 1	105	*0.05

Copying formulas is one of the most frequently used techniques in spreadsheet work and can be accomplished very easily.

The same principle applies for all the formulas we need for the $x(t)$-column. Since these formulas could be colloquially expressed as "just add the numbers above and above and to the right", for all $t > 1$ the formula is no more than a copy of the formula producing $x(1)$. Therefore, we can simply copy this formula down as far as we need. Additionally, all the formulas in the $\Delta x(t)$-column are also copies of the formula for $\Delta x(1)$. Therefore we can copy the formulas in both columns down like this:

t	x(t)	Δx(t)
0	100	5
1	105	5.25
2	110.25	5.5125
3	115.7625	5.788125
4	121.5506	6.077531
5	127.6282	6.381408
6	134.0096	6.700478

If we had thought about copying in the beginning, we would have noticed that for all three columns all the formulas for $t > 1$ are just copies from the row for $t = 1$. This structure is expressed even better by the following diagram:

t	x(t)	Δx(t)
0	100	5
1	105	5.25
2	110.25	5.5125
3	115.7625	5.788125
4	121.5506	6.077531
5	127.6282	6.381408
6	134.0096	6.700478

So the method of solving our difference equation consisted of three steps:

1. setting up the initial conditions for the variables
2. calculating the values for the variables one step later
3. copying down the formulas from step 2.

Step 3 only works when the formulas constructed in step 2 are valid not only for step 2 but give the general method for calculation of the values at any given time from the values "one step earlier". The basic spreadsheet concept of relative references is a very important ingredient in the construction of these formulas. One even might say that we are not just building a formula, but that we are building a prototype for many formulas at the same time.

Of course, it is possible to copy down all the formula much further than displayed on the diagrams in this book. Spreadsheet programs nowadays have a few thousand rows. For example, the version of Excel current at the time of writing (Excel 97) has 65,536 rows – four times more than previous versions.

Parametrized Models

Now, let us refine our model. The growth rate of $c = 0.05$ per time step is hard-wired in the model we just built. If we want to study the same model for $c = 0.07$, we have to change all the formulas in the column for $\Delta x(t)$. This can be done entering the new formula in the first row (below the header row) of this column and then copying this formula down far enough.

Now again we want to study our growth model:

$$\Delta x(t) = x(t) - x(t-1) = c.x(t-1),$$

but we want our model to be more user-friendly. We will set up the sheet in a way that one cell contains the value for c, and by changing the value in this cell we will immediately get a new "model run".

In this new sheet (already including the initial values) we could set up the formula for $\Delta x(t)$ in the following way:

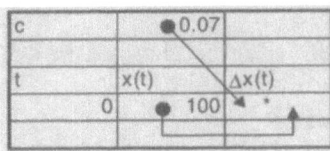

This formula, however, will not work if it is copied into the next row:

c			0.07	
t		x(t)		$\Delta x(t)$
	0		100	

The relative reference starting at the value 0.07 would be moved along with the copy of the formula, and this would produce an incorrect formula. Therefore, we need a way of "pinning or anchoring" the reference to the formula, or making this reference an *absolute reference*. We will indicate this the following way:

c		0.07	
t	x(t)	Δx(t)	
0	100	*	

The starting point of the reference is "pinned down" indicating it will not move along with the formula when the formula is copied. So our formula now has a relative reference to *x(t)* and an absolute reference to *c*. Looking at these model components from a more abstract point of view we see that *x(t)* is a model variable and *c* is a model parameter or a model constant. It turns out that in a spreadsheet model quite often variables are indicated by relative references and parameters or constants are indicated by absolute values.

But how do we create an absolute reference? Assuming that the upper left cell in our diagram is A1 we have to create the formula with the absolute reference in cell C4. In cell address notation, absolute references are indicated by dollar signs in front of the absolute reference component. So in C4 we need the formula =$B$1*B4. The two dollar signs are needed because it is also possible to have row absolute column relative references, so the absolute character of a reference has to be indicated separately for rows and columns. We could, of course, create this formula by typing it into the cell manually. But we already have seen the point and click method of creating formulas is much more comfortable than typing. Therefore we need a way of creating an absolute reference with point and click also. Here is how we can do it:

1. In cell C4, enter = from the keyboard.
2. Using the mouse or the cursor key move the "selected cell" to B1. Cell C4 now displays =B1.
3. Press function key F4 once; cell C4 will display =$B$1. We just changed the *relative reference* to an *absolute reference*.
4. Enter * from the keyboard, cell C4 displays =$B$1*
5. Again using the mouse or the cursor keys move the "selected cell" to B4. Cell C4 displays =$B$1*B4.

Press the Return key to finish the formula.

When creating a formula by pointing and clicking pressing the function key F4 directly after indicating a cell reference will change this reference to an absolute reference, pressing function key F4 more than once will cycle the reference through all possible combinations of absolute and relative references, so it is possible to create mixed type references (e.g. row absolute column relative) also with the point and click method.

Back to our model now. The formulas for $x(t)$ and t in row $t = 1$ are the same as in the simpler model:

c	0.07	
t	x(t)	Δx(t)
0	100	7
+ 1	+	

The formula for $\Delta x(t)$ can simply be copied down from the row for $t = 0$.

c	0.07	
t	x(t)	Δx(t)
0	100	7

This will work since we constructed the formula using an absolute reference to the cell containing the constant c, therefore all the copies will refer to the same cell.

All the formulas in the rows further down can be created by just copying the formulas from the row for $t = 1$

c	0.07	
t	x(t)	Δx(t)
0	100	7
1	107	7.49
2	114.49	8.0143
3	122.5043	8.575301
4	131.0796	9.175572

With this version of our model it is very easy to create a "model run" for a different parameter value. Just changing the value in cell B1 (currently 0.07) by entering a new value will automatically recalculate the whole model. Automatic bookkeeping of dependencies of formulas and automatic recalculation whenever necessary is also one of the most convenient features of spreadsheet programs. In almost all programming language based tools (which could be general programming languages, but also special purpose simulation languages) one would have to change a parameter value in the source code of the programs and then manually start a new model run. Spreadsheets will recalculate whenever a value used as input to a formula changes. Immediately after changing 0.07 in cell B1 to 0.09 we will see the worksheet next page:

c		0.09	
t	x(t)	Δx(t)	
0	100	9	
1	109	9.81	
2	118.81	10.6929	
3	129.5029	11.65526	
4	141.1582	12.70423	

When dealing with dynamical models we not only want the equations to be calculated, we also want graphical displays of the data derived from the model. Spreadsheet programs make it extremely easy to create graphs from columns or rows of data. To create a graph from the t and the $x(t)$-columns of our model we have to select the range with the data. Now we have to tell the program that we want to graph the data in the selected range. This can be done using the "Insert" menu and then the "Chart item". An easier way to accomplish our goal, at least

in Excel, is to click on the ChartWizard button ▨ on the standard toolbar. The cursor changes its shape to a crosswire indicating that we have to mark the region where we want the graph to be placed on the sheet. Similar to selecting cell ranges this is done by clicking at the location of the upper left corner of the graph, keeping the mouse button pressed, moving the cursor to the lower right corner of the graph, and releasing the mouse button. Then Excel asks a few questions (via dialog boxes) about the data range (which already is supplied with the selected cell range as the preliminary answer, and the graph type. When a "function plot" of data points is the kind of graph we want, it is very important to select the XY(Scatter)-plot and not the line graph. Line graphs tacitly assume that all the values for the x-scale are equally spaced and use the values in the cells only for the labels along the x-axis. XY-plots use the values in the cells of the first data column for the location of the points, and that is what is usually needed in scientific graphs. In a further step we will be asked about the subtype we want for our XY-graph, and among other things we will be offered choices of graphs with and without grids, of connected or isolated points, and of smoothed or straight connecting lines between the data points. Finally, we will be offered choices for graph headings and axes labels. If the data range we selected for the graph has text in the first row and numbers in all the other rows, the text will be used for labeling the line in the graph and for the heading of the graph.

If we create an XY-graph for our growth model (with t-values from 0 to 20), we will see the following chart:

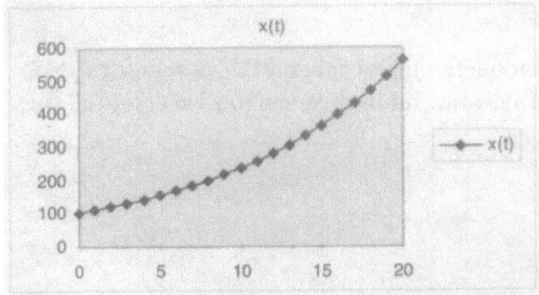

On the computer screen this chart is displayed side by side with the data table. And now we are ready to see an example of the conveniences of spreadsheets: if we type a new value for c in the respective cell, both the data and the graph will change immediately to display the values for this new model parameter.

A Slightly More Complex Model

Now let us refine our model a little bit. We will compare our simple model (which of course is an exponential growth model) with a model where the formula for the change from $x(t)$ to $x(t + 1)$ is slightly more complicated. The basic growth model we started from has the formula:

$$\Delta x(t) = x(t) - x(t - 1) = c.x(t - 1)$$

which also can be written as:

$$\frac{\Delta x(t)}{x(t)} = c$$

so we have constant relative growth. Such an equation cannot describe a situation in which there is a capacity limit since for any growth rate the value of $x(t)$ will exceed any bound. A first attempt at a model expressing limited growth can be described in the following way: let us assume there is an "ultimate" value U which $x(t)$ may never exceed. Let us furthermore assume that the growth rate is c when $x(t)$ is small compared with U, but that the growth rate gets smaller when $x(t)$ approaches U. We might call $\frac{x(t)}{U}$ saturation. Then $1 - \frac{x(t)}{U}$ could be called "percentage of unfilled space". If we calculate the relative growth (or growth rate) according to:

$$\frac{\Delta x(t)}{x(t)} = c(1 - \frac{x(t)}{U})$$

then we have a model where growth rate decreases with saturation.

This formula of course can be rewritten as

$$\Delta x(t) = c(1 - \frac{x(t)}{U})x(t)$$

We will now modify our original model so that it becomes this new model. To do so we insert a row just below the first row, and in this new row we enter the value for the capacity limit U:

c		0.09	
U		500	
t	x(t)		Δx(t)
0	100.00		9.00
1	109.00		9.81
2	118.81		10.69
3	129.50		11.66
4	141.16		12.70

To change or model to follow the new formula we only have to change the formulas in the $\Delta x(t)$-column. When we erase the formulas in the $\Delta x(t)$-column, the $x(t)$-column will display incorrect values since to correctly calculate values in this column we need the values from the $\Delta x(t)$-column which we just erased. Nevertheless we don't have to delete the formulas in the $x(t)$-column since they will "resume their work" as soon as we have created the new formulas in the $\Delta x(t)$-column.

After erasing these cells we can create the new formula for $\Delta x(t)$.

When this formula is entered, we just copy it down and the worksheet is adjusted to display the new model.

c		0.09	
U		500	
t	x(t)		Δx(t)
0	100.00		*(1-)*
1	100.00		
2	100.00		
3	100.00		
4	100.00		

Since the graph we displayed previously takes its data from the t- and the $x(t)$-column, the graph also was adjusted as soon as we completed the formulas in the $\Delta x(t)$-column.

Let us use our model to display another convenient feature of modern spreadsheet programs: solving equations numerically. Let us ask the question of how we would have to change the value of c to get a value of 150 for $x(4)$ (which currently is 131.07). We could use a trial and error method to find such a value. Each time we change the value in B2 (which is the "value cell for c) all the values in the worksheet immediately change to reflect this new model parameter, so we may just try values until the value for $x(4)$ is close to 150. But there is a better way.

To achieve our goal we select the cell with the value we want to be changed to 150 either by clicking on the cell or by using the cursor keys. Once the cell is selected from the "Tools" menu we select the "Goal Seek" item. The screen now displays a dialog box with three fields to be filled:

1. **Set cell:** already is tentatively filled with the address of the selected cell, so we accept it
2. **To value:** here we enter our target value, i.e. 150.
3. **By changing cell:** here the address of the cell which should be changed to create the target value is needed. This address can either be entered from the keyboard, or better by first clicking into the field to be filled in the dialog box and then into the field which should be changed directly in the worksheet.

We are really talking about two cells here: the cell containing the target value (which we have not yet achieved), which mathematically speaking is the dependent variable, and the cell which is allowed to be changed to achieve the desired value in the cell with the target value, which mathematically speaking is the independent variable.

Once we have entered all the three required inputs we press the OK button on the dialog box, and after a very short time the parameter value needed to achieve our desired goal is found and the worksheet is adjusted accordingly, and the value for $x(4)$ changes to 150.

Goal Seek is essentially a simple tool for solving equations with one unknown variable. It is important to understand that the formula for the dependent variable does not have to explicitly contain a reference to the cell with the independent variable. In a complex model there might be many cells calculating some intermediate results until finally the dependent variable is calculated. This is illustrated in our example already, $x(4)$ does not explicitly contain a reference to c, but depends on c through a longer chain of equations. Spreadsheet programs can handle these kind of complex dependencies quite well.

Most spreadsheet programs available today offer a "goal seek" function or something equivalent. Excel also offers the "Solver" tool (from the "Tools" menu) which is a much more sophisticated tool for solving multivariable equations and optimization problems with constraints (given as equalities and/or inequalities). Solver is quite a powerful tool and can handle problems up to 30 variables reasonably well. So for problems of medium size (more than just a few variables, less than hundred variables) Excel also can be used as an optimization program. The advantage of using Excel for optimization is that special optimization programs are specialized one-purpose tools, whereas Excel is general multipurpose tool, so after learning to use Excel (or any spreadsheet program) one has the potential to deal with a much wider class of problems. Additionally, in many circumstances Excel will be much easier available than special optimization programs.

Of course, goal seek is not only a tool for models, it is a general tool for numerical mathematics.

Using the Data Table Feature

Spreadsheets contain another important tool which is not very well known in scientific circles. This is the feature known as Data Table.

We will illustrate the its use using our growth model again. We will study the following problem. Given our extended growth model (which, as some readers might be aware of already, is called the logistic growth model), we want to derive a table which gives us the value of $x(10)$ for different values of c. So we want to see the state of our system after a certain time for a range of possible initial values.

Of course, just typing the different values in the c-cell immediately displays the corresponding value for $x(10)$ in the $x(10)$-cell. But we only see one value at a time. What we really want is a table. So we add the values for c we want to be used in our scenarios in a new column in our worksheet.

c		0.09				
U		500			0.01	
					0.02	
t	x(t)	Δx(t)			0.03	
0	100.00	7.20			0.04	
1	107.20	7.58			0.05	
2	114.78	7.96			0.06	
3	122.74	8.33			0.07	
4	131.07	8.70			0.08	
5	139.78	9.06			0.09	
6	148.84	9.41			0.1	
7	158.25	9.73				
8	167.98	10.04				
9	178.02	10.32				
10	188.34	10.57				

Our goal now could colloquially be described in the following way: we want these new values (in the fifth column of our worksheet) to be "plugged" into cell B2, and then we want the value from the $x(10)$-cell to be transferred to the cell immediately to the right of the cell with the "current c-value". Spreadsheet programs can do this, but it is a bit tricky. As a first step we have to add a formula to our worksheet. The formula goes in the header row of the columns directly to the right of the column with the parameter values, and the formula contains a reference to the cell we want to be "watched" for different parameter values.

c		0.09				
U		500			0.01	
					0.02	
t	x(t)	Δx(t)			0.03	
0	100.00	7.20			0.04	
1	107.20	7.58			0.05	
2	114.78	7.96			0.06	
3	122.74	8.33			0.07	
4	131.07	8.70			0.08	
5	139.78	9.06			0.09	
6	148.84	9.41			0.1	
7	158.25	9.73				
8	167.98	10.04				
9	178.02	10.32				
10	188.34	10.57				

With a setup like the one displayed in this diagram, cell E1 contains the formula =B15. Additionally, in this diagram we have indicated the cells to be used as the different parameter values. So far, our model worksheet has not been told in which cell to put the values in the shaded area. This last piece of information has to be entered in a dialog box. So to calculate our $x(10)$-value for different c-scenarios we have to take a series of steps:

At first we select the rectangular range containing the values to be used for the model runs, and additionally containing the reference to the cell to be watched:

c		0.09				
U		500			0.01	
					0.02	
t	x(t)	Δx(t)			0.03	
0	100.00	7.20			0.04	
1	107.20	7.58			0.05	
2	114.78	7.96			0.06	
3	122.74	8.33			0.07	
4	131.07	8.70			0.08	
5	139.78	9.06			0.09	
6	148.84	9.41			0.1	
7	158.25	9.73				
8	167.98	10.04				
9	178.02	10.32				
10	188.34	10.57				

The selected area has to include the row above the row with the first value for the parameter, and it also has to include a column with the reference to the cell to be watched directly to the right of the column with the parameter values. Otherwise what we are about to set up will not work.

In the next step we go into the "Data" menu and select "Table". Now we are prompted with a dialog box asking for a "row input cell" and a "column input cell". This dialog box does not state very clearly what it is needed for. To achieve what we want we have to indicate that we want the values from the first column of the selected range to be put into the cell with the c-value. Therefore, we have to indicate the call address B1 as the "column input range", and we can leave the "row input range" field empty. When this field is completed, we press the OK button on the dialog box, and all the necessary calculations are performed immediately, yielding the following worksheet:

So we see that for $c = 0.01$ our model yields $x(10) = 108.22$, $c = 0.02$ yields $x(10) = 116.86$ and so on.

c		0.09					188.34
U		500				0.01	108.22
						0.02	116.86
t		x(t)	Δx(t)			0.03	125.93
	0	100.00	7.20			0.04	135.41
	1	107.20	7.58			0.05	145.28
	2	114.78	7.96			0.06	155.54
	3	122.74	8.33			0.07	166.15
	4	131.07	8.70			0.08	177.10
	5	139.78	9.06			0.09	188.34
	6	148.84	9.41			0.1	199.85
	7	158.25	9.73				
	8	167.98	10.04				
	9	178.02	10.32				
	10	188.34	10.57				

Why is "Data Table" so important? If the cell to the right of each c-value contains the formula for how to calculate the value we are interested in, then we could just use the basic spreadsheet technique of copying the formula and we would not need the "Data Table" command at all. Our example clearly demonstrates why this method cannot solve the problem we just studied. We don't have a formula which calculates $x(10)$ directly from c. We only have a moderately complex worksheet using quite a few cells with intermediate results which calculates $x(10)$ once c is given. We might think of our sheet as a complicated input-output machine which takes inputs in the c-cell and then puts the output in the $x(10)$-cell. "Data Table" then acts as a "scribe". It allows us to supply a "tape" with a series of values, all to be put into the input hopper one after the other, and producing a protocol giving the output value associated with each of our supplied input values. Using this mechanism we can set up very complicated models and study the dependence of certain system variables on parameter values. If we have to find parameter values to achieve a certain goal in a complex model, we might use Data Table first to get an overview on how the parameter influences the goal variable, and after having collected this information we might use Goal Seek (or even the Solver) to achieve the goal we set.

The models we've studied so far were rather simple difference equations. It is not difficult to expand the techniques we applied to the numerical solutions for differential equations. The Euler-Cauchy method is essentially the method we were just using. All our models so far were of the basic type:

$$f(x + 1) = f(x) + D(f(x))$$

where D is the function describing the difference equations, so we had a fixed step width of 1. If we change this equation to:

$$f(x + h) = f(x) + h.D(f(x))$$

Then we already have the Euler-Cauchy method, and all we have to do is to change the formula in the t-column and in the formula in the $x(t)$-column. A worksheet using this somewhat generalized method could look like this:

In this diagram the arrows are used to indicate two formulas, the formula for the t-column and the formula for the $x(t)$-column. These two formulas are related to the solution method of our difference or differential equation, they are not related to the specific equation under consideration. The specific equation is located in the $\Delta x(t)$-column, and we did not have to change this column when we changed the solution method.

Multivariable Systems

Our basic method of using a spreadsheet to solve difference equations numerically is not restricted to one-variable problems. To elaborate the material a little let us study a simple two-variable system described by a Lotka-Volterra equation.

Our system has the following equations:

$$\Delta x_1(t) = x_1(t)\,(a_1 - a_2 x_2(t))$$

$$\Delta x_2(t) = x_2(t)\,(-b_1 + b_2 x_1(t))$$

In a certain sense this model is an extension of our previous one-variable-model. We can see this more explicitly if we look at it in a rewritten form:

$$\frac{\Delta x_1(t)}{x_1(t)} = a_1 - a_2 x_2(t)$$

$$\frac{\Delta x_2(t)}{x_2(t)} = -b_1 + b_2 x_1(t)$$

These equations show that $x_1(t)$ just follows a simple exponential growth model with constant relative growth if $x_2(0) = 0$. Similarly, $x_2(t)>$ just follows a simple exponential growth model if $x_1(0) = 0$. In out logistic growth model we changed the constant growth rate to a growth rate linear in state. In our new model we doing something similar. The important difference is that the growth rate depends linearly *on the state of the other system variable*. So we have a system of 2 coupled variables.

Let us set up our spreadsheet for this model:

a_1	0.03			
a_2	0.0002			
b_1	0.02			
b_2	0.0003			
t	$x_1(t)$	$x_2(t)$	$\Delta x_1(t)$	$\Delta x_2(t)$
0	100	1000		

Translating the equation for the difference for the first variable into the "arrow notation" yields:

a_1	0.03			
a_2	0.0002			
b_1	0.02			
b_2	0.0003			
t	$x_1(t)$	$x_2(t)$	$\Delta x_1(t)$	$\Delta x_2(t)$
0	100	1000	*(▲ - ▲ ▲)	

and similarly the equation for the second variable yields:

a_1	0.03			
a_2	0.0002			
b_1	0.02			
b_2	0.0003			
t	$x_1(t)$	$x_2(t)$	$\Delta x_1(t)$	$\Delta x_2(t)$
0	100	1000		*(-▲ + ▲ ▲)

In the next step we calculate $x_1(1)$:

a_1	0.03			
a_2	0.0002			
b_1	0.02			
b_2	0.0003			
t	$x_1(t)$	$x_2(t)$	$\Delta x_1(t)$	$\Delta x_2(t)$
0	100	1000	-17	10
	+			

As in the one-variable model this simple arrow diagram illustrates the Euler-Cauchy method of integrating our differential system. In the next step we complete the row for $t = 1$ by copying the formulas for Δx_1 and Δx_2 down and the formula for x_1 to the right.

a_1	0.03			
a_2	0.0002			
b_1	0.02			
b_2	0.0003			
t	$x_1(t)$	$x_2(t)$	$\Delta x_1(t)$	$\Delta x_2(t)$
0	100	1000	-17	10
+ 1	83			

Now again, like in the one variable model, we copy the last row down. For this model it is advisable to do this a few hundred times. In fact, to study what we will do now we will need this model up to $t = 2000$.

To study the behavior of this model we can graph our variables $x_1(t)$ and $x_2(t)$. Selecting the three columns with t, $x_1(t)$ and $x_2(t)$ for an x-y-graph we get the following chart:

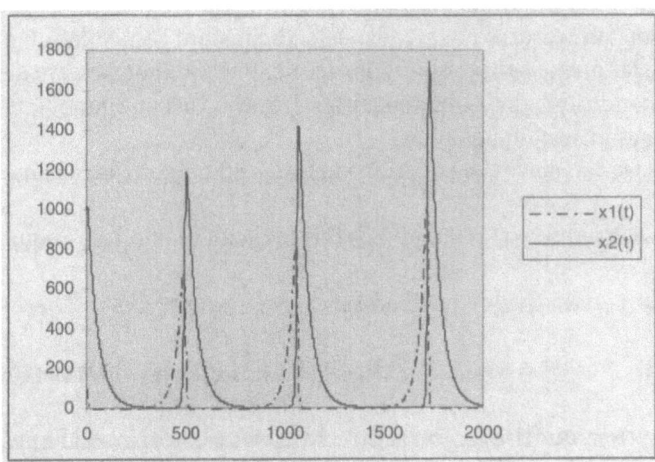

It is known from the theory of differential equations that this model can be understood better by means of another type of graph, namely the phase diagram. To display this diagram, we select $x_1(t)$ and $x_2(t)$ and create an XY-graph with these two variables.

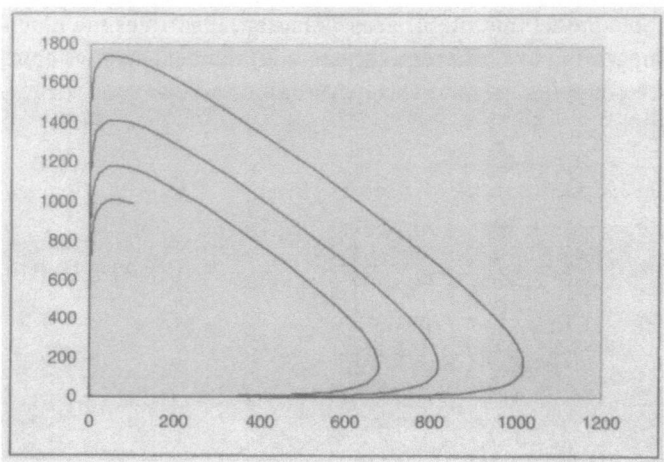

This graph seems to indicate that out model is cycling around some point close to $x_1 = 60$ and $x_2 = 200$. When we enter these values into the spreadsheet we immediately see how the graphs change, and we also see that the oscillations of $x_1(t)$ and $x_2(t)$ become much narrower. So the behavior of the system seems to indicate that there is a stable state. Changing our initial values a little bit more we see that for $x_1 = 65$ and $x_2 = 150$ the system is almost stable already. Doing so we are using one of the important features of spreadsheets: the instant recalculation of the worksheet whenever one or more values in it are changed. In our case we are changing the initial conditions for a dynamical system, and we can immediately see the consequences in system behavior.

In our special case we can solve the equations to find a stable state using algebraic methods also:

For a stable state we need $\Delta x_1(t) = 0$ and $\Delta x_2(t) = 0$. Using the defining equations we see that these conditions are fulfilled for $x_1 = \frac{b_1}{b_2}$ and $x_2 = \frac{a_1}{a_2}$. Therefore, in our case we get $x_1 = \frac{200}{3} = 66.6667$ and $x_2 = 150$. Using these values as the initial values of our system our first graph suddenly shows two horizontal lines, and the phase diagram shows only one point. So these values really do give us a stable state.

Higher-order Systems: Newtonian Dynamics and Planetary Motion

We can extend the range of dynamical systems once more and study a two-variable second order system. We will investigate a very well known example: a planet orbiting around a sun, following Newton's laws of gravity.

The sun is placed in the origin of our coordinate system, and the planet will have coordinates (x, y) and velocity (v_x, v_y). The Newtonian laws tell us that the acceleration vector (a_x, a_y) points in the opposite direction of (x, y) and has length $\frac{c}{r^2}$ where $r = \sqrt{x^2 + y^2}$. This can be rewritten algebraically: $(a_x, a_y) = \frac{c}{r^3}(-x, -y)$

Let us start building a spreadsheet model for this system:

c		120						
t	x(t)	y(t)	v$_x$(t)	v$_y$(t)	r	a$_x$(t)	a$_y$(t)	
0	100	0	0	1	SQRT(▲^2+▲^2)			

The equation for the acceleration columns then looks like this:

c	120							
t	x(t)	y(t)	v$_x$(t)	v$_y$(t)	r	a$_x$(t)	a$_y$(t)	
0	100	0	0	1	100 ▲*▲/▲^3			

There is a new symbol in the representation of this formula. The arrow in the r-column has a "circle on rails" at its tail. These rails indicate that this reference will move when the formula is copied down, but it will not move when the formula is copied horizontally. The formula content of this cell (it is cell G4) is =$B$1*B4/$F4. The reference $F4 is a mixed reference, row-relative and column-absolute, and since we have designed it this way we can copy it to the right to get the formula for a_y. Using our standard method of getting the next state of the system we enter the following formulas:

c		120						
t	x(t)	y(t)	v$_x$(t)	v$_y$(t)	r	a$_x$(t)	a$_y$(t)	
0	100	0	0	1	100	-0.012	0	
+ 1	+ ▲							

Setting up our sheet this way we used the equations:

$$(x(t+1), y(t+1)) = (x(t), y(t)) + (v_x(t), v_y(t))$$

for the location of our planet. The formula for velocity is similar:

$$(v_x(t+1), v_y(t+1)) = (v_x(t), v_y(t)) + (a_x(t), a_y(t))$$

Therefore we can extend the worksheet (again creating a formula just once and then copying it):

c	120						
t	x(t)	y(t)	$v_x(t)$	$v_y(t)$	r	$a_x(t)$	$a_y(t)$
0	100	0	0	1	100	.012	0
+1	100	1					

To create the formulas for $t = 1$ for the last three columns we copy down the formulas directly above. To study the system behavior we copy the formulas down for values up to $t = 1000$. Then we create an XY-graph, using the x- and y-columns of our data. This graph displays the orbit of the planet.

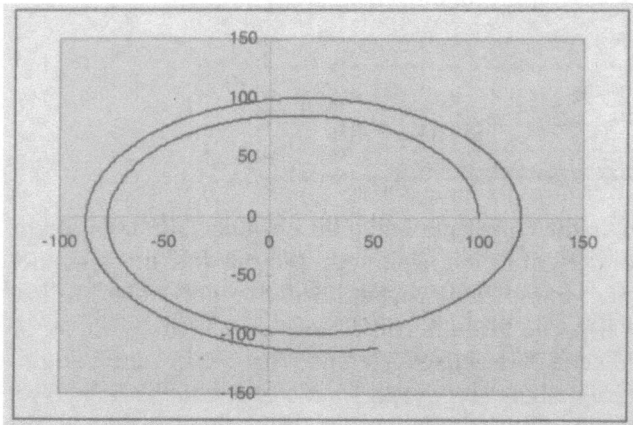

The graph looks reasonable, but it is not completely in agreement with what we know about planetary orbits: the orbit is not closed. The reason is that our simplistic model of integrating the underlying differential equation is not good enough. We were using:

$$(x(t + 1), y(t + 1)) = (x(t), y(t)) + (v_x(t), v_y(t)) \text{ and}$$

Using the same basic idea we also have:

$$(v_x(t + 1), v_y(t + 1)) = (v_x(t), v_y(t)) + (a_x(t), a_y(t))$$

So we were using the velocity at $t = 0$ to calculate the location at $t = 1$ from the location at $t = 0$. Just using common sense shows it would be more reasonable to use the velocity at $t = \frac{1}{2}$. Doing this we arrive at the equation:

Adjusting our spreadsheet according to these modifications of the equation we get:

c	120							
t	x(t)	y(t)	$v_x(t+\frac{1}{2})$	$v_y(t+\frac{1}{2})$	r		$a_x(t)$	$a_y(t)$
0	100	0	0	1	100		-0.012	0
1								

What happened is that in the first version (using the Euler-Cauchy method of integration) the increments for x, y, v_x, and v_y were taken from the row above. using the modified integration method where v_x and v_y are taken at times between the times for x, y, a_x, and a_x the increments for v_x and v_y are taken from the same row. So we changed the integration method in the spreadsheet by turning a diagonal arrow indicating a formula reference into a horizontal arrow.

It is well known that this method, the half step method, works much better than the classical Euler-Cauchy for second order dynamical systems where acceleration depends only on location and not on velocity.

After making this technically small but conceptually large change in the sheet the graph of the orbit changes, and the orbit closes.

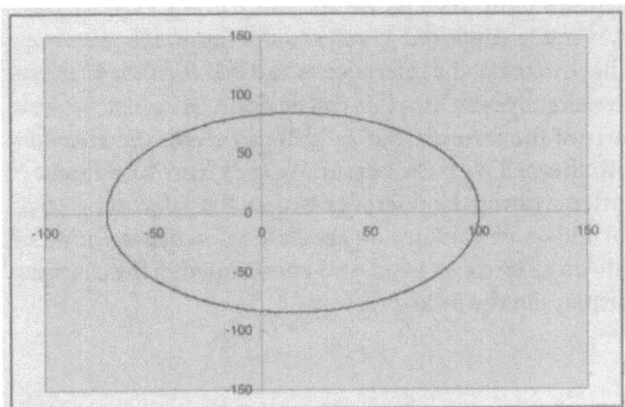

All the models developed so far are contained in the file Mathmod.XLS on the CD-ROM accompanying this book.

Simple Stochastic Models

So far we have only studied deterministic models. Spreadsheets can also be used to study stochastic models, and we even can do Monte-Carlo simulations with spreadsheets. Let us illustrate this with a simple example. We will simulate flipping a coin ten times and counting the number of heads. We will set up our model in a way that we can simulate biased and unbiased coins, in other words our

coin will have a parameter which is the probability of throwing a head. The basic
tool we will be using is the RAND function available in any modern spreadsheet.
RAND yields a floating point random number between 0 and 1. For our coin flipping model we will use 1 for head and 0 for tails. How can we create a formula
which will give us 1 with probability p and 0 with probability $-p$? The answer: by
using the IF function. IF(RAND()<=p,1,0) calculates a random number between
0 and 1, and if this random number is less than or equal to p we get 1, otherwise
we get 0. So we set up our sheet this way:

p		0.5
1	IF(RAND()<= ,1,0)	
2		
3		
4		
5		
6		
7		
8		
9		
10		

The shading in this diagram indicates that we are going to copy the formula
down. All the copies, however, are supposed to refer to the same cell containing
the p-value, therefore we have to make the reference to this cell absolute when we
create the formula. We have already seen how this can be done. Of course, we have
to type all the written parts of the formula, but we still can create the reference
to the p-cell by pointing to this cell with the cursor. We only have to remember
to press function key F4 after creating the reference to turn the reference into an
absolute one. Since the formula now contains an absolute reference to the p-cell
we can copy the formula down as far as we need, and as indicated in the diagram.

On screen this sheet displays in the following way:

p		0.5
1		1
2		0
3		0
4		1
5		0
6		0
7		0
8		1
9		1
10		0

This display corresponds to one series of ten coin flips. If we want to calculate
another series, we have to start recalculation. In Excel this is done by pressing the

F9 function key. Each time this key is pressed, we will get a new series of ten coin tosses, and each time the values will be different.

Originally we set out to count the number of heads in our series. This is modeled easily, we have to add the cells in the column with the 1 s and 0 s.

This diagram indicates that the cell with SUM in it contains the sum of all the cells directly above.

This cell displays the number of heads, and each time we press F9 we will "perform a new random experiment". What we want next is a protocol of performing 100 of these experiments. We already know the tool for doing this: the Data Table command. The only difference is that we do not have a parameter which directly (in a mathematically predictable way) influences the outcome here. But we have something similar: the p-value of our coin. So next to our table we set up a column with the p-values for the coins we want to be used. We want these p-values to be equal. So we set up the following sheet:

p	0.5			
			0.5	
1	1			
2	0			
3	0			
4	1			
5	0			
6	0			
7	0			
8	1			
9	1			
10	0			
	4			

Since we want to perform a hundred of these experiments the worksheet has to have many more rows, but the basic structure is fully given by this diagram. Since we want a protocol of the number of heads we have to add the formula indicated by the following arrow diagram in our worksheet:

p	0.5			
			0.5	
1	1		0.5	
2	0		0.5	
3	0		0.5	
4	1		0.5	
5	0		0.5	
6	0		0.5	
7	0		0.5	
8	1		0.5	
9	1		0.5	
10	0		0.5	
	4		0.5	

The grayshading in this diagram indicates the area to be selected for the Data Table command. When this area is selected and the Data Table command is selected from the menus, the last thing to be done is indicating cell B1 as the column input cell, and then the Monte-Carlo simulation will start running and after a very short time we will have a sheet like this:

p	0.5			4
			0.5	5
1	1		0.5	3
2	0		0.5	3
3	0		0.5	6
4	1		0.5	5
5	0		0.5	4
6	0		0.5	3
7	0		0.5	8
8	1		0.5	7
9	1		0.5	6
10	0		0.5	7
	4		0.5	8

giving the outcome of simulated repeated series of ten coin flips in the rightmost column. Changing the p-value in the shaded cell to 0.6 will immediately give us a new run of 100 series of 10 coin flips each, this time with a biased coin if we are using Excel. Excel puts ranges created with the Data Table command in the "recalculation loop", meaning that these areas will be recalculated automatically whenever anything in the sheet changes. Some other spreadsheet programs (especially Lotus 1-2-3) behave somewhat differently. They only put the values into the Data Table range. To update the table one has to go to the Data Table menu item again. In our example this would imply that to get the results for an unfair coin with such a program one would have to change the p-value and then additionally rerun the Data Table option manually to get the new simulation run.

Since all spreadsheet programs also have tools for counting frequencies, these simulation runs can be used as input for further formulas giving the frequencies of the different possible outcomes of our simulated experiments. The basic idea behind what we have just set up was to use Data Table for a purpose for which it was probably not originally designed. A file named CoinFlip.XLS containing our example with the simulated coin flips can be found on the CD ROM accompanying this book.

Instead of using the very simple formula simulating flipping the coin we also could set up more complicated formulas and do some rather complicated Monte-Carlo simulations.

Let us briefly mention a more complex example. Using this technique we can create random paths of a particles and set up a model containing the path of a single particle following some random mechanism. If we set up a model for simplified Brownian motion with a random step of fixed small length in a direction with equal distribution we can calculate the maximal distance from the origin of one such simulated particle. Using the Data Table technique we can repeat this simulation e.g. 1000 times and then calculate the frequency distribution of the maximal distance for these 1000 particles. A worksheet, Brown.XLS illustrating this example can be found on the CD-ROM accompanying this book. Having set up an example of this kind it is very easy to change the underlying random phenomenon, for example one might replace the directionally uniform random distribution for the particle step by another distribution and immediately see the distribution of the maximal distance from the origin for another kind of random motion. Generally speaking, the combination of a model containing random components, Data table, and the Frequency function allows us set up a Monte Carlo simulation for an extremely wide range of systems. The advantage of doing this with a spreadsheet compared to, say, special purpose simulation programming languages is at least twofold. Spreadsheets are a more general tool, knowledge of how to use a spreadsheet is applicable in many more different situations than knowledge of special purpose tools. And spreadsheets have the invaluable auto recalculation feature. When setting up a model step by step there is always immediate feedback on how the parts of the model designed so far, work. It is much more difficult to get feedback about the functioning of model components when the model is implemented in a classical multipurpose or special purpose programming language. With a language-based tool even if an interpreter is available and not only a compiler, one has to actively start "test runs". In a spreadsheet, each time a formula is entered, the sheet is recalculated, giving us test runs at every moment during the process of model implementation.

Similar arguments also are valid when we are modeling with difference and differential equations. When we use a spreadsheet as our modeling tool, the fact of having test runs at each moment of our modeling process helps in the stepwise refinement of the model. When using tools of a lesser degree of interactivity, quite often we try to finish the conceptual analysis of the model before we start writing code. With a spreadsheet, the conceptual modeling process and the model implementation tend to be much more interwoven. One main advantage of spreadsheet programs with dynamic equations is the ease of changing model parameters. In the examples discussed earlier in this chapter we have seen how to set up a model with parameters "to play with." The newer releases of Microsoft Excel allow us to go even one step further. We can connect "sliders" with parameters and then we can literally "turn the knobs" on our model. The CD-ROM bound into this book has two examples using this mechanism. The first model Logistic.XLS is the logistic iteration $x_{n+1} = \lambda x_n(1 - x_n)$ which leads to the famous Feigenbaum diagram. In this example, the parameter λ is controlled by a slider. Moving the slider shows the iteration path for different parameter values, and it helps to develop a "feeling" for the sensitivity of the model depending on the parameter range. Setting up a slider control is not too difficult. We only have to activate the Forms toolbar. This toolbar contains a tool called a "scroll bar". Clicking this tool we can draw a scroll bar anywhere on the worksheet. Then we have to click this scroll bar with the right mouse button (at least in Microsoft Windows, on an Apple Macintosh we have to press the Control key and then click the bar) and the dialog box appearing on screen allows us to connect the slider to a cell. Once the slider is connected, moving the slider will change the cell value. The CD-ROM contains a second example named Pense.XLS using this technique in a much more elaborate model. This second model is taken from social science and deals with the ratio of the number of people in retirement age and the number of people in workforce age. Sliders allow the retirement age to be modified, and using this one can see the numerical consequences of changing the laws governing retirement. Also included on the CD-ROM is a Word for Windows file named Pense.doc giving more details on the "mechanics" of the model. This and many more examples can be found on the Web site at URL:

http://sunsite.univie.ac.at/Spreadsite

This site, at the University of Vienna (managed by the author of this chapter), is devoted to computer support for mathematics, statistics, and science education, and it puts strong emphasis on spreadsheets as tools for these subject areas. Incidentally, the file Pense.DOC mentioned above was converted from the HTML file describing the worksheet Pense.XLS on this Web site. The original HTML version of the document, Pense.HTM is also provided on the disk. Non-Word users will, of course need a Web Browser to view it from there.

At this point we leave it to the reader to use the ideas presented here to study models of perhaps much higher complexity. The main purpose of this chapter was not to give "large scale" examples of deterministic and stochastic mathematical models. We wanted to demonstrate how spreadsheet techniques can help understanding models better while they are developed and refined. Additionally, we wanted to show that spreadsheet programs can be used as calculation engines for a rather wide range of mathematical models.

Applications of Spreadsheets in the Earth Sciences

J. P. Le Roux and R. D. O'Brien

Introduction

It is impossible to cover the full scope and capabilities of spreadsheets as applied to all branches of the earth sciences in a single chapter. For this reason, only a limited number of typical applications are discussed in some detail, but the reader is referred in each section to supplementary material in the literature. The content is structured in such a way that it is accessible to those with only a very basic knowledge of spreadsheet handling, but it covers most of the essential concepts and provides a foundation for more advanced operations.

The majority of spreadsheet templates published in geoscience journals and books over the last decade were for Lotus 1-2-3, which can usually be adapted with minor changes for similar spreadsheets such as Quattro, Twin or As-Easy-As. Microsoft Excel has also become very popular in recent years. Most spreadsheets are integrated software packages combining database management systems, graphics software, and built-in programming languages utilized as keyboard macros. They are thus ideally suited for solving problems in the earth sciences. Excel is particularly convenient in that it can open other types of spreadsheet files as well. Before opening such files in Excel, select *Tools – Options* from the menu. In the Options dialog box select the Transition tab and make sure that the Transition Formula Evaluation option is checked. This will ensure that the file is opened correctly without losing any data. The Transition Formula Entry option allows you to enter a formula in Excel in the Lotus syntax, as Excel will automatically convert the formula to its own syntax. [*Note:* Some Excel functions which have no Lotus equivalents do not operate as expected while the Transition Formula Evaluation option is checked. For example, the Excel function STDEV(list), which should return the standard deviation of a sample, will return the standard deviation of a population if this option is checked.]

How to Use this Chapter

For each of the examples discussed below two different files are provided, one for the raw data sets and another for the completed programs (Fig. 1). The names of files containing only data start with the prefix D, in contrast to the program file names which commence with P. The data (D) files are intended for use in conjunction with the step-by-step instructions in the text. Working through the latter should help to familiarize the reader with the most commonly used spreadsheet procedures. Answers can subsequently be checked in the program (P) files. All these exercise worksheets can be found in the file ErthSci.XLS on the disk.

Data Management and Presentation

Spreadsheets can be very useful for data storage and management, as they have many features in common with more advanced database systems. They can also provide a low-cost alternative to commercial software packages designed for specific tasks. Dahl (1990), for example, provided three Lotus macro programs for the acquisition, management and reduction of inductively coupled plasma spectrometry data, which can also be modified for use with diverse analytical instrumentation. This yields a distinct advantage over conventional PC-based data stations, which are commonly linked to specific instruments and cannot be customized to perform the additional routines required by analysts.

High-quality graphs such as scatter- and triangular plots can be produced by utilizing the graphic capabilities of spreadsheets. Excel programs which simplify the production of such diagrams, for example, were devised by Christie and Langmuir (1994) and Marshall (1996). An interesting application of graph functions was also demonstrated by Girard (1992), who developed a routine in Lotus to produce stereographic projections of planar and linear fabric measurements. The first five ranges (A-E) of the Y-axis on XY graphs are used to plot various selections of structural elements in different symbols, whereas the sixth range (F) is employed to draw the circular projection contour, the central cross-hairs and the north indicator according to the methods of Holm (1988b) or Berge (1991).

Database Functions

Two specialized data management commands of Excel are particularly convenient. *Data Sort* commands are used to order unsorted data either alphabetically or numerically, whereas *Data Filter* commands are applied to locate records in the database which conform to certain specified conditions.

In the example below, an introduction to some of the database management and graphics features of Excel is provided. The worksheet D-Cleavage in the accompanying diskette contains structural data (S1- and S2-cleavage orientations) of three hypothetical formations. The data are unsorted, as would be the case if they were entered directly from field notebooks.

As a first step in the data analysis, it may be necessary to separate the structural data of the three different formations. To do so, first select the data range to be sorted (A3:C83) and choose *Data Sort* from the menu bar using your mouse. The Sort dialog box will appear. As the column titles were included in the range selection they will appear in the list boxes. Sort the data according to the FORMATION. If the strike directions are to be sorted as well, select STRIKE from the second box. In both cases the sort order can also be selected: ascending (ABC, 123) or descending (CBA, 321). Click OK to sort the data.

While recording the strike and dip of cleavage in the field, it may have become apparent that two sets of cleavage prevail: an older S1-cleavage striking more or less east-west, and a younger S2-cleavage with a north-south orientation. To extract either one from the database, two options are available: AutoFilter and Advanced Filter. AutoFilter does not actually extract the data that evaluate true but hides those values that evaluate false. The Advanced Filter copies those records that evaluate true to another location. To use AutoFilter in the present example, first select the range to be queried, including the column titles (A3:C83). From the menu bar select *Data Filter AutoFilter*. The column titles will appear as drop down lists (buttons on the titles). Click on the arrow in the STRIKE cell and select Custom, which causes the custom AutoFilter dialog box to appear. Enter the filter criteria (in this case, strike directions exceeding 245 degrees) as: >=245 into the text box and click OK. The rows containing records that do not satisfy the criteria will be hidden from view. The cells that remain visible (A42:C57) and the column titles can now be selected and copied (*Edit Copy* or Ctrl + C) to another location. For example, with the active cell on D3, press Enter. As rows 4 to 41 are hidden only the column titles will be visible. In order to view all the data again click on the arrow in the STRIKE cell and select All. Cancel the AutoFilter by selecting *Data Filter AutoFilter* from the main menu bar.

The result shows that the older S1-cleavage is confined to the Butte Formation, indicating that this stratigraphic unit was affected by the first phase of compression before deposition of the Angel and Coyote Formations.

Data Presentation

To present the cleavage data in picture format, the graphics capabilities of Excel can be employed. For example, one might wish to show the strike directions or dip angles as bar graphs, which requires that the data be grouped first into class intervals. In this case, strike directions for the Angel Formation, which vary between 151 and 202 degrees, will be grouped into 10 degree sectors. In cells G3, H3 and I3 enter CLASS INTERVAL, FREQUENCY and RELATIVE PERCENTAGE (or appropriate abbreviations) respectively. The 10-step increment values between 160 and 210 are entered in cells G4:G10. In cell G4 enter 160 and with G4 as the active cell select *Edit Fill Series*. In the fill dialog box select Columns, enter 10 as the step value and 210 as the stop value, click OK. To calculate the frequencies select the output range (H4:H10). In the edit box (i.e. within the range of cells selected by dragging the mouse) enter the formula =**FREQUENCY(B4:B25,G4:G9)**. As this is an array function, press the [Ctrl]+[Shift]+[Enter] keys simultaneously. (To make use of the Function Wizard, click the wizard button and from the function category select Statistical and then Frequency from the function name list. Then click Next>. In the data_array text box enter the range B4:B25 or select the cells with the mouse, and press the [Tab] key to move to the bins_array text box, entering the range G4:G10. Complete the command by selecting the Finish button followed by [Ctrl]+[Shift]+[Enter]). The data will be sorted automatically into the appropriate 10 degree sectors between 150 and 210. In cell H10 enter =**SUM(H4:H9)** to add the values of H4:H9. In cell I4 enter =**ROUND(H4*100/$H$10,3)**, rounding the value of H4*100/H10 to the third decimal. The $ sign fixes the next symbol, so that it remains unchanged when the contents of cell I4 is copied to other cells. This is referred to as an absolute reference. Copy cell I4 by dragging the bottom-right corner of the active cell through I4:I10.

To construct a bar chart of the strike directions, select the range containing the relevant data (G4:I9) and click the ChartWizard button. Next, outline the area where the graph is to be viewed by dragging the cross hair with the chart symbol. In Step Two of the ChartWizard, choose the Column Graph, and in Step Three Chart Format 1. In Step Four, make sure that the data series is set to Columns and that the Category Axis is set to 1. Enter the graph titles in step five and select the Finish button to view the graph.

In addition to normal column charts, Excel has commands for producing 3-D column charts, pie charts and line charts. It is also possible to create more sophisticated diagrams by adapting the available graph functions. For example, histograms are commonly used in geology instead of bar graphs. To produce a conventional histogram (in which the bars are adjacent to one another, in contrast with the isolated bars of bar graphs) of the percentage of cleavage strike directions within each 10° sector, click the ChartWizard, select an area for the chart and define the range as G4:G9, I4:I9. On completion of the bar graph, double-click it to edit, click on the columns to select the data series and then click the right mouse button. From the shortcut menu select Format Column Group. In the dialog box select the Options tab and set the Gap width to 0. Fig. 1 shows the printed version.

In Excel, a formatted graph can be saved so that a new graph can later be formatted automatically with the same settings. Double-click the graph and then

Fig. 1. Histogram of cleavage strike directions in Angel Formation

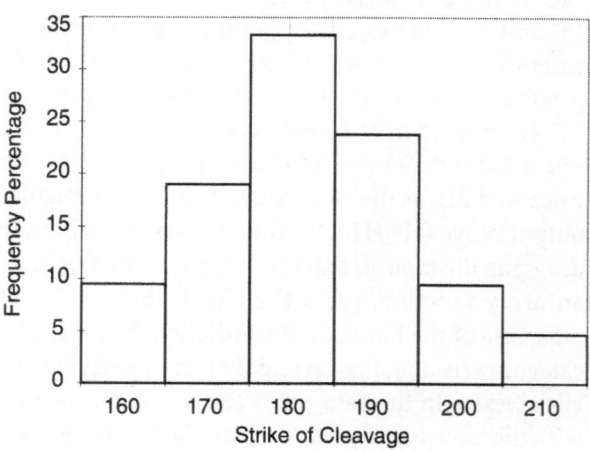

click the right mouse button. From the shortcut menu select Autoformat, User Defined, Customize, Add. Enter the name of the custom format (e.g. Histogram) and a description, then click OK and close the graph. To use this format again double-click the chart and from the shortcut menu select Autoformat, User Defined. From the list of user-defined formats select the name of the format.

To automate routine or repetitive tasks, for example when producing graphs, spreadsheet macros can also be very useful. Excel provides a macro recorder which will record all your actions so that you can run them later. The starting conditions must be set before you begin recording (for example, if the macro applies to a range, select the range first). From the *Tools* menu select *Record Macro –* Record New *Macro*. In the Record New Macro dialog box enter the name of the macro, select the Options button and check the Store in Personal Macro Workbook option. The macro can also be added to the Tools menu and/or assigned to a shortcut key. Carry out the required actions and then stop the macro recorder by clicking the Stop button. Macros that were recorded to Excel's Personal Macro Workbook (which is created when you first record a macro to it) are found in the file Personal.XLS. To open this file, select Unhide from the Window menu. The worksheet REE Macro contains a macro written in Visual Basic for applications, that will create and format a rare earth element diagram (Fig. 2). Select the worksheet REE Macro and copy it to the personal workbook as a new module by selecting *Insert – Macro – Module*. To assign a shortcut key to the macro, select *Tools – Macro*, choose the macro name (ReeGraph) and press the Options button. In the Assign To field enter the shortcut key or the text to appear on the Tools menu.

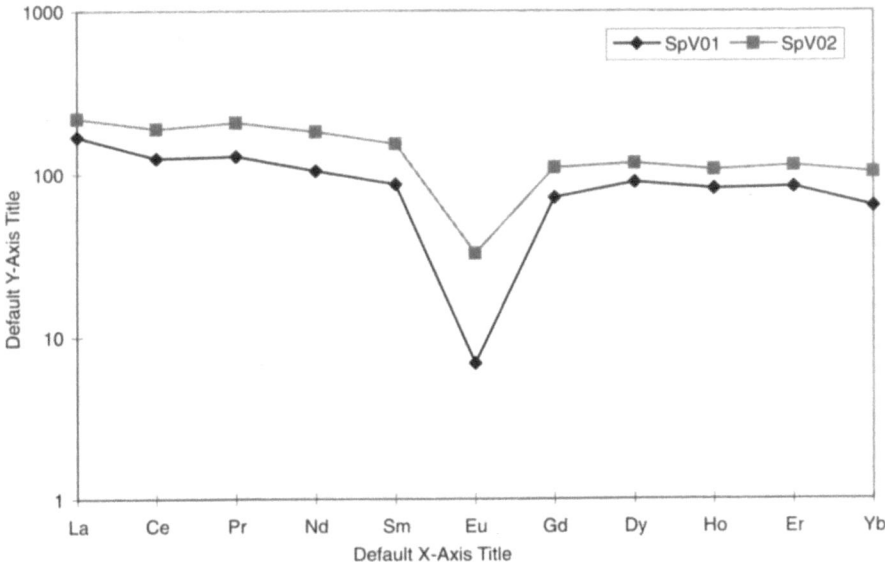

Fig. 2. Rare element diagram created with a macro written in Visual Basic

As this macro has been loaded into the Personal.XLS workbook it is available to any worksheet that you may have open. To draw the graph select the cells A3:L5 from the ReeGraph worksheet. Run the macro by pressing the shortcut key or selecting the macro name from the Tools menu. The macro program listing is provided in the Appendix at the end of this chapter.

Mathematical Analysis

The scope of numerical analysis in the earth sciences is virtually unlimited, and spreadsheets are used increasingly for a wide variety of problems. Excel provides a choice of mathematical, statistical and financial functions (always preceded by =), which are in effect shorthand methods of entering formulae. For example, the formula (B3+B4+B5+B6+B7) can be entered as =SUM(B3:B7). Extensive use is made of functions in the examples below.

Elementary Mathematical Applications

All branches of the earth sciences make use of mathematical analysis, which may involve elementary to very complex calculations. Spreadsheets are used mainly to expedite such tasks. The traditional approach to converting sample depths to ages, for example, requires meticulous hand-plotting of a depth/age curve based on known stratigraphic datum points, which is time-consuming and prone to subjectivity. Davies et al. (1992) outlined a simple Excel routine obviating the need for hand-plotting, as it automatically converts depths to ages and vice versa. Other applications in this category include that of Le Roux (1992b), who designed a Lotus spreadsheet to facilitate the numerous calculations required to predict the behavior of particles in fluids, and Rao (1995), who demonstrated the use of an interactive, macro-driven Lotus program for biotite-garnet thermometry. Martin (1996) constructed an Excel worksheet to perform thermodynamic calculations, create phase diagrams and retrieve thermodynamic data from phase equilibria experiments, whereas Reche and Martinez (1996) published a related Excel program to perform thermobarometric calculations in metapelitic rocks.

An example from hydrogeology is illustrated in the worksheet D-Permeability. To determine the permeability of a sandy soil, an auger hole was drilled to below the water table (Fig. 3).

The water was pumped out and the depth of the recovering water table monitored at 15 minute intervals, with the results as tabulated in the worksheet. The coefficient of permeability (k) can be obtained using, for example, the Hooghoudt (1936) method, which applies to variable head situations. The Hooghoudt equation is given by:

$$k = [aL/t(2H + a)] \ln (y_1/y_2)$$

where a is the radius of the auger hole, L is the empirical length over which head loss y occurs (given by aH/0.19 in this situation), y_1 and y_2 are the heads at the

Fig. 3. An example from hydrogeology: relationship between auger hole and water table

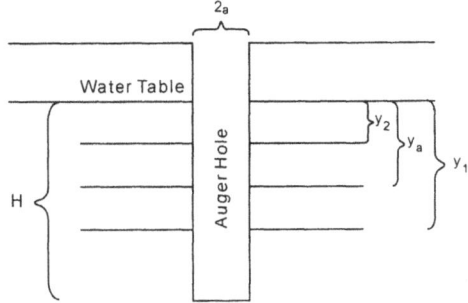

start and end of the experiment, t is the time in seconds, and H is the permanent head (the distance between the bottom of the hole and the original water table).

To calculate the permeability coefficient of the soil, use the following formula in cell B9:

= (((B4/2)*((B4/2)*(B5-B6)/0.19))/(2*((B5-B6)+(B4/2))*D7))*LN((B5-B6)/E7).

The permeability coefficient can also be determined using the equation of Ernst (1950), which is given by:

$$k = \{40/[(20 + H/a)(2-y_a/H)]\}(a/y_a)(\Delta y/ \Delta t)$$

where Δy is the rise in the water table during any time interval Δt, and y_a is the average head during the specific time interval. Write the formula for this equation in cell B8 and compare the results. (The permeability coefficient determined by these methods are very similar, with the Ernst equation yielding a slightly higher value of 2.352 x 10^{-5} against 2.241 x 10^{-5} m/s for the Hooghoudt equation).

Two-Dimensional Applications

Earth scientists often work with two-dimensional or spatial data such as geological maps, drillhole grids and geochemical or geophysical sample locations and data stations. As a result, there are many examples of two-dimensional data analysis in the literature. For example, Le Roux and Rust (1989) developed a procedure to combine an unlimited number of different types of maps, for which a Lotus template was provided by Le Roux (1991b). An iterative trigonometric technique to determine the transport patterns of sediments in the shallow marine environment from the spatial trends of their grain-size parameters, was proposed by Le Roux (1994b). As the iterative procedure is extremely time-consuming, the only practical approach is by computer, for which a spreadsheet such as Lotus is well suited (Le Roux, 1994c). Fourie and Odgers (1995) published a Lotus template for the processing and interpretation of seismic refraction data, which uses its plotting function to draw the interpreted structural model in two dimensions.

In some applications, spreadsheets can be used directly as orthogonal coordinate systems on which data can be recorded in their relative positions, as illustrated for example in the later section on Geological Modeling. In the example below, the use of Excel for the analysis of directional data is discussed. The program is updated from a Lotus 1-2-3 template of Le Roux (1991a) to incorporate an improved method of calculating the range of palaeocurrent directions as applied to the determination of channel sinuosities (Le Roux, 1994a). In the worksheet D-Palaeocurrent, palaeocurrent directions as recorded in the field are shown in column A. These have to be sorted into 10° sectors so that the conventional method of vector analysis for grouped data (Curray, 1956) can be applied. In this procedure, the vector mean azimuth (θ) is given by:

$$\tan \theta = \Sigma(n \sin \theta)/ \Sigma(n \cos \theta)$$

where n is the frequency percentage of records in each 10° sector. The vector magnitude in percent (r) is obtained by:

$$r = [\Sigma(n \sin \theta)^2 + \Sigma(n \cos \theta)^2]$$

The sinuosity value (P) can be calculated from r by using the method of Le Roux (1992a):

$$P = (\Omega/2)/\sin(\varnothing/2)$$

where Ω and $\varnothing$ are the directional range in radians and degrees, respectively, as obtained from the graph of $\varnothing$ against r provided by Le Roux (1994a). The values of $\varnothing$ as related to r are entered as a table (columns H and I) in the worksheet.

In the worksheet D-Palaeocurrent, enter =SUM(C4:C39) in cell C40. In cells D4 and E4 enter =C4/\$C\$40 and =D4*100 and copy to D5:D39 and E5:E39, respectively. In cell E40 enter =SUM(E4:E39). In column F4:F39 the sine of the sector median values are multiplied with their corresponding frequency percentages. For example, in cell F4 enter =E4*(0.087), i.e. (frequency percentage in sector 001°–010°) x (sin 5°), and in cell F33, =E33*(–0.906), i.e. =E33 x [sin (360°-295°)]. Note that the sine values in cells F4:F21 are positive, but negative in cells F22:F39. In cell F40 enter =SUM(F4:F39). The same procedure is repeated in column G4:G39 to multiply the frequency percentage of each sector with the cosine of its median value, e.g. in cell G4 enter =E4*(0.996). The cosine values are positive in cells G4:G12 and G31:G39, but negative in G13:G30. The values of G4:G39 are added in cell G40 using the =SUM function. Columns H4:H103 and I4:I103 have been used to tabulate the vector magnitude (VMT) against the corresponding operational range (twice the angular deviation) of directions.

Calculation of the vector mean is performed in cell J4 with the formula:

=ROUND(IF(ATAN2(G40,F40)*57.29577>0,ATAN2(G40,F40)*57.29577,
ATAN2(G40,F40)*57.29577+360),0)

The vector magnitude is given in cell J6 by: =ROUND(SQRT(F40^2+G40^2),0).
From the value of r, the corresponding value of the operational range $\varnothing$ can be

found in the table of columns H and I. In cell J8 enter =ROUND(VLOOKUP (J6,H4:I103,2),0) to find Ø. To calculate the sinuosity value P, enter =ROUND (J8*0.00873/SIN(J8*0.00873),2) in cell J10.

To sort the palaeocurrent directions in column A into 10° sectors, use the Frequency function. Select cells C4:C39 and enter the formula =FREQUENCY (A4:A41,B4:B39) in the formula bar followed by [Ctrl] + [Shift] + [Enter] to recalculate the worksheet. For this data set, the vector mean is 288°, the vector magnitude 87% and the sinuosity value 1.10.

Three-Dimensional Applications

In structural geology, it is often necessary to rotate inclined strata to some new orientation. For example, the analysis of structures below an angular unconformity overlain by tilted beds requires the latter to be rotated to their original attitude. This problem of secondary tilt can be solved geometrically or by the use of stereographic projections (Phillips, 1971). When a large number of data are involved, however, both these methods can be too time-consuming and it is quicker to use a computer.

A number of spreadsheet methods for solving such 3-dimensional problems have been published. Ozkaya (1995), for example, designed two Excel macros to calculate the vertical and true thickness of strata in deviated boreholes. Hinman (1993) presented a Lotus template for orientating diamond drillcore in regions where one planar structural element can be assumed to have a reasonably constant strike. In a combination of two- and three-dimensional applications, Le Roux (1991a) provided a 1-2-3 template to facilitate palaeocurrent analysis. In two separate subroutines, the dip azimuths of tilted crossbeds are rotated to obtain their original directions, followed by a vector analysis of the latter, as modified in the worksheet P-Palaeocurrent.

The worksheet D-Untilt contains structural data on the attitude of bedding planes (B), joints (JN) and cleavage (CV), as recorded in an older folded succession (O) of strata which are overlain unconformably by younger inclined beds (Y). It is assumed that the younger succession was horizontal at the time of deposition. To restore the younger strata to horizontal while tilting the older beds by the same amount, the method of Parks (1970) can be used. This involves computing the angle XYZ in the spherical triangle formed between the pole positions of the older planar surface, the horizontal plane and the younger beds on a Schmidt or Wulff stereonet.

The amount of rotated dip DIPR is given by:

$$\cos DIPR = \cos DIPX \times \cos DIPY + \sin DIPX \times \sin DIPY \times \cos DIFFS$$

where DIPX is the dip of the required planar surface (bedding, joint or cleavage), DIPY is the bedding dip of the younger succession, and DIFFS is the difference between the strike of the younger beds and the strike of the older planar surface.

Angle XYZ is used to calculate the direction of the rotated dip according to the following relationship:

cos XYZ = (cos DIPX – cos DIPR x cos DIPY)/(sin DIPR x sin DIPY).

The rotated dip azimuth of the older planar surface is given by:

RDA = APY – XYZ

when APX is greater than APY, or

RDA = APX + XYZ

when APX is less than APY, where APX and APY are the pole positions of the older planar surface and the younger beds, respectively.

In the worksheet D-Untilt, a calculating area can be created in block H1:T4, labeling the columns as follows: H1: APY; I1: APX; J1: DIFFS; K1: COSDIP; L1: XYZ; M1:T1; RDA1:RDA8. Enter the following formulae in Row 2.

- H2: = IF(B3>0,IF(B3<=90,B3+270,B3–90),0)
- I2: = IF(B5>0,IF(B5<=90,B5+270,B5–90),0)
- J2: = ABS(B3-B5)
- K2: = IF(B3>0,COS(B6*0.01745)*COS(B4*0.01745)+SIN(B6*0.01745)
        *SIN(B4*0.01745)*COS(J2*0.01745),0)
- L2: = IF(H2<>I2,ACOS((COS(B6*0.01745)-
        K2*COS(B4*0.01745))/(SIN((ACOS(K2)
        *57.29577)*0.01745)*SIN(B4*0.01745)))*57.29577,0)
- M2: = IF(K2>0,IF(AND(H2<I2,ABS(H2-I2)<180),H2-
        L2,IF(AND(H2>I2,ABS(H2–I2) <180),H2+L2,0)),0)
- N2: = IF(K2>0,IF(AND(H2<I2,ABS(H2-
        I2)>180),H2+L2,IF(AND(H2>I2,ABS(H2–I2)>180),H2-L2,0)),0)
- O2: = IF(K2<0,IF(H2<I2,H2-L2-180,IF(H2>I2,H2+L2-180,0)),0)
- P2: = IF(B3>0,IF(AND(H2=I2,B4<B6),H2-180,
        IF(AND(H2=I2,B4>B6),H2,0)),0)
- Q2: = IF(AND(M2>0,M2<360),M2,IF(M2<0,M2+360,
        IF(M2>360,M2-360,0)))
- R2: = IF(AND(N2>0,N2<360),N2,IF(N2<0,N2+360,IF(N2>360,N2-360,0)))
- S2: = IF(AND(O2>0,O2<360),O2,IF(O2<0,O2+360,IF(O2>360,O2-360,0)))
- T2: = IF(AND(P2>0,P2<360),P2,IF(P2<0,P2+360,IF(P2>360,P2-360,0)))

Complete cells H3:T4 by adapting the formulae in H2:T2. In the output area (block C1:D9), the following formulae are entered in column D:

- D3: = ROUND(IF(MAX(Q2:T2)>=90,MAX(Q2:T2)-
        90,270+MAX(Q2:T2)),0)
- D4: = ROUND(ACOS(K2)*57.29577,0)
- D5: = IF(D3<=270,D3+90,D3–270)
- D6: = ROUND(IF(MAX(Q3:T3)>=90,MAX(Q3:T3)-

        90,270+MAX(Q3:T3)),0)
- D7: = ROUND(ACOS(K3)*57.29577,0)
- D8: = IF(D6<=270,D6+90,D6–270)
- D9: = ROUND(IF(MAX(Q4:T4)>=90,MAX(Q4:T4)-
        90,270+MAX(Q4:T4)),0)
- D10: = ROUND(ACOS(K4)*57.29577,0)
- D11: = IF(D9<=270,D9+90,D9–270)

The input and output values in this program assume that the strike direction is recorded on the left of the observer while looking down-dip.

Statistical Analysis

Although statistical analysis of geological data is commonly performed using specialized software packages such as Statgraphics, spreadsheets provide an alternative approach. Koch (1990), for example, discussed the use of Lotus 1-2-3 as applied to confidence intervals, frequency distributions and probability calculations based on fictitious as well as real geological and geochemical data sets. In stratigraphy, Chi-square tests have been used for many years in Markov analyses of cyclicity in sedimentary successions. Modern techniques are based on quasi-independence, with the expected frequencies in the independent trials matrix being generated by an iterative method (Powers and Easterling, 1982). Le Roux (1994d) developed a modified approach based on the methods of Carr (1982) and Powers and Easterling (1982), for which a Lotus template was provided. [*Note:* This program as published contains an error, in that the row totals were used accidentally instead of the column totals in generating the expected transition frequencies.]

   In the following selection of some typical applications, it is assumed that the reader is familiar with basic statistical principles. Useful textbooks which may be consulted in this regard include Krumbein and Graybill (1965), Kock and Link (1971), Koosis (1977), and Davis (1986).

Elementary Statistical Applications

The worksheet D-Pebbles can be used to illustrate the application of Excel's around twenty statistical functions in the analysis of pebble data. At twenty data stations along a braided stream, the relative percentage and mean intermediate diameter of quartz pebbles have been recorded. Station numbers correspond to the distance from the first station (0) in km.

   It is useful to summarize the pebble data so that broad inferences may be drawn. Values typical of a distribution of measurements, known as *measures of central tendency*, include the mean, the median and the mode, for which Excel provides statistical functions.

   The mean value for the relative percentage of quartz pebbles over the entire 14.6 km stretch of stream channel, is determined by entering =**AVERAGE(B4:B23)** in any unused cell. For later reference, however, it is convenient if all cells are iden-

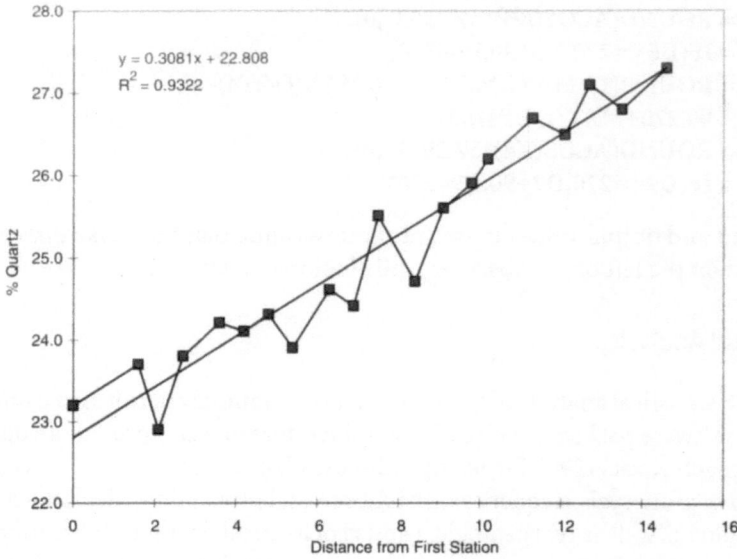

Fig. 4. Percentage of quartz pebbles as a function of distance

tified. As a number of summary statistics will be determined here, label cells A24: A28 as MEAN, RANGE, STANDARD DEVIATION, VARIANCE and NUMBER OF OBSERVATIONS (or n), respectively. In cell B24, enter =**AVERAGE(B4:B23)**. The range of percentage distributions is given by =**MAX(B4:B23)-MIN(B4:B23)**, which is entered in cell B25. Other summary statistics, known as *measures of variability*, are the standard deviation and the variance, which can be entered in cells B26 and B27 as =**ROUND(STDEV(B4:B23),3)** and =**ROUND(VAR(B4:B23),3)**, respectively. [Note: These are sample statistics; the standard deviation and variance of a population are given by the functions =STDEVP(list) and =VARP(list)]. The value of n can be obtained by =**COUNTA(B4:B23)**, which simply returns the number of nonblank cells in the specified list in cell B28.

The next step in the pebble data analysis would be to examine whether there is any statistical correlation between the relative percentage of quartz pebbles at any station and the distance from the first station. This requires regression of the data points on a scattergram. Regression analysis in Excel is performed by selecting the Tools – Data Analysis command and Regression from the Data Analysis dialog. Enter the X and Y ranges of the data (A3:A23, B3:B23), checking the Labels option as the column titles are included. In the output options section, select the Output Range (E1) and click OK. The regression statistics will appear in block E1:F8, while the graph will be produced in a separate worksheet. This can be copied onto the data sheet using the cut or copy commands.

A regression line can also be inserted onto the existing chart. Double-click the chart and then click the first data point. Select the Insert – Trendline command from the menu. From the Trendline dialog select the type of regression line (Linear), click the Options button and check the Display Equation on Chart and the Display R-squared Value. The regression line will be displayed on the graph with a text box containing the equation and R^2 value. To reposition the text box on the chart select it and holding it by the border drag it to a new location (Fig. 4).

The value of the correlation coefficient r (squared as R^2) can be found by entering =SQRT(0.9322) in cell B32 (label A32 as r). For 18 degrees of freedom, a correlation coefficient of 0.965 is significant at the 99.5% confidence level (see standard statistical tables, e.g. Rohlf and Sokal, 1969; Neave, 1981).

The regression equation y = ax + b can now be used to predict the relative percentage of quartz pebbles at any distance from the first station. The values of a and b in this case are 0.3081 and 22.808, respectively. Label cells A29 and A30 as DISTANCE and % QUARTZ, using cell B29 to enter any distance from the origin. In cell B30, type =ROUND((0.3081*B29+22.808),1)

For practice, set up a template to determine the mean pebble size at any distance from the first station.

Analysis of Variance

A somewhat more advanced statistical technique with wide application in the earth sciences, is the analysis of variance (see, e.g. Koch, 1990; Le Roux et al., 1994). This method uses the ratio between two sample variances (known as the F-ratio) to test the null hypothesis that a number of samples all come from populations with the same mean. Differences between sample means are used to calculate the variance of the population, which is compared with an estimation of the population variance based only on the difference between individual observations. These are known as "between groups" and "within groups" variances, respectively.

Load the worksheet D-Anova for an illustration of a one-way analysis of variance to solve a sedimentological problem: the influence of energy conditions on bioturbation patterns. It is suspected that the activity of certain burrowing organisms varies according to current velocities in a tidal inlet. The null hypothesis is that there is no significant change in activity under different energy conditions. To test the hypothesis, 4 groups of experiments, each involving the same number of sand prawns, are carried out. The prawns are placed in a sand-bottomed flume for 24 hours, with current velocities fixed at 10, 30, 50 or 70 cm/sec for each group of experiments. After every 24 hour period the intensity of bioturbation is determined on a relative scale and the sand is homogenised for the next experiment. The X-values in the worksheet D-Anova represent the degree of bioturbation.

Excel provides three types of ANOVAS in the Data Analysis toolpack on the Tools menu. (If this has not been installed yet, choose the Add-Ins command from the Tools menu and then click the check box next to Data Analysis). From the Data Analysis dialog select Single Factor Anova. In the Input Range text box enter the data range (A5:D16), and under Output options select Output Range (E1). Click OK to complete the analysis.

From F-tables for 3 and 39 degrees of freedom, the critical region is established as F >5.0 (rounded) at the 99.5% confidence level. As the F-ratio determined above (5.07) exceeds this value, the results are significant and the null hypothesis can be rejected. Energy conditions do have a significant influence on bioturbation patterns.

Chi-Square Tests

There are numerous references in the geological literature on applications of the Chi-square (χ^2) test, including Strahler (1950) on valley-wall slope angles, Ferm (1955) on the radioactivity of fresh and weathered rocks, Harrison (1957) on particle orientation data, and Cadigan (1962) on the homogeneity of heavy mineral occurrences. Chi-square tests are used to determine the probability that a given sample was drawn from a population with a specific distribution. The method has the advantage that it can be applied even where observations do not have normal distributions or equal variances, which would rule out t-tests or analyses of variance. As the data are grouped into categories or class intervals, the measured attributes may also be expressed on interval, ratio, nominal or ordinal scales.

The worksheet D-Chi-square can be used to perform a χ^2 test for data involving two variables. A palaeontologist is investigating the possibility that certain burrowing organisms occurred preferentially in specific types of sediments. A random sample of 500 observations is taken and the results are summarized in a two-way contingency table or frequency matrix.

The first step is to predict a frequency for each cell by considering the number of fossils and rock types in each category. These numbers are entered on the edges of the matrix by adding the row and column values. So, in cell E5, enter =SUM(B5:D5) and copy to E6:E8. Similarly, enter =SUM(B5:B7) in cell B8 and copy to C8:D8. In Row 9, the column totals are expressed as relative frequencies of the grand total by entering =(B8/$E$8) in cell B9 and copying to C9:E9.

As sandstone comprises 24% of all the rock types investigated, about 0.24 × 150 (36) brachiopods would be expected in sandstone if they had no preference for any specific sediments. Using similar reasoning, a second matrix of predicted frequencies can be constructed. Employing the same format as in the original matrix, the formula =B9*$E$5 is entered in cell B13 and copied to C13:D13. Similarly, =B9*$E$6 and =B9*$E$7 are copied from cells B14 and B15 to C14:D14 and C15:D15. Next, a χ^2 matrix is constructed by entering =ROUND(((B5-B13)^2/B13),2) in cell B19 and copying to B19:D21. In cell A23 enter the label χ^2-VALUE and in cell B23 the formula =ROUND(SUM(B19:D21),2). The number of degrees of freedom is labeled in A24 as D.F. and calculated in B24 by =(COUNTA(B13:D13)-1)*(COUNTA(B13:B15)-1). In cell A25, enter the label P. RANDOM (the probability that the sample was drawn from a random distribution) and in cell B25 the function =CHIDIST(B23,B24). The value of χ^2 (173.53) exceeds 13.28 (the 1% significance level for 4 degrees of freedom, as shown by the fact that B25 returns a value of less than 0.01), so that the null hypothesis (that the results reflect only random variations) can be rejected. It can be concluded that the organisms did prefer certain sediments.

Geostatistics

The application of the Theory of Regionalised Variables (Matheron, 1971) to the estimation of ore reserves is well established, but it is not always appreciated that the same techniques can be used to solve problems involving any sample values affected by their location relative to that of their neighbors. In the example discussed below, it is assumed that the reader has a basic knowledge of what is generally known as "Geostatistics". Introductory texts on this subject include Royle (1977), Rendu (1978) and Clark (1979).

The estimation of ore reserves rely on certain assumptions about the relationship between sample values and their relative spatial orientation, which is described by a semi-variogram. The latter is obtained by fitting a theoretical model to experimental or measured data, as explained in the section on Geological Modeling. Figure 5 and the worksheet D-Geostatistics contain the relevant information to illustrate the approach in solving a typical two-dimensional problem related to ore reserve estimation.

A development drive parallel to the sides of a planned stoping panel in a sulphide orebody has been closely sampled so that its average grade is known. An estimate of the standard deviation around the mean is required for the panel, to assess the financial risk involved in developing the stope. Some of the symbols used in the worksheet D-Geostatistics are explained in Fig. 5. T* is the average grade of the development drive, whereas C and a are the sill and range of influence of the spherical semi-variogram, respectively. A separate input area for values which have to be looked up in tables, has been created in block A11:B14, and a working area in block C3:D14. For convenience, the labels have been supplied in column C.

Three values are required to calculate the standard error (square root of the extension variance), which is necessary to solve the problem. These are known

Fig. 5. Stopping panel and development drive in sulphide orebody

Stoping Panel

Development Drive

b

d

as gamma-bar terms. $\bar{\gamma}$(A,A) is the average semi-variogram value between every point in the mine panel and every other point in the panel, $\bar{\gamma}$(S,S) is the average semi-variogram value between every point in the drive and every other point in the drive, and $\bar{\gamma}$(S,A) is the average semi-variogram value between the drive and every point in the panel. The values of $\bar{\gamma}$(A,A) and $\bar{\gamma}$(S,S) must be removed from the system if only the average grade of the panel is to be considered, so that the extension variance (σ^2) is given by:

$$\sigma^2 = 2\bar{\gamma}(S,A) - \bar{\gamma}(S,S) - \bar{\gamma}(A,A)$$

To calculate $\bar{\gamma}$(A,A), the F(l,b) function must be evaluated. As published F-function tables apply to a standardized spherical model with C = 1 and a = 1, the measurements l and b must be divided by the range a of the semi-variogram. In cells D4 and D5, enter =ROUND(B6/B10,2) and =ROUND(B7/B10,2), respectively. The converted values of l and b in the auxiliary F-function F(3.26, 2.5) are now used to find the variance of grades within the panel using the correct tables (e.g. Table 3.5 of Clark, 1979). In this case, linear interpolation of the values in the table for F(3.26, 2.5) gives an F-value of **0.9367** when C = 1. This is entered in cell B12. As the F-value must be multiplied with the actual value of the sill to find the gamma-bar term $\bar{\gamma}$(A,A), enter **(B12*B9)** in cell D12.

For a spherical model, the value of $\bar{\gamma}$(S,S) is given by:

$$\bar{\gamma}(S,S) = (C/20)(20 -15(a/l) + 4(a^2/l^2))$$

This is entered in cell D13 as **(B9/20)*(20–15*(B10/B6)+4*(B10^2/B6^2))**.

The term $\bar{\gamma}$(S,A) is obtained by making use of the auxiliary function χ(l,b). The latter applies only to situations where a line is on one side of a panel, so that the panel in this case has to be divided into an upper and a lower part, separated by the drive. The sum of the semi-variogram values between the drive and every point in the upper panel is given by 135 × 235 x χ(135; 235), and that of the lower panel by 45 × 235 x χ(45; 235). These χ(l,b) functions also have to be converted to apply to the standardised spherical model, so that the following formulae are entered in D6:D9. D6: =ROUND(B8/B10,2); D7: =ROUND(B5/B10,2); D8: =ROUND(((B7-B8)/B10),2); D9: =ROUND(B5/B10,2). The converted χ(l,b) functions, χ2(1.88; 3.26) and χ1(0.63; 3.26) are used to find the corresponding c-values in c-tables applicable to spherical models (e.g. Table 4.2 in Clark, 1979). These χ-values (**0.8662 and 0.9576**) are entered in cells B14 and B13, respectively.

As the two sub-panels represent fractions of the complete panel, enter =(B7-B8)*B6/(B7*B6) and =(B8*B6)/(B7*B6) in cells D10 and D11. The gamma-bar term $\bar{\gamma}$(S,A) is calculated in cell D14 by the formula: =(B13*B9*D11)+(B14*B9*D10). The final calculation is executed in cell B16 with the formula =ROUND(SQRT (2*D14-D12-D13)*2,2). This is twice the standard error for the complete panel, which gives the estimated "true" range of grades around the average grade of the panel (equal to the average grade of the drive) at the 95% confidence level (assuming normality for the error distribution). In this case, T = 125±9.38 kg/ton. If the cut-off grade is set at 115 kg/ton, minor risk can be assumed for the operations.

Financial Analysis

In the earth sciences, financial analysis is mostly focused on the economic evaluation of mineral or hydrocarbon deposits. Koch (1990), for example, illustrated the use of Lotus to estimate block grades and tonnages from development workings and drillhole data. Some textbooks giving a background to financial analysis include Parks (1973) and Annels (1991).

Economic Evaluation

The economic evaluation of mineral deposits is an orderly and planned analysis of all available information to determine the probable value of a project. It can be based on several different investment decision techniques, most of which are facilitated by the use of spreadsheets. Excel provides some of the basic financial functions (which are all related to the time value of money) required in such analyses.

The worksheet D-Finance compares the cash flow (CF) characteristics of two mutually exclusive petroleum investment opportunities. To select the best proposition, a number of discounted cash flow (DCF) selection criteria can be used, all of which have certain inherent advantages and disadvantages. In this exercise, three of the most commonly used criteria are illustrated.

Firstly, the net present value (NPV) of the anticipated time distribution of cash flows for the specific period (5 years) has to be calculated. The discount rate is taken as 8% or 12%, respectively, which are entered in cells C5, D5 (0.08, 0.12) and F5, G5 (0.08, 0.12). In cell C7 enter =ROUND($B$6+NPV(C$5,$B$7),0) and copy to D7. In cell C8 enter =ROUND($B$6+NPV(C$5,$B$7:$B8),0), copy to C11 and then to D8:D11. Repeat this procedure for project 2. A positive NPV indicates the maximum amount that can be invested in each project to still obtain a minimum acceptable rate of return at different projected discount rates. At a DCF of 8% in this example, project 1 with a NPV of $214 954 after 5 years is a marginally better proposition than project 2, which would yield $212 254 in the same period. (If different capital investments were required for the two projects, they could be compared by using the present value ratio, which is obtained by simply dividing the total NPV by the initial investment capital.)

The second selection criterion examined here is the growth or reinvestment rate of return (GROR), which determines the future value of an investment if all positive cash flows were reinvested to the end of the final year of operation. In effect, this means that compound interest is calculated for each separate cash flow, the totals are added, and the present value of the investment is determined. The GROR is given by:

$$GROR = \sqrt[n]{(CF/PV)} - 1$$

where n is the number of investment periods (years in this case) for each cash flow reinvestment, and PV is the present value. In cells C14 to C17 enter =ROUND($B7*(1+$C$5)^4,0) to =ROUND($B10*(1+$C$5),0), reducing the number of years from 4 to 1. Repeat for cells D14:D17, F14:F17 and G14:G17, using an interest rate of 8% for F14:F17 and 12% for C14:C17 and G14:G17. In cell C18, enter =$B$11 and copy to D18. In F18, enter $E$11 and copy to G18. Add the values in cells C19, D19, F19 and G19, by entering, for example =SUM(C14:C18) in C19. The GROR at 8% and 12% are entered in cells C20, F20, C21 and F21 using the formula: =ROUND(((C19/ABS(B6))^(1/5)-1)*100,2) in C20, for example. In this case, the GROR at a reinvestment rate of 8% is 11.93% for project 1 and 11.88% for project 2, again favouring the former option.

A third investment criterion is the internal rate of return (IRR), also known as the discounted cash flow rate of return, internal yield, profitability index, or the interest rate of return. This is the average percentage return that an investment opportunity is expected to yield over its lifetime. In effect, this means that the discount rate which equates the present value of the positive cash flows with the present values of the negative cash flows must be found. In Excel, this iterative procedure is carried out using the =IRR function, which requires an initial guess (usually between 0 and 1) to serve as a starting point. In cells C22 and F22, enter =ROUND(100*IRR(B6:B11,C5),2) and =ROUND(100*IRR(E6:E11,F5),2) using 8% as the initial guess. The results indicate that project 1 would yield a much better internal rate of return of 17.97% against 13.76% of project 2, confirming this as the preferred investment opportunity.

Geological Modeling

One of the major advantages of spreadsheets is that formulae are not separated from data as in conventional high level programming languages, so that any modifications are reflected immediately in the output. This makes them ideally suited for geological modeling, i.e. the process of fitting theoretical simulations to observed data, in order to formulate a model which can be used as a predictive device.

The literature abounds with examples of geological, geochemical and geophysical modeling based on spreadsheets. Holm (1988a, 1990) described the use of Lotus to perform petrogenetic modeling of continuous melting and open-system fractional crystallization in magmas. Friberg (1989) proposed a Lotus template for garnet stoichiometry, which can also be modified to process other minerals such as pyroxene, amphibole and chlorite. An Excel template for the simultaneous solution of geobarometers and geothermometers, which can also be adapted for mixing models or calibrations, was published by Boyle (1991). Dexter and Avery (1991) explained the convenience of Lotus 1-2-3 and Excel for teaching and applying water-balance modeling techniques. Sprenke (1991) demonstrated the use of Lotus for interactive gravity modeling, the purpose of which is to determine the shapes and physical parameters of the geological sources of gravity anomalies. Forward modeling of electrical sounding experiments using Quattro Pro, was demonstrated by Sheriff (1992). Biddle et al. (1995) described the use of spreadsheets in assessing mineral stability during water-rock interactions, in order to define element mass balances.

One example of a predictive model is the semi-variogram, which forms the basis of all geostatistical calculations. It can be defined as a formula or graph portraying the expected difference in value between sample pairs with a given relative orientation. If the orientation and distance between sample pairs are expressed as h, the mean difference between their ore grades as m(h) and the variance between these differences as $2\gamma(h)$, the experimental value (denoted by *) of $m^*(h)$ can be determined for any specific distance by:

$$2\gamma(h) = (1/n)\Sigma[g(x)-g(x+h)]$$

where n is the number of data pairs, g is the ore grade, x is the position of one sample in the data pair, and (x+h) is the position of the other sample. The term $\gamma^*(h)$, known as the experimental semi-variogram, is expressed as:

$$\gamma^*(h) = (1/n)\Sigma[g(x)-g(x+h)]^2$$

In the graph, the distance h is plotted against the x-axis (abscissa) and the corresponding value of $\gamma^*(h)$ against the y-axis (ordinate). Various theoretical models (e.g. spherical, exponential, linear and De Wijsian) of the semi-variogram $(\gamma(h))$ can be tested to obtain the best possible fit.

In the worksheet D-Variogram, the spreadsheet represents an orthogonal drilling grid with the ore grades entered in their relative geographic locations. Cells are considered to be 100 m apart. For this example, an experimental semi-variogram will be constructed for the east-west direction only, but in practice different semi-variograms may be constructed for different directions. First, data stations 100 m apart along the E-W direction will be considered. In cell K1 enter: =IF(AND(A3>0.001,B3>0.001),(A3-B3)^2,0) and copy to K1:S8. This takes all the data pairs 100 m apart into account over the complete drilling area. To find the value of n (the number of data pairs 100 m apart for which actual data exist), type =IF(OR(AND(AND(A3>0,B3>0),A3=B3),K1>0),1,0) in T1 and copy to AB8. Similarly, to calculate the semi-variogram for distances between pairs of 200 m, 300 m and 400 m, respectively, enter the following formulae in the first cell mentioned, copying them to the rest of the cells as indicated:

- AC1:AJ8: = IF(AND(A3>0.001,C3>0.001),(A3-C3)^2,0)
- AK1:AR8: = IF(OR(AND(AND(A3>0,C3>0),A3=C3),AC1>0),1,0)
- AS1:AY8: = IF(AND(A3>0.001,D3>0.001),(A3-D3)^2,0)
- AZ1:BF8: = IF(OR(AND(AND(A3>0,D3>0),A3=D3),AS1>0),1,0)
- BG1:BL8: = IF(AND(A3>0.001,E3>0.001),(A3-E3)^2,0)
- BM1:BR8: = IF(OR(AND(AND(A3>0,E3>0),A3=E3),BG1>0),1,0)

Tabulate the distances h and corresponding values of the semi-variogram in cells A14 to B17, after entering the labels h in A13, $g^*(h)$ in B13 and g(h) in C13.

- A14: **100**; B14: = SUM(K1:S8)/(2*SUM(T1:AB8))

- A15: **200**; B15: = SUM(AC1:AJ8)/(2*SUM(AK1:AR8))
- A16: **300**; B16: = SUM(AS1:AY8)/(2*SUM(AZ1:BF8))
- A17: **400**; B17: = SUM(BG1:BL8)/(2*SUM(BM1:BR8))

In cells A19 and A20 enter the labels **a** and **C**, respectively.

The experimental semi-variogram can be viewed using the ChartWizard and defining the Range as A14:B17. As the $\gamma^*(h)$ values level off at a sill of about 112, a spherical model would probably provide the best fit. This model is derived theoretically as follows:

$$\gamma(h) = C[3\,h/2a)-(h^3/2a^3)]$$

for h < a, and:

$$\gamma(h) = C$$

where h ≥ a.

In cell E14, enter:

=IF(A14<$B$19,$B$20*(((3*A14)/(2*$B$19))-(A14^3/(2*$B$19^3))),$B$20).

This is copied to E15:E17. Various values of C and a can now be entered in cells B20 and B19 to obtain the best fit, in this case defining the Y-range as C14:C17. In the case of a spherical model, the value of C can be estimated by examining the graph of the experimental semi-variogram, while a line through the first few plots on the graph will cut the sill at a value of about (2/3)a. This can be used as a first estimate of a. Fine-tuning these values for C and a and plotting both the experimental and theoretical semi-variograms on the same graph to find the best possible agreement, gives a value of about 345 m and 10.8^2 for the range of influence and sill, respectively.

In Excel, Solver (from the Tools menu) can also be used to perform the iteration process automatically. In cell B21, enter **=ABS(B14-E14)+ABS(B15-E15)+ABS(B16-E16)+ABS(B17-E17)**. Use $B$21 as the Target cell in the Solver dialog box, and choose the minimize command. The changing cells are $B$19 and $B$20. In the Constraints box, enter **$B$19 >=0, $B$19 >=300, $B$19 <=400, $B$20 >=10**, and **$B$20 <=11.5**. Click the Options button and check Quadratic. Clicking Solve starts the iteration process and gives a value of 363.33 for a and 11.0 for C. The final theoretical semi-variogram for the east-west direction is compared with the experimental one in Fig. 6.

For practice, determine the semi-variogram in the north-south direction for the given data set. Calculate also the mean semi-variogram for the east-west and north-south directions by weighting the $\gamma^*(h)$ values for each distance with the number of data pairs.

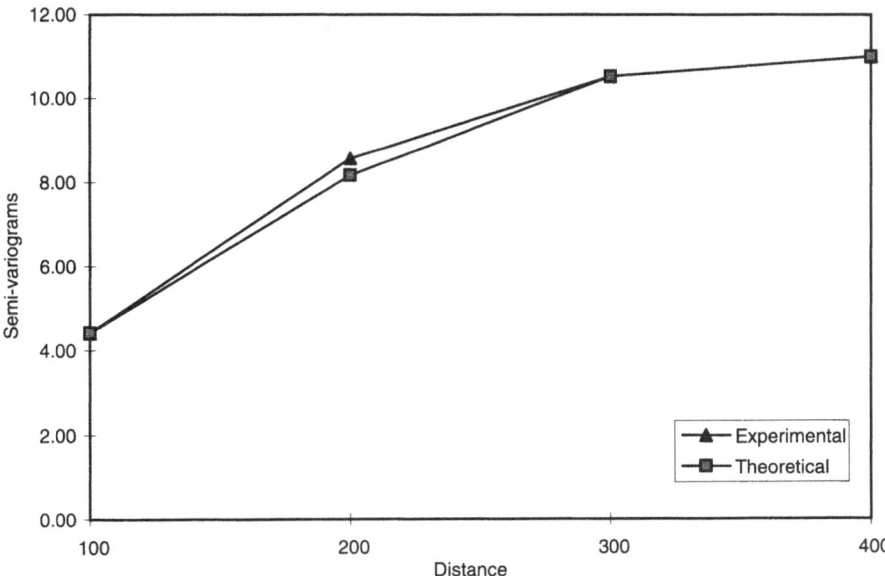

Fig. 6. Comparison of theoretical and experimental semi-variograms for east-west direction

Conclusions

The data management and graphical capabilities of spreadsheets make them ideally suited for applications in the earth sciences. Within the Windows environment spreadsheets offer the added advantage of data sharing between different programs, e.g.: extracting data from a database (for example Microsoft Access) for analysis in Excel and inserting the results directly into a word processor.

To date relatively few Excel macros have been published in the scientific literature. With the capability to write user defined functions or macros in Excel which are accessible to all spreadsheets we should see an increase in the development of Excel applications in the future.

References

Annels, A.E. (1991). Mineral Deposit Evaluation: A Practical Approach. Chapman & Hall, London, 436 pp

Berge, T.M. (1991). Stereographic projections using Lotus 123. Geobyte, 5, 36–39

Biddle, D.L., Percival, H.J. and Chittleborough, D.J. (1995). An interactive spreadsheet for graphing mineral stability diagrams. Computers and Geosciences, 21, 175–185

Boyle, A.P. (1991). Simultaneous solution of geobarometers and geothermometers using a microcomputer spreadsheet. Computers & Geosciences, 17, 1473–1479

Cadigan, R.A. (1962). A method for determining the randomness of regionally distributed quantitative data. Journal of Sedimentary Petrology, 32, 813–818

Carr, T.R. (1982). Log-linear models, Markov chains and cyclic sedimentation. Journal of Sedimentary Petrology, 52, 905–912

Christie, D.M. and Langmuir, C.H. (1994). Automated XY plots from Microsoft Excel. Computers and Geosciences, 20, 47–52

Clark, I. (1979). Practical Geostatistics. Applied Science Publishers Ltd, London, 129 pp

Curray, J.R. (1956). The analysis of two-dimensional orientation data. Journal of Geology, 64, 117–131

Dahl, P.S. (1990). A PC- and LOTUS-based data acquisition/reduction system for an ICP spectrometer. Computers and Geosciences, 16, 881–896

Davies, T.A., Baldauf, J.G. and Kidd, R.B. (1992). A simple spreadsheet routine for calculating depth/age relations. Computers and Geosciences, 18, 579–585

Davis, J.C. (1986). Statistics and Data Analysis in Geology (2nd ed.). John Wiley & Sons, New York, 646 pp

Dexter, L.R. and Avery, C.C. (1991). Using spreadsheet software in water-balance modeling. Computers and Geosciences, 17, 527–536

Ernst, L.F. (1950). Een nieuwe formule voor de berekening van de doorlaatfactor met de boorgatenmethode. Rap. Landbouwproefsta. en Bodemkundig Inst. T.N.O., Groningen, Netherlands

Ferm, J.C. (1955). Radioactivity of coals and associated rocks in Beaver, Clearfield and Jefferson Counties, Pennsylvania. U.S. Department of the Interior, TEI-468

Fourie, C.J.S. and Odgers, A.T.R. (1995). Spreadsheet interpretation of seismic refraction data. Computers and Geosciences, 21, 273–277

Friberg, L.M. (1989). Garnet stoichiometry program using a LOTUS 1-2-3 spreadsheet. Computers & Geosciences, 15, 1169–1172

Girard, R. (1992). Spreadsheet routine for the management of structural data with a microcomputer. Computers and Geosciences, 18, 29–45

Harrison, P.W. (1957). New techniques for three-dimensional fabric analysis of till and englacial debris containing particles from 3 to 40 mm in size. Journal of Geology, 65, 98–105

Hinman, M. (1993). A LOTUS 1-2-3 diamond drillhole structural manipulation spreadsheet: drillcore structural data generation. Computers and Geosciences, 19, 343–354

Holm, P.E. (1988a). Petrogenetic modeling with a spreadsheet program. Journal of Geological Education, 36, 155–156

Holm, P.E. (1988b). Triangular plots and spreadsheet software. Journal of Geological Education, 36, 157–159

Holm, P.E. (1990). Complex petrogenetic modeling using spreadsheet software. Computers and Geosciences, 16, 1117–1122

Hooghoudt, S.B. (1936). Bijdragen tot de kennis van eenige natuurkundige grootheden van den grond, 4. Versl. Lambd., Ond., 42(13), B: 449–541. Algemeene Landsdrukkerij, The Hague

Koch, G.S. (1990). Geological Problem Solving with LOTUS 1-2-3 for Exploration and Mining Geology. Computer Methods in the Geosciences, Vol. 8, Pergamon Press, New York, 208 pp

Koch, G.S. and Link, R.F. (1971). Statistical Analysis of Geological Data, Vol. 2. John Wiley & Sons, New York, 438 pp

Koosis, D.J. (1977). Statistics (2nd ed.). John Wiley & Sons, New York, 282 pp

Krumbein, W.C. and Graybill, F.A. (1965). An Introduction to Statistical Models in Geology. McGraw-Hill, New York, 475 pp

Le Roux, J.P. (1991a). Paleocurrent analysis using LOTUS 1-2-3. Computers and Geosciences, 17, 1465–1468

Le Roux, J.P. (1991b). A spreadsheet model for integrating stratigraphic and lithofacies maps. Computers and Geosciences, 17, 1469–1472

Le Roux, J.P. (1992a). Determining the channel sinuosity of ancient fluvial systems from paleocurrent data. Journal of Sedimentary Petrology, 62, 283–291

Le Roux, J.P. (1992b). Behavior of spherical grains in fluids: a convenient spreadsheet template for engineers and sedimentologists. Computers and Geosciences, 18, 1255–1257

Le Roux, J.P. (1994a). The angular deviation in circular statistics as applied to the calculation of channel sinuosities. Journal of Sedimentary Research, A64, 86–87

Le Roux, J.P. (1994b). An alternative approach to the identification of net sediment transport paths based on grain-size trends. Sedimentary Geology, 94, 97–107

Le Roux, J.P. (1994c). A spreadsheet template for determining sediment transport vectors from grain-size parameters. Computers and Geosciences, 20, 433–440

Le Roux, J.P. (1994d). Spreadsheet procedure for modified first-order embedded Markov analysis of cyclicity in sediments. Computers and Geosciences, 20, 17–22

Le Roux, J.P., Grobler, L. and Smit, P.H. (1994). Monoclines and palaeochannels: evidence for syntectonic sedimentation in the Beaufort Group of the Karoo Basin, South Africa. Journal of African Earth Sciences, 18, 219–226

Le Roux, J.P. and Rust, I.C. (1989). Composite facies maps: a new aid for palaeo-environmental reconstruction. South African Journal of Geology, 92, 436–443

Marshall, D. (1996). TernPLot: an Excel spreadsheet for ternary diagrams. Computers and Geosciences, 22, 697–700

Martin, J.D. (1996). EQMIN, a Microsoft Excel spreadsheet to perform thermodynamic calculations: a didactic approach. Computers and Geosciences, 22, 639–650

Matheron, G. (1971). The Theory of Regionalised Variables and its Applications. Cahier No. 5, Centre de Morphologie Mathematique de Fontainebleau, 211 pp

Neave, H.R. (1981). Elementary Statistics Tables. Alden Press, London, 48 pp

Ozkaya, S.I. (1995). Two EXCEL macros for tracing deviated boreholes using cubic splines and calculation of formation depth and thickness. Computers and Geosciences, 21, 851–858

Parks, J.M. (1970). Computerized trigonometric method for rotation of structurally tilted sedimentary directional features. Geological Society of America Bulletin, 81, 537–540

Parks, R.D. (1973). Examination and Valuation of Mineral Property, 4th ed. Addison-Wesley, London, 507 pp

Phillips, F.C. (1971). The Use of Stereographic Projection in Structural Geology. Edward Arnold, London, 90 pp

Powers, D.W. and Easterling, R.G. (1982). Improved methodology for using embedded Markov chains to describe cyclical sediments. Journal of Sedimentary Petrology, 52, 913–923

Rao, D.R. (1995). BGT – the macros driven spreadsheet program for biotite-garnet thermometry. Computers and Geosciences, 21, 593–604

Reche, J. and Martinez, F.J. (1996). GTP: an Excel spreadsheet for thermobarometric calculations in metapelitic rocks. Computers and Geosciences, 22, 775–784

Rendu, J-M. (1978). An Introduction to Geostatistical Methods of Mineral Evaluation. Monograph of the South African Institute of Mining and Metallurgy, 100 pp

Rohlf, F.J. and Sokal, R.R. (1969). Statistical Tables. W.H. Freeman & Co., San Francisco, 353 pp

Royle, A.G. (1977). A Practical Introduction to Geostatistics. Course Notes of the University of Leeds, Department of Mining and Mineral Sciences

Sheriff, S.D. (1992). Forward modeling of electrical sounding experiments using convolution and a spreadsheet. Computers and Geosciences, 18, 75–78

Sprenke, K.F. (1991). Gravity modeling with Lotus 1-2-3. Computers and Geosciences, 17, 719–725

Strahler, A.N. (1950). Equilibrium theory of erosional slopes approached by frequency distribution analysis, I. American Journal of Science, 248, 673–696

Appendix: Some Example VBA Code

The following section of VBA code is used to plot the rare earth diagram described in the Data presentation section.

```
Sub GraphPlot()
' This simple plot macro, provided with appropriate data, will produce the rare
earth element plot in Fig. 2
```

```
On Error GoTo ErrorHandler 'Enable error-handling routine.
        DataRange = Selection.Address 'assign range to variable
'Create New Chart
        ActiveSheet.ChartObjects.Add(30, 30, 300, 200).Select
        Application.CutCopyMode = False
        ActiveChart.ChartWizard Source:=Range(DataRange), Gallery:=xlLine, _
        Format:=1, PlotBy:=xlRows, CategoryLabels:=1, SeriesLabels _
        :=1, HasLegend:=1, Title:="Default Title", _
        CategoryTitle:="Default X-Axis Title", ValueTitle:="Default Y-axis Title"
'Select chart and apply formats
        GraphName = Selection.Name
        ActiveSheet.ChartObjects(GraphName).Activate
        ActiveChart.PlotArea.Select
        Selection.Interior.ColorIndex = xlNone
        ActiveChart.Axes(xlValue).Select
        With ActiveChart.Axes(xlValue)
                .ScaleType = True 'xlLogarithmic
                .MinimumScaleIsAuto = True
                .MaximumScaleIsAuto = True
                .MinorUnitIsAuto = True
                .MajorUnitIsAuto = True
                .CrossesAt = .MinimumScale
                .ReversePlotOrder = False
        End With
        ActiveChart.Axes(xlCategory).Select
        Selection.TickLabels.Orientation = xlHorizontal
'Position and Size Graph Objects
        ActiveChart.Axes(xlValue).AxisTitle.Select
        Selection.Left = 1
        ActiveChart.PlotArea.Select
        Selection.Top = ActiveChart.ChartTitle.Top + ActiveChart.ChartTitle.
                        Font.Size * 2
        Selection.Left = ActiveChart.Axes(xlValue).AxisTitle.Font.Size * 2
        Selection.Width = ActiveSheet.ChartObjects(GraphName).Width - _
                Selection.Left - ActiveChart.Axes(xlValue).AxisTitle.Font.Size
        ActiveChart.Legend.Select
Selection.Position = xlBottom
        ActiveSheet.Select
        Exit Sub
' Error-handling routine.
ErrorHandler:
        Select Case Err ' Evaluate Error Number.
        Case 1005
                MsgBox Err & " " & Error() ' Property cannot be set.
```

```
        Case Else
                MsgBox Err & " " & Error() ' All other errors.
                Exit Sub ' Exit macro
        End Select
        Resume Next ' Resume execution at next line.
End Sub
```

Spreadsheet Applications in Aquatic Chemistry

S. Leharne

Introduction

The power of modern spreadsheets coupled with their almost transparent procedures for treating and transforming data provide them with an important role in many quantitative aspects of chemistry. No doubt all experimental scientists can recognise the usefulness of a software package that enables one to store data, analyse it, mathematically transform it, and, if desired, present the results graphically. The purpose of this chapter is to demonstrate how spreadsheets can be used as tools for a variety of practical and pedagogic purposes and for the solution of complex problems. These uses include:

- the computation and visualisation of thermodynamic equilibria
- fitting experimental data to model equations
- simulating instrumental output so that the underlying physical processes giving rise to particular signals are properly understood
- modelling in environmental chemistry.

The text will confine itself therefore to applications that arise in physical, analytical and environmental chemistry as they impinge upon the chemistry of aqueous systems. The applications outlined in the following text have all been developed using Excel 5.0a. There are many features in the latest versions of Excel that make it particularly useful for application development. These include the provision of the macro programming language Visual Basic for Applications (VBA) that enables developers to write user defined procedures and functions in Visual Basic. In fact Visual Basic will also allow the developer to control many aspects of the software package. In addition there is the inclusion of data analysis tools which can provide statistical analyses of the data; and finally Excel provides an extremely powerful parameter optimisation program – called Solver – that may be used for fitting data to model equations using least squares techniques. Indeed the ability to define complex multi-equation models using VBA, which can be incorporated into a user defined function; coupled with the parameter optimisation capabilities of Solver; and in turn coupled with the variety of statistical analyses that may be undertaken with the output, makes for an exceptional all round data analysis package.

Some Working Applications

Visual analysis of chemical equilibria

Frequently problems involving thermodynamic equilibria may be made more accessible if the problem can be visualised. This is especially true from a pedagogic perspective. Graphical displays customarily involve an elaboration of the pertinent equilibrium expressions which have been written in such a way that they all contain a common independent variable. In most cases this variable will be pH, redox potential – which may be expressed as electron activity (Sillén 1959) – or ligand concentration. The dependent variable will be the concentrations of various chemical species which are involved in the equilibrium.

Consider the following: How does the speciation of a monoprotic acid change with pH? The mass action expression for the dissociation of a monoprotic acid is given by:

$$K_a = \frac{\left[H^+\right]\left[A^-\right]}{\left[HA\right]}\tag{1}$$

where K_a is the acid dissociation constant. If C_a is the total acid concentration then the corresponding mass balance expression may be written:

$$C_a = \left[HA\right]+\left[A^-\right]\tag{2}$$

If we consider pH to be the independent variable it then becomes possible to follow the manner in which the concentrations of the acid anion and the undissociated acid alter with pH. By rearrangement and appropriate substitution the following expressions for the acid species may be obtained.

$$\left[HA\right]=\frac{C_a\left[H^+\right]}{K_a+\left[H^+\right]}$$

$$\tag{3}$$

$$\left[A^-\right]=\frac{C_a K_a}{K_a+\left[H^+\right]}$$

It is a reasonably straightforward task to design the spreadsheet application necessary to compute the concentrations of HA and A⁻ as a function of pH. The steps required to produce such a spreadsheet application will be dealt with below. Figure 1 shows the characteristic output obtained for a monoprotic weak acid – in this case ethanoic acid. Sillén (1959) and Stumm and Morgan (1996) have developed some important themes in their descriptions of the value of these diagrams. Among the advantages is an ability to visually identify the most important species

Fig. 1. Acid speciation diagram for a 10^{-2} M solution of ethanoic acid

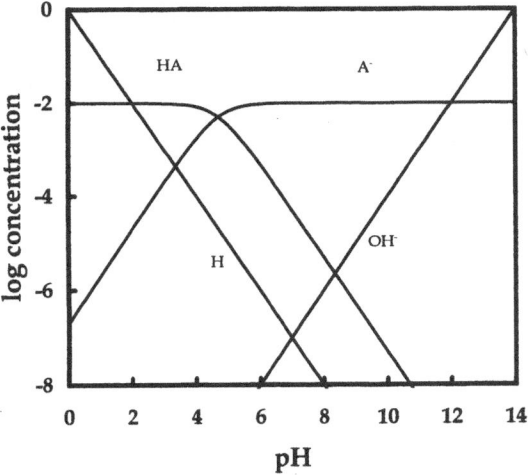

under a particular set of conditions. These graphs can be extended quite easily to view: the speciation patterns of polyprotic acids, the fate of volatile acids, the distribution of such species between air and aqueous solution and the distribution of acid species between non-aqueous and aqueous solvents. As such they are of extensive benefit in analytical chemistry and environmental chemistry.

The mathematical and chemical development of the subject, so far, has ignored the important thermodynamic distinction between concentration and activity. Activity is the product of concentration and activity coefficient. The Debye Hückel theory is commonly used to calculate *dilute* solution activity coefficients. Several modifications are found to be useful for solutions of *relatively high* ionic strength. (Stumm and Morgan 1996). Several important equations, which may be used for the calculation of single ion activity coefficients, are presented below (Stomm and Morgan 1996):

$$\log f \quad = \quad -Az^2\sqrt{I} \qquad \qquad \text{Deby Hückel}$$

$$= \quad -Az^2 \frac{\sqrt{I}}{1+\sqrt{I}} \qquad \qquad \text{Güntelberg}$$

$$= \quad -Az^2\left(\frac{\sqrt{I}}{1+\sqrt{I}} - 0.2I\right) \qquad \text{Davies}$$

$$\text{where}$$

$$I \quad = \quad \sum C_i z_i^2$$

f is the single ion activity coefficient, I is the ionic strength, z is the ionic charge and $A \approx 0.5$ for water.

It is common to specify the ionic strength by using an approximately ten-fold excess of an appropriate electrolyte solution. Under these circumstances the following expression for the corrected equilibrium constant can be provided for acid dissociation (Stumm and morgan 1996):

$$K' = K \frac{f_{HA}}{f_{A^-}} \tag{4}$$

where K' is the corrected acid dissociation constant, K is the acid dissociation constant for all species under standard conditions, f_{HA} and f_A are the activity coefficients for the acid species HA and A^- respectively. We do not require the activity coefficient for H^+ since it is assumed that the measurement of pH is based upon hydrogen ion activity.

Consideration of the arithmetic that describes the speciation pattern for a polyprotic acid $H_n A$ such as phosphoric acid (H_3PO_4) permits the development of general expressions for acid speciation diagrams. Such acids are characterised in same way as the monoprotic acid by mass action expressions describing dissociation and by a mass balance. The following scheme describes these species completely using a combination of mass action equilibrium expressions and formulated mass balances. The following mass action expressions are used:

$$K_1 = \frac{\left[H^+ \right]\left[H_{n-1}A^- \right]}{\left[H_n A \right]}$$

$$K_2 = \frac{\left[H^+ \right]\left[H_{n-2}A^{2-} \right]}{\left[H_{n-1}A \right]} \tag{5}$$

$$\vdots$$

$$K_n = \frac{\left[H^+ \right]\left[A^{n-} \right]}{\left[HA^{(n-1)-} \right]}$$

as well as the mass balance:

$$C_a = \left[H_n A \right] + \left[H_{n-1}A^- \right] \ \ldots \ + \left[A^{n-} \right] \tag{6}$$

substitution into and rearrangement of the mass balance expression provides the following:

$$\left[A^{n-}\right] = \cfrac{C_a}{1 + \sum_{i=1}^{n} \left[H^+\right]^i \times \cfrac{1}{\prod_{j=1}^{i} K_{n+1-i}}}$$

(7)

and

$$\left[H_i A^{(n-i)-}\right] = \left[A^{n-}\right] \times \cfrac{\left[H^+\right]^i}{\prod_{j=1}^{i} K_{n+1-i}}$$

It is a none too demanding task to produce a spreadsheet application which shows the change in species concentration for the polyprotic acid as a function of pH. The Acidspec.XLS worksheet (and Figure 2) shows one way in which this may be executed.

If the reader examines Figure 2 it will be noted that the acid dissociation constants and the water dissociation constant are arranged in a suitable part of the worksheet. In our application the cell range chosen is B4:B7. In the range D4:E7 we place the ionic charges of the acid species involved in the corresponding equilibria. z_{HP} refers to the initial acid species (i.e. the species losing the proton in the equilibrium) and z_P refers to the final acid species. Thus in row 4 we have information about the first dissociation constant. Here z_{HP} refers to the ionic charge on H_3PO_4 – which is of course zero; and z_P is the ionic charge on $H_2PO_4^-$ – which is 1. Using the Güntelberg formula we can calculate single ion activity coefficients and the corresponding ratio of the two which is used as our activity correction

Fig. 2. Spreadsheet layout for computing the speciation diagram for 10^{-3}M solution of phosphoric acid. The solution has an ionic strength of 0.025 M

	A	B	C	D	E	F	G
1		Speciation diagram for phosphoric acid					
2							
3				z_{HP}	z_P		Activity correction
4	K₁	7.94E-03		0	1	1.170209	
5	K₂	6.31E-08		1	2	1.602473	
6	K₃	5.01E-13		2	3	2.194409	
7	K_w	1.00E-14		0	1	1.170209	
8							
9	K₁	9.30E-03	C_a	1.00E-03			
10	K₂	1.01E-07					
11	K₃	1.1E-12	I	0.025			
12	K_w	1.17E-14					
13							
14							
15	pH	[H⁺]	[OH⁻]	[P]	[HP]	[H₂P]	[H₃P]
16		1	1E-14	2.49E-25	4.97E-13	7.88E-06	0.000992
17	0	1	1E-14	2.49E-25	4.97E-13	7.88E-06	0.000992

to the equilibrium constants. In range F4:F7 we use the following equation to calculate the activity correction:

$$\frac{f_{HA}}{f_A} = 10^{\left(z_P - z_{HP}\right)A\frac{\sqrt{I}}{1+\sqrt{I}}}$$

The formula inserted, for instance, in cell F4 is written in the following way:

=10^(E4-D4)*0.5*SQRT($E$11)/(1+SQRT($E$11)

It should be appreciated that all formulae entered in EXCEL are preceded by the '=' sign. The cell references D4 and E4 refer to the numerical contents of those cells. SQRT is an EXCEL function which returns the square root of the value placed between the function brackets. Finally the reference to the ionic strength in cell E11 appears as $E$11 in the formula. The purpose of this becomes clear when one copies the formula in F4 into cells F5 to F7. As elsewhere in this book the action of the '$' sign can be seen as anchoring the cell reference following.

In the cell range B9:B12 we place those formulae – based upon equation 4 – which will be used to calculate the corrected acid dissociation and water self ionisation equilibrium constants. For example in cell B9 the formula will be =B4*F4. In the worksheet below these entries the species concentrations will be computed as a function of pH. Thus in a suitable part of the worksheet column labels – outlining the contents of the columns – are placed. These are not necessary for the software but are usually indispensable for the human developer. Under each label the appropriate speciation formula is entered. pH is the independent or 'master' variable and values for pH are placed in the leftmost column – in our case column A. The formulae in equation 7 however use [H$^+$] the proton concentration. Thus in cell B16 we place the applicable conversion formula. Since for this application use is made of the Data Table procedure to fill the spreadsheet with acid species' values, the formula used refers to the cell location A14. The Data Table procedure requires the user to specify an empty cell location into which input values from the data table may be inserted and formulae evaluated. The conversion formula used is therefore =10^-A14. Into cells C16 to G16 we place those formulae which will be used to compute the appropriate species' concentration. The reader should refer to Acidspec.XLS on the disk for details of the formulas. To generate the table of values we enter in column A – starting at cell A17 the chosen input pH values (0 to 14 in steps of 0.5). Using the mouse select the cell range A17:A45. This means placing the cursor on cell A17, depressing the left hand button and moving the mouse and cursor to cell A45 at which point the button is then released. Next select Edit from the MenuBar followed by Fill followed by Series. A dialogue box will pop-up. Since we want a series of values to appear n the data range selected which increase linearly in steps of 0.5 we check the Columns and Linear options and enter in the Step Value box 0.5. The next step is to select the rectangular range – again by using the mouse – which contains the entire range of input values and the formulae we wish to calculate. This range being A16:G45. The cells in row 16 are included because they contain the formulae, the cells in range A17:A45 are included because they contain the Input values. Finally we select Data from the MenuBar followed by Table. A dialogue box will appear prompting the user to enter the input cell reference. Since we wish to generate columns of data we place in the box entitled Column Input Cell the cell location A14 and then select the button labelled OK. A table of values will then be created. The data produced by this procedure can be shown graphically. However more useful visual information can be obtained by creating a plot of log species concentration as a function of pH. Figure 3 shows this for phosphoric acid. The additional columns containing values for log species concentration are easily entered on row 16 in columns H to M. Refer to Acidspec.XLS for details of the formulae used.

One important use of these diagrams is to answer such questions as: What is the pH of a 10^{-3} M solution of Na_2HPO_4? Such questions can arise frequently in titrimetry where for example a 10^{-3} M solution of H_3PO_4 is titrated with an aqueous solution of sodium hydroxide. The end-point of the titration is reached with the formation of a 10^{-3} M solution of Na_2HPO_4. It is possible to solve this problem algebraically though this may involve solving a complex polynomial expression. To make matters difficult the normal numerical methods employed to solve such problems such as the Newton-Raphson or secant methods can sometimes fail because of the nature of the curve in the vicinity of x axis (see Figure 4). The speciation diagram in Figure 3 can however be used to furnish a good and quick estimate of the pH required. To formulate the answer we need to generate the appropriate proton condition that characterises the disodium hydrogen phosphate solution. The proton condition is a mathematical statement that identifies and equates those acid species that have gained protons and those that have lost them based upon some reference condition. The reference condition is merely our Na_2HPO_4 solution consisting of Na^+ and HPO_4^{3-} ions plus water. The for-

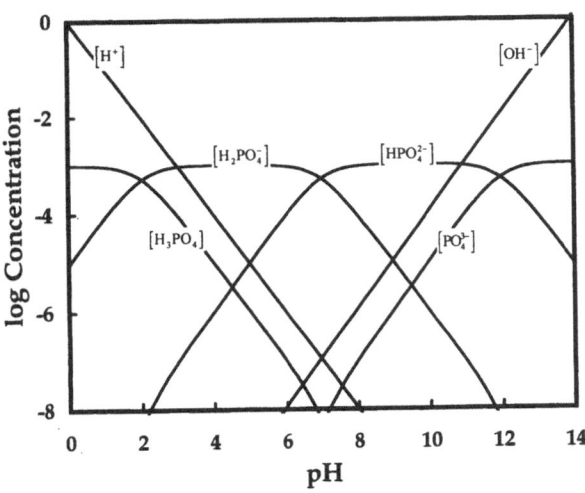

Fig. 3. Acid speciation diagram for a 10^{-3} M solution of phosphoric acid

Fig. 4. Proton condition
equation derived from
equation 9 shown as a
function of pH

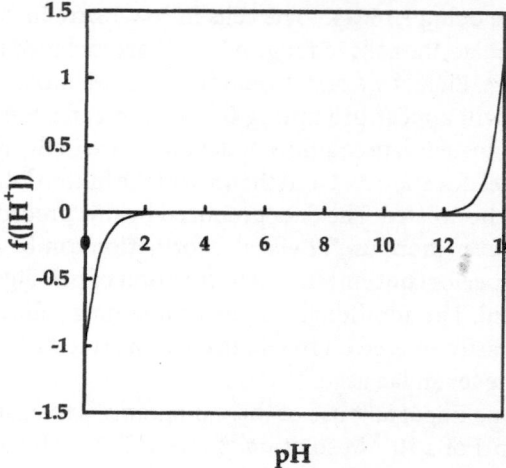

Fig. 4. Proton condition equation derived from equation 9 shown as a function of pH

mation of species such as $H_2PO_4^-$, H_3PO_4 and H_3O^+ represents the production of species having an excess of protons. On the other hand the species PO_4^{3-} and OH^- are proton deficient. For the disodium hydrogen phosphate solution the appropriate proton balance equation is given by:

$$\left[H_2PO_4^-\right]+2\left[H_3PO_4\right]+\left[H_3O^+\right] \qquad = \qquad \left[PO_4^{3-}\right]+\left[OH^-\right] \qquad (8)$$

Proton excess   Proton deficit

It is easily seen by inspection of Figure 3 that this equation is satisfied by the intersection of the dihydrogen phosphate ion and hydroxide ion lines. This intersection occurs at a pH of about 9.0. The concentrations of the other species at this pH being small enough to be disregarded.

A more direct way of tackling the problem – when using Excel – is to use Solver. It is worth noting that Quattro Pro for Windows contains a similar parameter optimisation program called Optimizer which works in fundamentally the same way. In both spreadsheet programs this particular feature may be obtained from the Tools sub-menu. To obtain an estimate of the hydrogen ion concentration for the solution in question the problem is basically reformulated by substituting the various species concentration terms shown in equation 7 into the proton condition expression equation 8 to give the following equation:

$$P+\frac{K_w}{\left[H^+\right]}-P\frac{\left[H^+\right]^2}{K_3K_2}-2P\frac{\left[H^+\right]^3}{K_3K_2K_1}-\left[H^+\right]=0$$

where

$$P = \cfrac{C_a}{1 + \cfrac{\left[H^+\right]}{K_3} + \cfrac{\left[H^+\right]^2}{K_3 K_2} + \cfrac{\left[H^+\right]^3}{K_3 K_2 K_1}} \quad (9)$$

Figure 5 shows a functional spreadsheet arrangement that may be used for numerically solving the problem. Into cell C5 we enter our initial estimate of the pH. In cell D5 we enter the formula =10^-C5, to convert pH to a hydrogen ion concentration. Finally in cell E5 we enter our expression (based upon equation 9). The more accurate the pH entered into C5 the closer the function evaluation in E5 will be to zero. It is convenient for our purposes to produce a user defined function which will calculate the value of equation 9. This can be done using Visual Basic for Applications (VBA). To produce a user defined function we select from the MenuBar the Insert option followed by Macro followed by Module. A new sheet with the default name Module1 will appear in which the following VBA function may be written.

```
Function f(H, K1, K2, K3, Kw, Ca)

    P = Ca / (1 + H / K3 + H ^ 2 / (K2 * K3) + H ^ 3 / (K1 * K2 * K3))
    f = P + Kw / H – P * H ^ 2 / (K2 * K3) - 2 * P * H ^ 3 / (K1 * K2 * K3) - H

End Function
```

H is the hydrogen ion concentration, K1, K2 and K3 are the acid dissociation constants for phosphoric acid, Kw is the water dissociation constant and Ca is the acid concentration. Note that the values for the equilibrium constants in cell range B7:B10 are corrected for activity.

Once the spreadsheet has been set up as shown in Fig. 5 we are ready to use Solver. From the MenuBar select Tools followed by Solver. A dialogue box will

	A	B	C	D	E	F	H
1	What is the pH of a 10^{-3} M solution of Na_2HPO_4?						
2							
3		Solver gives					
4			pH	[H$^+$]	Function		
5		Guess	8.941567	1.14E-09	-3E-16		
6							
7	K_1	9.30E-03		C_a	1.00E-03		
8	K_2	1.01E-07					
9	K_3	1.1E-12					
10	K_w	1.17E-14					

Fig. 5. Spreadsheet design for the calculation of the pH of a 10^{-3} M solution of Na_2HPO_4

appear. In the box labelled Set Target Cell enter E5. On the next line labelled Equal to, check the option Value to and in the box enter zero. Finally in the box labelled By Changing Cells enter C5. Finally select Solve. Basically we have indicated our wish to find a value for pH located in cell C5 which will permit the function in cell E5 to equal zero. The pH is found to be 8.94 - which is of course very close to the graphically obtained value of 9.0. Solver should generally converge to a value very quickly. If it fails to reach a solution then this may indicate that a new initial guess is required.

An alternative method for those not wishing to use the black-box method – Solver – but who wish to obtain a more accurate value for the pH than that obtained from the speciation diagram is to plot equation 8 as a function of pH. The pH can be identified by observing the position at which the line crosses the x axis as shown in Fig. 3. Clearly reducing the amount of data viewed makes this a successively simpler task. In fact, inspection of the cell contents containing the data to be plotted will show where the cross-over point is located and trial and error could be used to produce a reasonably accurate value of the pH at this point.

Other Visual Methods for Outlining Species Distribution

Figure 6 shows two alternative ways in which phosphoric acid species distribution may be shown, graphically, as a function of pH. Both graphs were produced using the same data namely the species concentration data – located in the Sheet labelled Speciation Calculations in Acidspec.XLS. The pH values were copied and pasted to a new Sheet (named Other Speciation Diagrams in the application). The species' concentrations were likewise copied and pasted to the same sheet. Both plots are concerned with showing the fraction or proportion of acid present as a particular species as a function of pH. This, therefore, requires the species' concentration data to be divided by the total acid concentration. For the Acidspec.XLS application an easy way of performing this was to copy the species concentration values to a 'clean' Sheet. In the Acidspec.XLS application this is named Other Speciation Diagrams. This was executed in the following way. The cursor was placed over the cell located at the top left hand corner of the relevant cell range, the left button was depressed, the cursor was moved to the cell located at the bottom right hand corner of the cell range and the button was then released. From the MenuBar Edit was selected followed by Copy. A new sheet was then selected (named Other Speciation Diagrams) by clicking on the Tab of this sheet. These tabs are located at the bottom of the screen. A convenient cell in the sheet was selected to be the top left hand corner of the cell range and then from the MenuBar, Edit followed by Paste Special was selected. Immediately a dialogue box appeared and the option labelled "Values" was checked (place the cursor over Values and click the left hand mouse button) and the button labelled OK was finally checked. The numerical values of the copied cells were automatically positioned in this sheet. The acid concentration term (located in the Speciation Calculations sheet) was next copied. Following this the Other Speciation Diagrams sheet was

Fig. 6. Fractional speciation curves for phosphoric acid

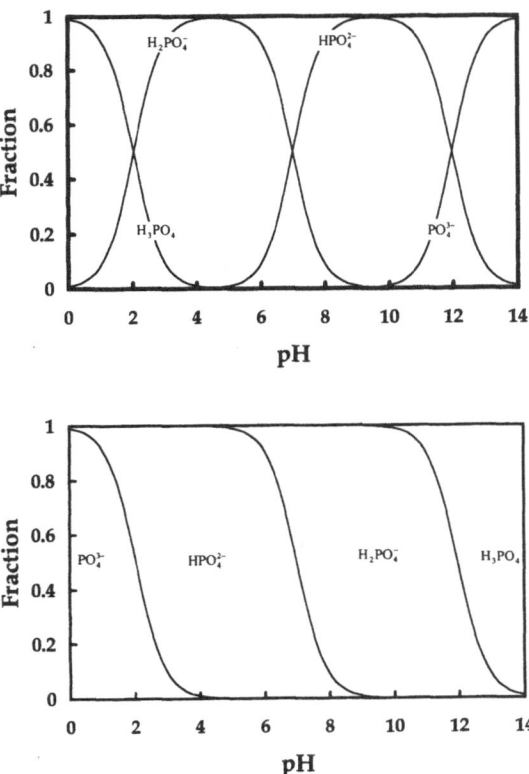

selected and the data range containg all the acid speciation concentrations was selected by placing the cursor over the top left hand cell, depressing the left mouse button and moving to the bottom right hand cell. The entire data range was surrounded by a thick line. From the MenuBar Edit followed by Paste Special was again selected and on the Paste Special dialogue box both Values and Divide were checked followed by OK. The concentration data was immediately transformed to fraction values. It was this fraction data that is shown in Figure 6a.

For Figure 6b the following data is actually plotted as a function of pH: line 1 is the fraction of H_3PO_4, line 2 is the fraction of H_3PO_4, + the fraction of $H_2PO_4^-$, line 3 is is the fraction of H_3PO_4, + the fraction of $H_2PO_4^-$ + the fraction of HPO_4^{2-}. These data sets are easily created using a similar procedure to that outlined in the previous paragraph. The pH data is copied and pasted to another part of the sheet. In the right three adjacent columns the H_3PO_4 fraction data is copied and pasted. Into the last two columns the $H_2PO_4^-$ fraction data is pasted but this time using the Paste Special procedure. From the dialogue box Values and Add are checked. Finally in the last column the HPO_4^{2-} fraction data is added – again using Paste Special. The output is shown in Figure 6b. Sillén (1959) has discussed in some detail the use of these latter distribution diagrams to which the reader is invited to refer. However basically the diagram is used to delineate the proportions of each species at a particular pH. This is readily achieved by drawing a vertical line at a pH of interest and then reading of from the y axis the proportions of phosphoric acid in a particular species form.

Following the Course of an Acid Base Titration

Consider an acid base titration between sodium hydroxide and a mono-protic acid HA. The following electroneutrality equation can be written:

$$\left[Na^+\right]+\left[H^+\right]=\left[OH^-\right]+\left[A^-\right] \tag{10}$$

The electroneutrality expression describes the physical necessity of all aqueous solutions the sum total of positive charges must equal the sum total of negative charges. This equation can be appropriately rearranged to provide an expression which indicates the functional relationship between the volume of base added in the course of a titration and the measured pH.

In equation 10 sodium ions are conservative – that is, there are no processes in which these ions are created or destroyed. As such the concentration of sodium ions is given by:

$$\left[Na^+\right]=\frac{C_{Base}\times V_{Base}}{V_{Acid}+V_{Base}} \tag{11}$$

C_{Base} is the concentration of base, V_{Acid} and V_{Base} are the volumes of acid and base placed in the titration vessel respectively. For a titration in which a base of known concentration is added to a known amount of acid of unknown concentration the term:

$$\frac{V_{Base}}{V_{Acid}+V_{Base}}$$

is a correction factor to account for the dilution of the base as it is added the acid solution. The relationship shown in equation 11 can be substituted into equation 10. We can further substitute for [A⁻] in equation 10 using the relevant expression shown in equation 3 and multiply it by the applicable dilution correction and substitute too for [OH⁻] using the expression:

$$\frac{K_W}{\left[H^+\right]}$$

The final expression becomes:

$$\left[H^+\right]^3 + \left(\frac{C_{Base} \times V_{Base}}{V_{Acid} + V_{Base}} + K_a\right)\left[H^+\right]^2 + \dots$$

$$\dots \left(\left(\frac{C_{Base} \times V_{Base} - C_{Acid} \times V_{Acid}}{V_{Acid} + V_{Base}}\right)K_a - K_w\right)\left[H^+\right] - K_w K_a = 0 \tag{12}$$

It is possible to solve this type of equation numerically using Newton's method. The Acidtitr.XLS workbook provides a VBA user defined function which may be used to compute a pH given input values for C_{Acid}, C_{Base}, V_{Acid}, V_{Base}, K_a and K_w. The function is readily defined. Recall that to write a VBA function we select Insert from the MenuBar followed by Macro followed by Module. The following function based upon a similar program by Miller (1987) can then be written.

```
Function pH(Cacid, Vacid, pKa, Cbase, Vbase, guess_pH)

  Ka=10^(-pKa)
  Kw=10^(-14)

  Atotal=Cacid*Vacid/(Vacid+Vabse)
  OHadded=Cbase*Vbase/(Vacid+Vbase)
  delta=OHadded-Atotal

  guess=10^(-guess_pH)
  counter=1
  tol=0.000001
Do
  counter=counter+1
  If counter>100 Then Exit Do
  f=guess^3+(OHadded+Ka)*guess^2+(delta*Ka-Kw)*guess-Kw*Ka
  df=3*guess^2+2* (OHadded+Ka)*guess+delta*Ka-Kw
  new_guess=guess-f/df
  If Abs(f/df)<=Abs(tol*new_guess) Then Exit Do
  guess=new_guess
Loop
  pH= -log10(guess)
End Function
```

This and the next function are to be found on the Sheet named pH definition. Since VBA doesn't recognise the function log10 this must be defined too. The following definition provides the pertinent function.

```
Static Function log10(x)
  log10=Log(x)/Log(10#)
End Function
```

The first two lines of the pH definition convert the pKa value – the common way an acid dissociation constant is quoted – into a K_a value. The second line provides a definition for K_w. The next three lines are included to tidy up the mathematical expressions encountered later on in the function. For the Newton method to work we need an initial guess to substitute into equation 12 $f([H^+])$ and its first derivative $f'([H^+])$ The new refined guess is obtained from the equation:

$$New\ guess = guess - \frac{f\left(\left[H^+\right]\right)}{f'\left(\left[H^+\right]\right)} \tag{13}$$

Line 6 converts the initial guess value for the pH into a hydrogen ion concentration. It sometimes happens that Newton's method does not converge or possibly converges very slowly. To obviate any problems of non or slow convergence we define a counter which adds up the number of iterations performed. If this becomes greater than 100 the function terminates the Do Loop. The tolerance value helps define the number of significant figures required for each estimate of the hydrogen ion concentration. The smaller the tolerance the larger the number of significant figures. The values of equation 12 and its first derivative are calculated and the refined value of the guess is computed using equation 13. We terminate the Do Loop if the difference between the old and new guesses is less than a prescribed value. Since we are only interested in the magnitudes of the differences, not the signs, we calculate absolute values. The absolute value of the difference between the two guesses in equation 13 is in the absolute value of $f([H^+])/ f'([H^+])$. For the Do Loop to terminate this value should be smaller than the tolerance multiplied by the value of the guess. This ensures that despite a possible difference in initial and final hydrogen ion concentrations – in a titration – of some 10 to 12 orders of magnitude the Do Loop will always provide the same degree of accuracy for all calculations (Miller 1987).

We can use the described functions in Excel to simulate a titration. All that is necessary is to provide the necessary input values required by the function. Normally on a spreadsheet all of these input values would be constant except V_{Base} which would be used as the independent variable. Solution pH would be the dependent variable. We can however, go further and use the pH function coupled with the power of Solver to fit potentiometric titration data to this model. Figure 7 (and the Sheet labelled Data Fitting Application in Acidtitr.XLS) outline one way in which this model fitting process may be achieved. The diagram shows the data obtained for the titration of approximately 0.1 M acetic acid with 0.1 M NaOH. For the model fitting process we shall assume that C_{Acid} and pK_a are unknown. Estimates of these values are placed in cells B1 and B3 respectively. V_{Acid} and C_{Base} are known and their values are entered into cells B2 and D1 respectively. The potentiometric titration data set provides values for V_{Base} and pH. This data is placed in range A12:B29. In range C12 we enter our user defined pH function to provide an estimated pH given the input data. The formula will be entered as

Fig. 7. Model fitting of data obtained for an approximately 0.1 M ethanoic acid solution titrated with 0.1 M NaOH solution

	A	B	C	D	E
1	C_{acid}		0.102	C_{base}	0.1	
2	V_{acid}		25.00			
3	pK_a		4.61			
4						
5			Sum of			
6			squared			
7			deviations	0.8364		
8						
9						
10	Vbase	Measured	Calculated	Squared	
11	mL	pH	pH	Deviation	
12	0.0	2.89	2.80	0.0073	
13	5.0	3.8	4.00	0.0407	
14	10.0	4.33	4.42	0.0084	
15	15.0	4.73	4.77	0.0013	
16	20.0	5.19	5.17	0.0004	
17	22.0	5.43	5.41	0.0005	
18	24.0	5.84	5.81	0.0010	
19	24.5	6.05	5.99	0.0035	
20	25.0	6.38	6.29	0.0081	
21	25.5	7.69	7.63	0.0041	
22	25.8	10.3	10.73	0.1886	
23	26.0	10.64	10.97	0.1086	
24	26.5	11.02	11.28	0.0663	
25	27.0	11.2	11.45	0.0640	
26	30.0	11.66	11.91	0.0627	
27	35.0	11.92	12.20	0.0775	
28	40.0	12.05	12.35	0.0886	
29	50.0	12.19	12.51	0.1048	
30					

=pH($B$1,$B$2,$B$3,$D$1,A12,2) - where 2 is our guess of the pH value. In cell D12 we enter the formula =(B12-C12)^2. This computes the square of the deviation between the measured pH and the model pH. This formula is copied into the range D13:D29. In cell C7 we calculate the sum of the square of the deviations using the formula =SUM(D12:D29). Model fitting is now straightforward. From the Tools Menu on the Excel MenuBar select Solver. The target cell is identified as C7. We wish to minimise its value, so we check Min on the next line. And on the following line we indicate that the sum of the square of the deviations should be minimised by changing the values in cells B1 and B3 (entered as B1,B3). Finally press the Solve button. In our application the optimised values for C_{Acid} and pK_a are found to be 0.102 M and 4.61 respectively. The value for the pKa value compares well with a quoted figure of 4.7 (Budevsky 1979). No attempt has been made in this application to correct for activity which may explain the discrepancy. The time taken for a model fitting process is quite modest.

Fig. 8. Spreadsheet design to show the data and best fit line for the titration of approxiamtely 0.1 M ethanoic acid solution with 0.1 M NaOH solution

	A	B	C	D
1	Cacid	0.102098	Cbase	0.1
2	Vacid	25		
3	pKa	4.611772		
4				
5				
6				
7	Vbase	pH	Fitted	
8	0	2.89		
9	5	3.8		
10	10	4.33		
11	15	4.73		
12	20	5.19		
13	22	5.43		
14	24	5.84		
15	24.5	6.05		
16	25	6.38		
17	25.5	7.69		
18	25.8	10.3		
19	26	10.64		
20	26.5	11.02		
21	27	11.2		
22	30	11.66		
23	35	11.92		
24	40	12.05		
25	50	12.19		
26	0		2.8047377	
27	1		3.2801294	
28	2		3.5585678	
29	3		3.7444521	
30	4		3.8857383	
31	5		4.0016906	

Showing the data points and the best fit line is straightforward. Figure 8 shows a worksheet design which was used to generate Fig. 9. Copy the V_{Base} vs pH data onto a new Sheet (e.g. Sheet2) along with the input data in the range A1:D3. The data points are placed in the cell range A8:B25 and the input data is inserted in the cell range A1:D3. Next in the cells immediately below the last data entry for V_{Base} - i.e. cell A26 - enter a series of V_{Base} values which will be used to calculate the pH values of the best fit line. In Fig. 8 this series of values goes from 0 to 50 in steps of 0.5. A quick way of entering such a series of numbers is to put 0 in the first cell of the range followed by 0.5 in the next cell down. Place the cursor on the cell A26 which contains 0, depress the left hand mouse button and drag the cursor down to the cell A27 which contains 0.5. The two cells will be surround-

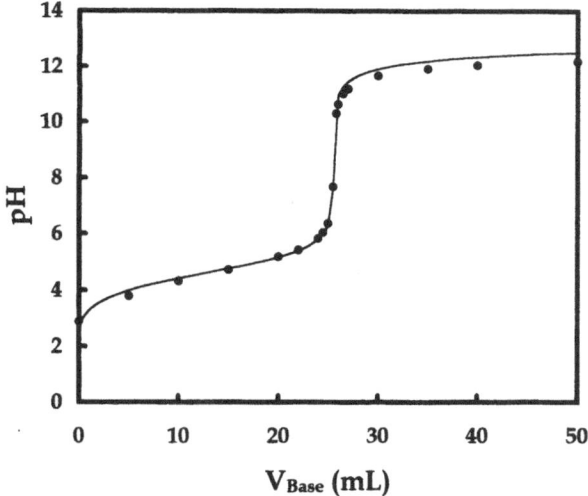

Fig. 9. The visual result of fitting data obtained for the titration of an approximately 0.1 M solution of ethanoic acid with 0.1 M NaOH to equation 12. Experimental data (·); line of best fit (–)

ed by a thick line on the bottom right hand corner of which will be found a little solid square. If the cursor is placed on this square the normal cursor cross will change to a thin cross. Press the left hand mouse button and drag the cursor down. As the cursor moves down the cells will automatically fill up with the other number in the series. Stop at 50! In cell C26 enter the user defined pH function. This can be copied to the rest of the cells in the range by placing the cursor on cell C26 and clicking the left-hand mouse button. The cell is surrounded with a thick line with the solid square in the bottom left-hand corner. Place the cursor over this square. The thick cursor cross will convert to a thin cursor cross. Press the left-hand mouse button and drag the cursor down. The formula will automatically be copied into the rest of the cell range. To create the graph select Insert and Chart from the MenuBar in the first dialogue box enter the cell range that contains the all the data to be graphed. In this application the range is A8:C126. Press the Next button. Double click the XY graph option followed by double clicking the Data markers only option. In step 4 press Next and finally in step 5 enter the axis titles followed by pressing the Finish button. Once the graph is displayed double click on the graph and then double click the data markers representing the line of best fit. From the dialogue box that appears select Patterns and from this menu check the option labelled Automatic under the section named Lines and check the option labelled None in the section named Markers and finally press OK. The experimental data will be shown as markers and the best fit curve is shown as a solid line.

The goodness of fit of the model to the data can be determined by calculating the coefficient of determination, which is given by the following formula:

$$\frac{\sum_{i=1}^{n}\left(y_{obs_i} - \overline{y}_{obs}\right)^2 - \sum_{i=1}^{n}\left(y_{obs_i} - y_{cal_i}\right)^2}{\sum_{i=1}^{n}\left(y_{obs_i} - \overline{y}_{obs}\right)^2} \tag{14}$$

The spreadsheet design used for the computation of the goodness-of-fit data is shown in Fig. 10. The data and the calculated (or fitted) pH values are held in the range A12:C29. The mean value of y is calculated in cell B3 by inputting the function =AVERAGE(B12:B29). In the range E12:E29 individual values for $(y_{obs}-y_{cal})^2$ (the square of difference between each measured pH value in the titration

Fig. 10. Spreadsheet layout for the computation of the goodness-of-fit statistics for the model fitting of the ethanoic acid sodium hydroxide titration data

	A	B	C	D	E	F
1	C of D	0.995549				
2						
3	Mean	7.961667				
4						
5						
6						
7						
8						
9						
10	Data		Fitted			
11	Vbase	pH	pH		$(y_{obs}-y_{cal})^2$	$(y_{obs-mean})^2$
12	0	2.89	2.804738		0.00727	25.7218
13	5	3.8	4.001691		0.040679	17.31947
14	10	4.33	4.421701		0.008409	13.189
15	15	4.73	4.766144		0.001306	10.44367
16	20	5.19	5.170814		0.000368	7.682136
17	22	5.43	5.407361		0.000513	6.409336
18	24	5.84	5.809087		0.000956	4.501469
19	24.5	6.05	5.990649		0.003523	3.654469
20	25	6.38	6.290178		0.008068	2.501669
21	25.5	7.69	7.625654		0.00414	0.073803
22	25.8	10.3	10.73429		0.188611	5.467803
23	26	10.64	10.96959		0.108632	7.173469
24	26.5	11.02	11.27742		0.066267	9.353403
25	27	11.2	11.45294		0.063978	10.4868
26	30	11.66	11.91048		0.06274	13.67767
27	35	11.92	12.19845		0.077535	15.6684
28	40	12.05	12.34772		0.088637	16.71447
29	50	12.19	12.51367		0.104763	17.8788
30					0.836394	187.9177
31						
32						

and the calculated pH value using the user defined pH function. In cell F30 the sum of the squared differences is calculated using the formula =SUM(F12:F29). In the range F12:F29 we calculate individual values for $(y_{obs}-mean)^2$ (the square of the difference between each measured pH value in the titration and the mean pH value). The sum of these squared differences is calculated in cell F30. The value of the coefficient is then determined in cell B1 using the formula =(F30-E30)/F30.

One interesting aspect of the titration experiment to pursue is to place the C_a and K_a data obtained from the model fitting into the Acidspec.XLS spreadsheet. The acid speciation diagram for ethanoic acid based upon these experimentally determined parameters can be produced. The question is immediately posed: what is the pH of the ethanoic acid solution at the beginning and the end-point of the titration? The speciation diagram provides the answer by indicating the pH values at which the relevant proton conditions are satisfied. At the beginning of the titration we have 0.102 M solution of ethanoic acid. The proton condition in this case is:

$$\left[H^+\right]=\left[OH^-\right]+\left[CH_3COO^-\right]$$

It can be seen by inspection of the chart in Acidspec.XLS that this equation is approximated to by the intersection of the H^+ and CH_3COO^- lines. This intersection occurs at a pH of 2.8. From the experimental data it an be seen that the initial pH is in fact 2.89. The pH of the end-point is determined by realising that we are interested in the proton condition that describes a 0.102 M solution of sodium ethanoate. The relevant equation in this case is:

$$\left[H^+\right]+\left[CH_3COOH\right]=\left[OH^-\right]$$

and it is approximated to by the intersection of CH_3COOH and OH^- lines. The pH in this case is 8.8.

Reading off these pH values is straightforward. The scale on the x axis can be altered so that the major tick marks appear for pH values from 0 to 14 increasing in steps of 1. The minor tick marks can also shown. These can also increase in steps of 0.1. To make these alterations one positions the cursor on the chart and 'double clicks' the left mouse button. The cursor is then placed on the x-axis and the mouse double clicked. From the Format Axis menu that appears Scale is selected and suitable entries made. Next Patterns is selected to ensure that the minor tick marks will be shown. Once the appropriate selections have been made the Format Axis may be closed followed by the closure of the Chart menu. It is possible that the graph may not be big enough to adequately show the tick marks on the x-axis. In this case the chart can be expanded by placing the cursor over the chart and clicking once. A box will appear around the chart. At each corner and at the centre of each side a small solid box will appear. If the cursor is placed over any of these boxes it will change to reveal a double arrowed line. The arrows point in the direction in which expansion or contraction of the chart may be achieved. The chart may in this way be expanded to any desired size. To aid the graph reading process the Drawing tool bar can be summoned and the Straight line button selected. A straight line of any desired length may then be drawn perpendicular to the x-axis and subsequently moved to any required point on the graph – e.g. the intersection of CH_3COOH and OH^- lines. The pH of the intersection point is then easily read off.

How does the Solubility of Aluminium Hydroxide change as a Function of pH?

In environmental chemistry the important task of assessing how changes in aqueous pH modify metal solubility is often encountered – especially in investigations of environmental acidification. Spreadsheet applications can be readily developed which aid the elaboration of the functional relationship between solubility and pH. A group of such applications are shown in the Alspec.XLS spreadsheet. For instance the problems posed by aluminium solubility in aquatic systems are of considerable interest. To analyse how we may develop spreadsheet applications in these area let us consider the solubilization of aluminium from the mineral gibbsite as a function of pH. The fundamental equation we require is the solubility product expression:

$$Al(OH)_{3\,(s)} \longrightarrow Al^{3+}_{(aq)} + 3OH^-_{(aq)}$$

Adopting the usual convention concerning the activities of solid phase materials in mass action expressions we may write:

$$\left[Al^{3+}_{(aq)}\right] = \frac{K_{sp}}{\left[OH^-_{(aq)}\right]^3} \tag{15}$$

Clearly equation 15 denotes that the concentration of Al^{3+} decreases rapidly as the hydroxide ion concentration increases – or as pH increases. It is important however to recognise that other aluminium species may be present in aqueous solution contributing to the overall aqueous solubility of aluminium. These species will include the following hydroxy species: $AlOH^{2+}$, $Al(OH)_2^+$, $Al(OH)_3$, $Al(OH)_4^-$ and $Al_3(OH)_4^{5+}$. There are, no doubt, further polymeric species that could be included. The concentrations of these species are given by the following equilibrium expressions:

$$\frac{\left[AlOH^{2+}\right]}{\left[Al^{3+}\right]\left[OH^{-}\right]}=\beta_{1} \qquad\qquad \left[AlOH^{2+}\right]=\beta_{1}\left[Al^{3+}\right]\left[OH^{-}\right] \quad (a)$$

$$\frac{\left[Al(OH)_{2}^{+}\right]}{\left[Al^{3+}\right]\left[OH^{-}\right]^{2}}=\beta_{2} \qquad\qquad \left[Al(OH)_{2}^{+}\right]=\beta_{2}\left[Al^{3+}\right]\left[OH^{-}\right]^{2} \quad (b)$$

$$\frac{\left[Al(OH)_{3}^{0}\right]}{\left[Al^{3+}\right]\left[OH^{-}\right]^{3}}=\beta_{3} \qquad or \qquad \left[Al(OH)_{3}^{0}\right]=\beta_{3}\left[Al^{3+}\right]\left[OH^{-}\right]^{3} \quad (c) \quad (16)$$

$$\frac{\left[Al(OH)_{4}^{-}\right]}{\left[Al^{3+}\right]\left[OH^{-}\right]^{4}}=\beta_{4} \qquad\qquad \left[Al(OH)_{4}^{-}\right]=\beta_{4}\left[Al^{3+}\right]\left[OH^{-}\right]^{4} \quad (d)$$

$$\frac{\left[Al_{3}(OH)_{4}^{5+}\right]}{\left[Al^{3+}\right]^{3}\left[OH^{-}\right]^{4}}=\beta_{5} \qquad\qquad \left[Al_{3}(OH)_{4}^{5+}\right]=\beta_{5}\left[Al^{3+}\right]^{3}\left[OH^{-}\right]^{4} (e)$$

To illustrate how the solubility of aluminium varies with pH for a system where crystalline gibbsite controls solubility is now a straightforward assignment. Figure 11 presents a suitable layout for this task. The relevant equilibrium data is presented in the top part of the spreadsheet. D2 holds the solubility product while the range D3:D7 contain the hydroxide formation constants which are referred to in equations 16 a to e. Cell G2 contains the water dissociation constant value. The computations are carried out below using the Data Table procedure as outlined earlier. The strategy used for managing the calculation is as follows. A pH range is selected over which Al solubility is to be investigated. For this particular application we have used the pH range 3 to 12. These pH values rising in steps of 0.1 are inserted in column A starting at cell A12. In cell B11 we insert the formula $=10^{-A9}$ which is used to calculate the hydrogen ion concentration. It may be recalled that for the Data Table procedure input values of the master variable are placed in an empty cell called the input cell to calculate formula values. For our application the input cell is A9. The formula in cell C11, = G2/A11, is used to calculate $[OH^{-}]$. In the next cell we determine $[Al^{3+}]$ using equation 15. As a spreadsheet formula this appears as $=D2/C11^{3}$. Furnished with the value of $[Al^{3+}]$ we are now in a position to compute the concentrations of the other hydroxy aluminium species using the formation constant data, and values for $[Al^{3+}]$ and $[OH^{-}]$. The formulae used are, of course, based upon equations 16 a to e. Finally in cell J11 we compute the log of the total concentration of soluble Al species. As outlined previously we can now generate a table of values based upon the formulae in the cell range B11:J11 using pH values in the range A12:A102 as the input

Fig. 11. Spreadsheet layout for the computation of aluminium solubility.

	A	B	C	D	E
1			Al/OH Data		
2			Ksp	1.29E-34	
3			β_1	1.00E+09	
4			β_2	5.01E+18	
5			β_3	1.00E+27	
6			β_4	1.00E+33	
7			β_5	1.26E+42	
8					
9					
10	pH	H$^+$	OH$^-$	Al^{3+}	AlOH^{2+}
11		1	1E-14	128824955	1288.2496
12	3	0.001	1E-11	0.128825	0.0012882
13	3.1	0.0007943	1.259E-11	0.0645654	0.0008128

	F	G	H	I	J
1					
2	Kw	1E-14			
3					
4					
5					
6					
7					
8					
9					
10	Al(OH)$_2^+$	Al(OH)$_3$	Al(OH)$_4^-$	Al$_3$(OH)$_4^{5+}$	log Al $_{Total}$
11	0.0645654	1.288E-07	1.288E-15	2.692E+10	10.907814
12	6.457E-05	1.288E-07	1.288E-12	2.692E-05	-0.885193
13	5.129E-05	1.288E-07	1.622E-12	8.511E-06	-1.184056

Fig. 12. Variation in the solubility of aluminium with pH for a system where aqueous solubility is controlled by Gibbsite

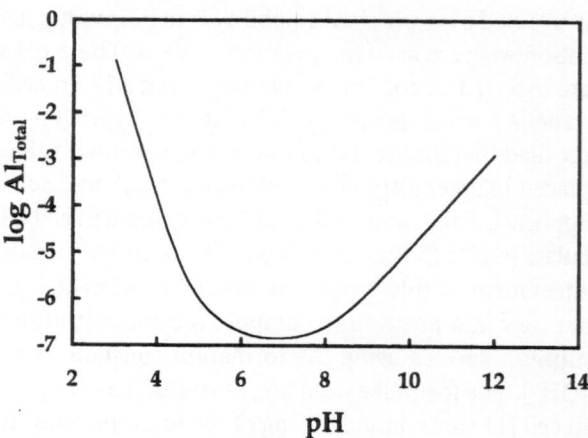

values. The resulting data may then be plotted. Figure 12 shows the chart of log Al_{Total} as a function of pH produced.

Figure 12 provides an explanation for the observation that in acidic waters fish mortality may be high because of clogging of the gills. In acid waters (pH about 4) aluminium solubility is about 10^{-4} M. At a pH of 7 to 8 this solubility figure drops to the range of $10^{-6.6}$ to $10^{-6.8}$ M. The transport of water across the gills of a fish gives rise to a change in pH in which it rises from about 4 (the pH of the surrounding water) to a pH in the range 7 to 8 (the physiological pH of fish blood). The net effect of this change is that aluminium solubility drops rapidly (it drops by nearly three orders of magnitude) and precipitates from solution at the blood/water interface and as a result brings about obstruction of the gills.

How is Aluminium Solubility altered by the Presence of Complexing Agents?

The spreadsheet application produced above (see Alspec.XLS) may be further expanded to address even more interesting questions. For instance in both inorganic and environmental chemistry it is recognised that the presence of chelating agents may have a profound influence on solubility. This may be demonstrated by considering the effect of EDTA upon aluminium solubility. Aluminium forms three species with EDTA the equilibrium expressions for which are given below. The ionic charges in these and subsequent equations are omitted for clarity:

$$[AlL] = \beta_1 [Al][L] \qquad\qquad (a)$$

$$[AlHL] = \beta_1 [Al][L][H] \qquad\qquad (b) \qquad (17)$$

$$[AlOHL] = \beta_1 [Al][L][OH] \qquad\qquad (c)$$

Computation of [L] can be completed by considering the following equations which outline the pertinent equilibrium expressions for the successive protonation of the fully dissociated EDTA ligand.

$$\frac{[HL]}{[H][L]} = \pi_1$$

$$\vdots \qquad\qquad\qquad (18)$$

$$\frac{[H_6L]}{[H]^6[L]} = \pi_6$$

The mass balance expressions for the aqueous concentrations of all forms of aluminium and EDTA are given by:

$$Al_{Total} = [Al] + \ldots [Al(OH)_4] + [Al_3(OH)_4] + \ldots \qquad (a)$$
$$\ldots [AlL] + [AlHL] + [AlOHL] \qquad\qquad\qquad (19)$$

$$L_{Total} = [L] + \ldots [H_6L] + [AlL] + [AlHL] + [AlOHL] \qquad (b)$$

If the mass balance expression for EDTA – shown in equation 19b – is divided through by [L] (the undissociated ligand concentration) and the ensuing expression rearranged and appropriate substitutions made, the following formula for the undissociated ligand concentration may be obtained:

$$L = \frac{L_{Total}}{1 + \sum_{i=1}^{6} \pi_i [H]^i + \beta_1[Al] + \beta_2[Al][H] + \beta_3[Al][OH]} \qquad (20)$$

Figure 13 reveals those additions that are necessary to make to Figure 11 to facilitate the estimation of the effect of EDTA on aluminium solubility. The protonation constants for EDTA π_1 to π_6 are inserted into the cell range L2:L7. The total acid concentration is placed in L8. The formation constants (ϕ_1 to ϕ_3) for the aluminium EDTA complexes are placed in cells O2:O4. The formula for the undissociated ligand concentration

```
=L8/(1+L2*B11+L3*B11^2 ...
+L7*B11^6+O2*D11+O3*D11*B11+O4*D11*C11)
```

Fig. 13. Computation of the effect of EDTA on aluminium solubility

	K	L	M	N	O
1	EDTA data				
2	π_1	2.19E+10		ϕ_1	1.35E+16
3	π_2	3.8E+16		ϕ_2	5.01E+18
4	π_3	2.14E+19		ϕ_3	1.58E+24
5	π_4	2.51E+21			
6	π_5	1E+23			
7	π_6	7.94E+23			
8	A_{Total}	1.00E-06			
9					
10	L^{6-}	AlL	AlHL	AlOHL	log Al $_{Total}$
11	1.54E-33	2.68E-09	9.96E-07	3.15E-15	10.90781
12	4.19E-22	7.28E-07	2.71E-07	8.56E-10	-0.88519

Fig. 14. The impact of EDTA on aluminium solubility for the system shown in Fig. 12

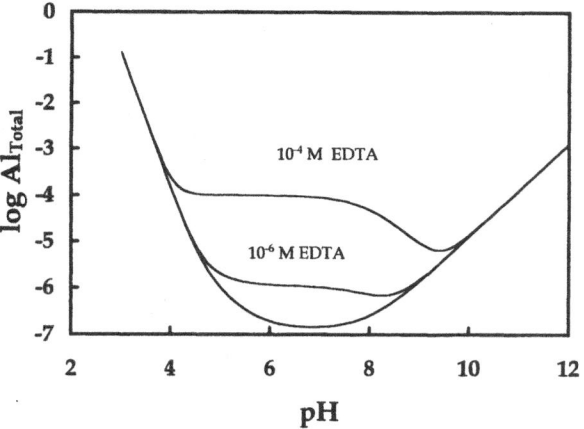

is located in K11. Once this is calculated the only remaining task is to determine the concentrations of the aluminium/EDTA complexes, AlL, AlHL and AlOHL using equations 17 a to c. In cell O11 the formula for the new total aluminium concentration in aqueous solution is entered.

The output of interest from the data table are the values in cell range J12:J102 the total aluminium concentration in the absence of EDTA, in the pH range 3 to 12; and the values in the cell range O12:O102 the total aluminium concentration in the presence of EDTA. To graph these values against pH it is a good idea to copy them to a new worksheet. The pH values may be copied and pasted. However the total aluminium concentrations should be copied and then pasted using the Paste Special option. The option object for Paste Special will appear and since we are only interested in the concentration values the Values option is 'checked'. The output from this exercise is shown in Figure 14. Clearly even a modest aqueous concentration of EDTA can have a profound effect on aluminium solubility.

It is worthwhile pointing out at this stage that the way in which the spreadsheet application has been designed should help students focus on the nature of the pertinent equilibria in aqueous solution. For instance, so long as gibbsite is present as a solid phase of unit activity then regardless of the many different equilibrium processes that may occur in solution the concentration of Al^{3+} is determined solely by the solubility product expression. Any process which removes Al^{3+} is compensated for by the release of more Al^{3+} from the solid phase – in compliance with Le Chatelier's principle.

How is the Solubility of Aluminium in Aqueous EDTA Solution Affected by Competing Equilibria?

In real environmental systems the reactions between EDTA and aluminium are likely to affected by the presence of competing reactions. The result of such competition is to reduce the undissociated ligand concentration and thus reduce the concentrations of aluminium EDTA complexes. In many natural aquatic systems

Fig. 15. Spreadsheet design for calculating the effect of calcium ion competition on the aluminium EDTA system

	Q	R	S	T	U
1	Calcium data				
2	Ca$_{Total}$	1.00E-03		ϕ_1	5.01E+10
3				ϕ_2	6.31E+13
4					
5					
6					
7					
8					
9					
10	L^{6-}	AlL	AlHL	AlOHL	log Al $_{Total}$
11	1.54E-33	2.68E-09	9.96E-07	3.15E-15	10.90781
12	4.19E-22	7.28E-07	2.71E-07	8.56E-10	-0.88519

the competition is most likely to be provided by calcium ions. Such competition may be readily incorporated into the spreadsheet model. In Figure 15 it is assumed that the total aqueous concentration of calcium is 10^{-3} M. The combination of the completely dissociated EDTA chelate with Ca^{2+} ions does not significantly reduce this concentration. The total aqueous calcium concentration is therefore set equal to the concentration of Ca^{2+} ions. The following mass action expressions characterise the complexation reactions that occur between calcium and EDTA.

$$\frac{[CaL]}{[Ca][L]} = \phi_1$$

$$\frac{[CaHL]}{[Ca][L][H]} = \phi_2$$

The calcium ion concentration is known as is the proton concentration in the second reaction. The formation constants are known. The concentration of fully dissociated EDTA is obtained from the following modification of equation 20.

$$L = \frac{L_{Total}}{1 + \sum_{i=1}^{6} \pi_i [H]^i + \beta_1 [Al] + \beta_2 [Al][H] + \beta_3 [Al][OH] + \phi_1 [Ca] + \phi_2 [Ca][H]} \quad (21)$$

The appropriate spreadsheet formula is placed in cell Q11 and the formulae used to compute the aluminium EDTA complex concentrations are inserted in cells R11, S11 and T11. The equation used to determine the log of the total aqueous aluminium concentration is placed in cell U11. The new data obtained is plotted against pH along with the data obtained from the previous section and is shown in Figure 16. It is interesting to note that the presence of a modest quantity of cal-

Fig. 16. The impact of calcium on aluminium solubility for the system shown in Fig. 14

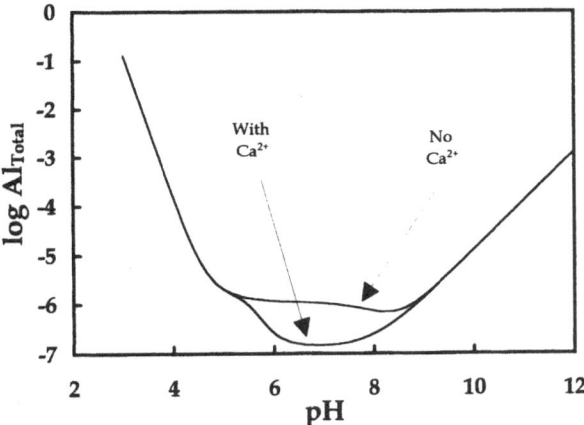

cium in aqueous solution can reduce the concentration of fully dissociated EDTA by such an amount that the increases in aluminium solubility in the pH range of 5 to 9 noted in the previous section are effectively counterbalanced.

Using Solver to Determine Speciation Patterns in more Complex Systems

One final question we shall study in this section on solving speciation problems in thermodynamics is reproduced from Stumm and Morgan (1996) and deals with the question of determining the extent of formation of hydrogen carbonato and carbonato complexes of magnesium in fresh waters. The following thermodynamic information is supplied. The total concentration of carbonic acid in its many forms is $4 * 10^{-3}$ M, the total concentration of magnesium is 10^{-3} M and the ionic strength is given as $4 * 10^{-3}$ M. Detailed below are the relevant mass action and mass balance expressions:

$$\frac{\left[MgHCO_3^-\right]}{\left[Mg^{2+}\right]\left[HCO_3^-\right]} = 4.1 \tag{a}$$

$$\frac{\left[Mg(HCO_3)_2\right]}{\left[Mg^{2+}\right]\left[HCO_3^-\right]^2} = 8.5 \tag{b}$$

$$\tag{22}$$

$$\frac{\left[MgCO_3\right]}{\left[Mg^{2+}\right]\left[CO_3^{2-}\right]} = 500 \tag{c}$$

$$\frac{\left[MgOH^+\right]\{H^+\}}{\left[Mg^{2+}\right]} = 10^{-11.52} \tag{d}$$

$$\frac{\left[HCO_3^-\right]\left\{H^+\right\}}{\left[H_2CO_3^*\right]} = 10^{-6.33} \tag{e}$$

$$\frac{\left[CO_3^{2-}\right]\left\{H^+\right\}}{\left[HCO_3^-\right]} = 10^{-10.27} \tag{f}$$

$$\tag{22}$$

$$\left[H_2CO_3^*\right]+\left[HCO_3^-\right]+\left[CO_3^{2-}\right]+\left[MgHCO_3^-\right]+\ldots$$
$$2\left[Mg\left(HCO_3\right)_2\right]+\left[MgCO_3\right] = 4\times10^{-3} \tag{g}$$

$$\left[Mg^{2+}\right]+\left[MgOH^+\right]+\left[MgHCO_3^-\right]+\left[Mg\left(HCO_3\right)_2\right]+\ldots$$
$$\ldots\left[MgCO_3\right] = 10^{-3} \tag{h}$$

Figure 17 outlines an application designed to simultaneously solve the above mass action and mass balance equations. The strategy adopted for these calculations is as follows. The pH of the system is, as in previous computations, the independent variable. Its value is inserted in cell C1. In the cell range E3:E10 we insert the initial estimates of the concentrations of the species involved in the reactions. In cell E11 we enter the formula for the conversion of the pH value to the hydrogen ion concentration – the formula being =10^-C1. The initial estimates of the concentrations of the carbonic acid species are determined by using equation 7. These equations are entered in the cell range G8:G10 The values for these acid species are then copied across to cells E8:E10 respectively using the Copy and Paste Special commands in which the Values option is checked. In the cell range E15:E22 we enter the accepted equilibrium values for the mass action expressions (modified for the given ionic strength) outlined in equations 22 a to f and the preset values for the mass balance expressions. In the cell range G15:G22 we enter spreadsheet formulae – based upon equations 22 a to h – which use the values in the cell range E3:E11 to calculate the values of these expressions. For instance in cell G15 we enter the formula =E3/(E7*E9). We now use Solver – selecting it in the usual way. The target cell is set as G19. Solver is instructed to change the values in the cell range E3:E10 (we don't change the hydrogen ion concentration in E11 since this is preset) such that G19 equals a value of 0.001. There are however er certain constraints on the range of acceptable values located in the cell range E3:E10. These constraints are that the values in the cell range E3:E11 are able to provide satisfactory numerical values for the other mass action and mass balance equations presented in equations 22 a to h. These constraints are entered in the following way. In the Solver dialogue box one observes a box labelled "Subject to the Constraints". To enter a constraint press the Add button. A new dia-

	A	B	C	D	E	F	G
1		pH =	7	Compounds			
2							
3				MgHCO₃⁻	1.33E-05		
4				Mg(HCO₃)₂	9.03E-08		Initial
5				MgCO₃	8.69E-07		carbonate
6				MgOH⁺	2.98E-08		guesses
7				Mg²⁺	9.86E-04		
8				H₂CO₃	7.02E-04		7.04E-04
9				HCO₃⁻	3.28E-03		3.29E-03
10				CO₃²⁻	1.76E-06		1.77E-06
11				H⁺	1E-07		
12							
13				Equations	Target values		Mass action values
14							
15				22 a	4.10E+00		4.10E+00
16				22 b	8.50E+00		8.50E+00
17				22 c	5.00E+02		5.00E+02
18				22 d	3.02E-12		3.02E-12
19				22 e	4.68E-07		4.68E-07
20				22 f	2.51E-17		2.51E-17
21				21 g	4.00E-03		4.00E-03
22				21 h	1.00E-03		1.00E-03

Fig. 17. Spreadsheet design for the computation of carbonato and hydrogen carbonato magnesium complex concentrations

logue box labelled "Add Constraint" will appear. In the box under the title Cell Reference enter the coordinates of cell whose value is to be restricted. Next select the list of constraints and choose =. Finally in the right hand box enter either the value you wish the constrained cell to equal or enter the cell reference which contains this value. In our spreadsheet application the accepted equilibrium constant values and mass balance values are located in the cell range E15:E22. Once all the constraints have been identified we eventually return to the main Solver menu by pressing OK. We finally kick Solver into action by pressing the Solve button. If the initial estimates in cell range E3:E10 are reasonable Solver will converge to a solution. If the estimates are not so good Solver may fail to converge. In this case new estimates are required.

It is possible to investigate how the speciation pattern and concentrations alter with changing pH. After every successful convergence the data in cells E3:E6 are copied to another Sheet using the Copy Paste Special commands. On the Paste Special menu we check the box entitled Transpose. This converts the vertically aligned data on the model sheet to horizontally aligned data on the other sheet. Thus on the fresh sheet we can have columns of data. In the first column we can have the pH values used. In the subsequent columns we insert the magnesium complex concentrations, each column containing the values for one particular complex. Aligning the data in this way makes it straightforward to inspect the data and to plot it. Figure 18 shows a graph of the data obtained. This spreadsheet application is outlined in Mgspec.XLS.

Fig. 18. The effect of pH on the speciation of carbonato and hydroxy magnesium species

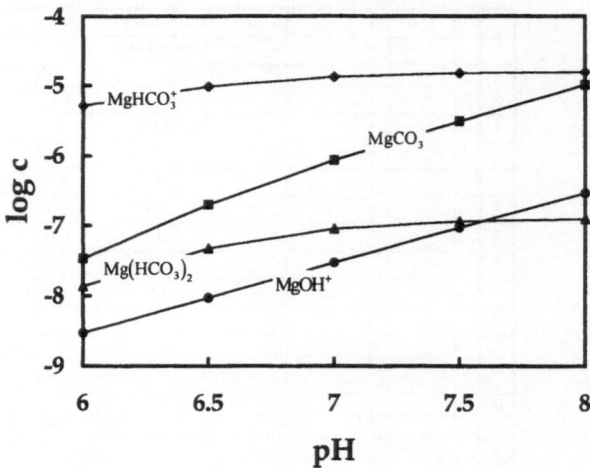

Concluding Remarks

The intention of this chapter has been to show the reader how thermodynamic relationships may be explored using the power of Excel. Against this backcloth the reader should have learnt how to use certain features of Excel possibly in novel ways. Most importantly it is hoped that the reader will feel capable of producing their own spreadsheet applications in their areas of expertise.

References

Budevsky O (1979) Foundations of Chemical Analysis. Ellis Horwood, Chichester
Miller AR (1987) Turbo Pascal Programs for Scientists and Engineers. Sybex, San Francisco
Sillén LG (1959) Treatise on Analytical Chemistry, Part 1, Vol 1. Kolthoff IM, Elving PJ (eds) Interscience, New York
Stumm W, Morgan JJ (1996) Aquatic Chemistry: Chemical Equilibria and Rates in Natural Waters, 3rd edn. Wiley-Interscience, New York

Using Spreadsheets in Chemical Engineering Problems

6

F.M. Julian

General

The row and column format of spreadsheets fits beautifully into the type of calculations which Chemical Engineers often make. In this chapter we will show that many spreadsheet functions can be used in Chemical Engineering. As a vehicle for this demonstration we will use the simple process flow diagram shown in Fig. 1, but the techniques used in this example have much wider applicability than just flowsheeting. In addition to simply drawing and converging the flowsheet, we will consider:

- Two methods of equation solving
- Table lookup procedures
- Linear and logarithmic interpolation
- Linear regression of data
- Flash calculations
- Damage Control
- Graphing
- Spreadsheet documentation

The examples in this chapter were created using Microsoft Excel, Version 7.0 (for Windows95). Readers still using earlier versions will notice slight though usually self-explanatory differences. With the exception of the code used in macros, anything shown in this chapter can be performed on any of the major spreadsheets with very little modification. Beginning in version 5.0, Excel altered its macro language to Visual Basic for Applications (VBA), making it entirely different from its own previous macro code and that of other spreadsheets. The code for two useful macros is shown in this chapter, but the rest of the examples can be performed perfectly well without using macros at all.

A reader who is interested in general Chemical Engineering can easily follow the steps outlined in this chapter to produce the final flowsheet and thereby use all its associated equation solvers, table lookups and interpolators. Whoever wishes to do so should start with a blank workbook containing seven worksheets. He can then follow the steps in this chapter exactly. A reader only interested in some of the procedures (for example, equation solving) can begin at that step and reproduce only the relevant portion of the spreadsheet. In some cases, where input values come from another part of the spreadsheet, the selective reader may have to create artificial input values.

The complete spreadsheet containing all the examples detailed in this chapter is included on the disk and named ChemEng.XLS

Flowsheeting

Let us suppose that a chemist at our company research facility has invented a marvelous new product which we will call "P". It is made from two raw materials, "R1" and "R2" in the gas phase according to this stoichiometry:

$$R1 + 2R2 \rightarrow 2P$$

Our job is to develop a flowsheet and material balance for this process. Apart from the chemistry, all we know is that we can expect 70% of the R1 fed to the Reactor to be converted to P. There is no excess R2. The Separator (we will assume for now) is a perfect separator which produces pure P in one stream and pure R1 for recycle to the Reactor in another.

Actual drawing of the flowsheet in Fig. 1 is done by clicking on the drawing button (circle, triangle and square icon) and putting the arrows and rectangles together to give the proper diagram. We will not dwell on the details of how this

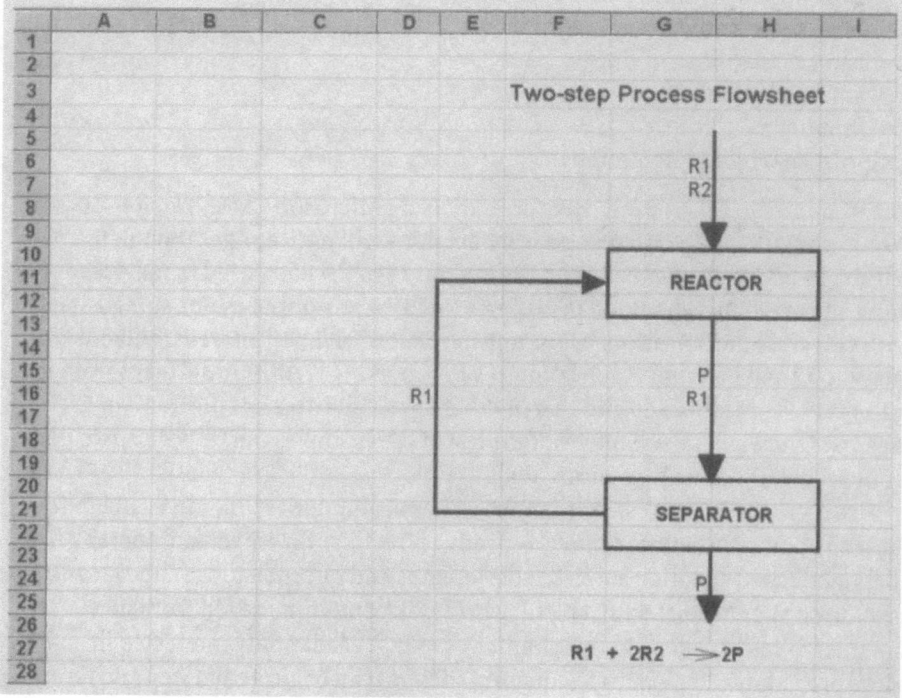

Fig. 1.

is done. The flowrates of each compound in lb-mols/hr (although they could equally well be in weight or volumetric units) will be shown directly on the flowsheet opposite the name of the compound. If we prefer, the flows could all be shown in a table to one side of the drawing, but the format shown here is convenient and easy to understand.

Throughout these examples we will give names to all cells of any importance to the calculations. We do this for two reasons. First, properly chosen cell names make formulas easy to read and understand. Second, use of cell names in macros allows us to move cells around on the sheet without rewriting the macros to change the cell addresses. Excel interprets all cell names as absolute addresses.

Before proceeding further, it is a good idea to set calculation to manual in the TOOLS Options menu. Also check Iterations and set the number of iterations to one. Since converging this flowsheet will be an iterative calculation, we must be able to recalculate at will, but not to let the computer recalculate at times when we may not be ready for it to be done. Until we are sure that the flowsheet is stable and all the formulas are correct, it is a good idea to see the results of each iteration before going on to the next one.

Now we are ready to put in the formulas for the various flows. The cell addresses will depend on where we locate the drawing on the sheet, but the ones below correspond with Fig. 1. We will start by assuming a feed rate of 100 mols/hr of R1 (shown as BASIS in Table 1).

After all the formulas have been entered, remember that we are on manual recalculation, so tap the F9 key a few times to iterate through the flowsheet until the numbers stop changing. The worksheet is now said to have converged. It should take five or six iterations to reach the "final result" figures shown below.

Just as we think our job is about done, the purchasing agent informs us that he has found a cheap source of R1, but that it is only 99% pure. It contains 1% R3. From previous work we know that R3 is an inert which passes right through the reactor unchanged. In the Separator it follows the R1. This means we must add to each stream a formula for the amount of R3 in it.

In addition, we must add a purge stream to remove the R3. We will take this purge from the recycle stream so we will not lose any P. We will, however, lose

Table 1.

Cell Name	Address	Contents	Formula	Final Result
BASIS	H1	Feed Rate of R1	100	100
CONV	H2	Conversion of R1 per pass	.7 (shown as %)	70%
FEEDR1	H6	Feed rate of R1	=BASIS	100.0
FEEDR2	H7	Feed rate of R2	=2*FEEDR1	200.0
EXITP	H15	Product from Reactor	=2*CONV*(FEEDR1+RECYR1)	200.0
EXITR1	H16	Unreacted R1	=(1-CONV)*(FEEDR1+RECYR1)	42.9
	H24	Product from Separator	=H15	200.0
RECYR1	E16	Recycle R1	=H16	42.9

some R1 since the composition of the purge is the same as the composition of the recycle. We can name the R1 flow in the purge PURR1 and the R3 in the purge PURR3. Since we are adding one mol/hr of R3 to the system, we will set PURR3 = 1.0. Then we can say:

$$\frac{PURR1}{PURR3} = \frac{RECYR1}{RECYR3} \qquad \text{or} \qquad PURR1 = PURR3\frac{RECYR1}{RECYR3}$$

This, however, is a risky formula to use. It contains a denominator which, under some circumstances, might equal zero. Spreadsheets are very intolerant of dividing by zero. To avoid this possibility, we can instead use the formula:

$$PURR1 = \frac{PURR3 * RECYR1}{MAX(.01, RECYR3)}$$

Now the denominator will have either the value .01 or the value RECYR3, whichever is larger. Since both are greater than zero, the denominator can never be zero.

The loss of some R1 in the purge means we must also change the formulas for the flows of R1, R2 and P in the flowsheet to allow for the R1 which is lost and no longer available to recycle to the reactor. Our worksheet will now appear as shown in Table 2.

At steady state, the value of PURR3 must equal the value of FEEDR3 (in this case 1.0). To start the flowsheet, however, we should set PURR3 at zero and recalculate once or twice to build up a R3 level in the circulating recycle stream. Then

Table 2.

Cell Name	Address	Contents	Formula	Final Result
BASIS	H1	Feed Rate of R1+R3	100	100
CONV	H2	Conversion of R1 per pass	.7 (shown as %)	70%
PURITY	H3	Purity of purchased R1	.99 (shown as %)	99%
FEEDR1	H6	Feed rate of R1	=BASIS*PURITY	99.0
FEEDR2	H7	Feed rate of R2	=2*(FEEDR1-PURR1)	163.1
FEEDR3	H8	Feed rate of R3	=BASIS*(1-PURITY)	1.0
EXITP	H15	Product from Reactor	=2*CONV*(FEEDR1+RECYR1-PURR1))	163.1
EXITR1	H16	Unreacted R1	=(1-CONV)*(FEEDR1+RECYR1-PURR1)	34.9
EXITR2	H17	R3 from Reactor	=FEEDR3+RECYR3-PURR3	2.0
	H24	Product from Separator	=EXITP	163.1
RECYR1	E16	Recycle R1	=EXITR1	34.9
RECYR3	E17	Recycle R3	=EXITR3	2.0
PURR1	C12	R1 in the Purge	=PURR3*RECYR1/MAX(.01, RECYR3)	17.5
PURR3	C13	R3 in the Purge	1.0	1.0

set PURR3 to 1.0. The values in the table below are obtained by setting PURR3 to zero, recalculating once, setting PURR3 to 1.0, and recalculating until convergence is reached. Recalculating more than twice with PURR3=0 will give a higher R3 concentration, but will be equally correct. In this case we still have one degree of freedom, in that we can have any inert (R3) level we want in the system.

In real life a plant operator has the option (within the limits of his equipment) of taking a large purge or a small one. The large purge (such as in this example) keeps the level of inerts (R3) very low in the system but loses a large quantity of the principal raw material (R1). The small purge loses very little R1, but at the expense of a high inert level in the process. The spreadsheet works exactly the same way. By pinching back on the purge valve, thus setting PURR3 to a value lower than FEEDR3, we can increase the inert level in the system as much as we want. If we set PURR3 greater than FEEDR3, the inert level will drop.

To illustrate this, pick a nearby empty cell, such as C19 and enter the formula for inert (R3) level in the recycle stream:

$$\frac{RECYR3}{\left(RECYR3 + RECYR1\right)}$$

Now experiment by setting PURR3 equal to 0.75 and recalculating several times. Watch the effect on the inert level shown in cell C19. As we approach the inert

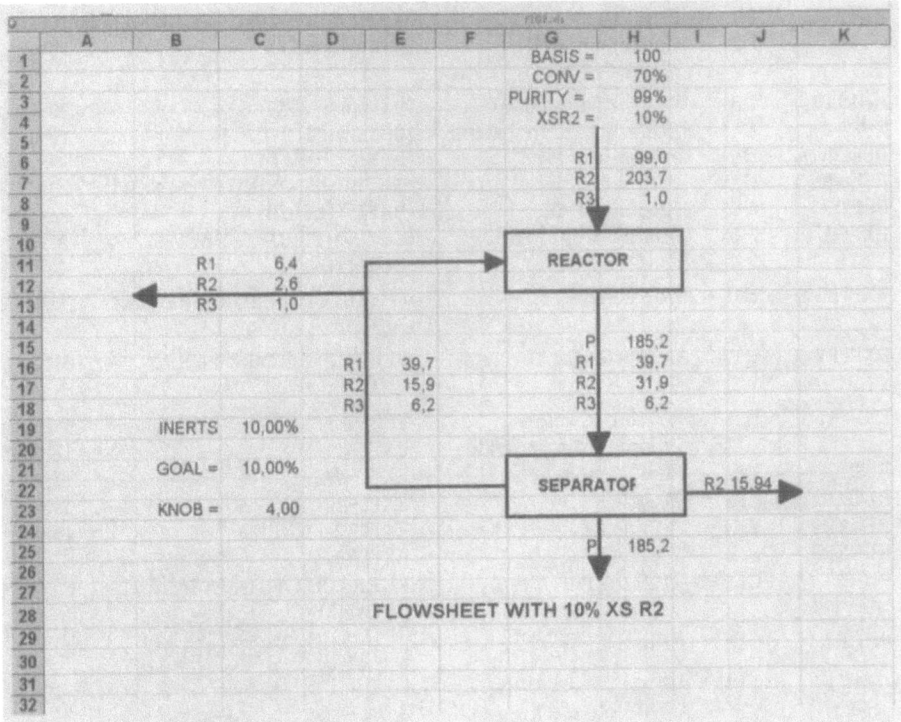

Fig. 2.

level we want, adjust PURR3 to be closer and closer to FEEDR3 (1.0 in our case). At final steady state conditions, it must be exactly equal to FEEDR3. If we expect to be adjusting the inert level often, we might use the Visual Basic macro shown in the Appendix to this chapter and titled sub CONVERGE(). This requires giving cell C19 the name INERTS and creating two nearby cells named GOAL and KNOB. In GOAL we enter the inert level we would like to have in the system. In KNOB we put a value for the sensitivity of the corrections the macro will make to the flowsheet. A good value for KNOB is 4.0. The macro works as a "proportional only" controller. Higher values of KNOB will result in larger changes to PURR3 and quicker but less stable solutions.

Meanwhile the research people have continued to work on this process and have discovered that there is value in having about a 10% excess of R2 in the reaction. This means there will now be R2 present in the Reactor exit stream, the recycle and the purge. Their work has shown that the Separator removes half the R2 as pure R2 in a new side draw stream, and the other half follows the R1. They have also shown that an inert level of 10% is optimum for the process.

Table 3.

Cell Name	Address	Contents	Formula	Final Result
BASIS	H1	R1+R3	100	100
CONV	H2	Conversion of R1 per pass	.7 (shown as %)	70%
PURITY	H3	Purity of purchased R1	.99 (shown as %)	99%
XSR2	H4	Excess R2 fed to Reactor	.10(shown as %)	10%
FEEDR1	H6	Feed rate of R1	=BASIS*PURITY	99.0
FEEDR2	H7	Feed rate of R2	=2*(FEEDR1-PURR1)*(1+XSR2)	203.7
FEEDR3	H8	Feed rate of R3	=BASIS*(1-PURITY)	1.0
EXITP	H15	Product from Reactor	=CONV*(FEEDR1+RECYR1-PURR1))	185.2
EXITR1	H16	Unreacted R1	=(1-CONV)*(FEEDR1+RECYR1-PURR1)	39.7
EXITR2	H17	Unreacted R2	=FEEDR2+RECYR2-PURR2-EXITP	31.9
EXITR3	H18	R3 from Reactor	=FEEDR3+RECYR3-PURR3	6.2
	H24	Product from Separator	=EXITP	185.2
PURER2	J20	Pure R2 from Seperator	=EXITR2/2	15.9
RECYR1	E16	Recycle R1	=EXITR1	39.7
RECYR2	E17	Recycle R2	=EXITR2-PURER2	15.9
RECYR3	E18	Recycle R3	=EXITR3	6.2
PURR1	C11	R1 in the Purge	=PURR3*RECYR1/MAX(.01, RECYR3)	6.4
PURR2	C12	R2 in the Purge	=PURR3*RECYR2/MAX(.01, RECYR3)	2.6
PURR3	C 13	R3 in the Purge	1.0	1.0
INERTS	C19	% Inerts in the Recycle	RECYR3/(RECYR1+RECYR2+RECYR3)	10.0%

Table 3 shows the cell formulas for the flowsheet with 10% inerts and 10% excess R2, and Fig. 2 shows how it will look on the spreadsheet. Note that the formula for FEEDR2 is the same as in the previous case but multiplied by 1.1.

Equation Solving

Another refinement to the flowsheet would be to add some details about the separation of the product (P) from the unreacted raw materials (R1, R2, R3). Until now we have assumed it to be a perfect separation, but in reality this is seldom true. Let us say it develops into a two-stage separation. The first is an isothermal flash conducted at 40°. The liquid from the flash is rich in P and is sent to a second stage, such as a distillation column, for further purification. The vapor from the flash is recycled to the reactor. Distillation columns, although they can be simulated on a spreadsheet, do not lend themselves to this type of calculation. Extensive physical property and vapor-liquid equilibrium data are required to calculate a distillation column accurately, and that can best be done using one of the many simulation packages designed to handle it. We will assume that the product compositions of the distillation step are known from experimental or calculated data.

The multi-component flash, however, can be handled neatly on a spreadsheet if the compounds do not engage in such non-ideal behavior as forming azeotropes. This will be another iterative calculation and requires that we know the K values of the various components at the conditions of the flash (say 3.5 atm and 40°). We will assume that we have the following experimental data on these K values:

Compound	K
R1	25
R2	0.2
R3	20
P	0.04

where:

$$K_i = \frac{y_i}{x_i}$$

and:

y_i = mol fraction of component i in the vapor
x_i = mol fraction of component i in the liquid
z_i = mol fraction of component i in the feed to the flash

The calculation of the amounts and compositions of the vapor and liquid streams from the flash can be performed by the method of Rachford and Rice (1952) by solving the following equation:

$$\sum_{1}^{n} \frac{z_i(1-K_i)}{1+V(K_i-1)} = 0$$

where V = fraction of flash feed which leaves as vapor and n = number of compounds in the mixture.

For spreadsheet purposes we will give certain cells names which correspond to the nomenclature shown above. The one exception to this is V which will refer to a cell containing the current guess at the value of V. For each trial, that cell address will be different, so for now we will continue to use V (though the actual formula will refer to a relative cell address). We can express the equation in spreadsheet terminology as:

ZR1*(1-KR1)/(1+V*(KR1–1))+ZR2*(1-KR2)/(1+V*(KR2–1))+ZR3*(1-KR3)/(1+V*(KR3–1))+ZP*(1-KP)/(1+V*(KP-1))

In Table 4 below, this equation will be referred to as "Eqn. for V".

The objective is to find the value of V for which this expression equals zero. Then, knowing V, we can calculate the compositions of the vapor and liquid by using the formulas:

$$x_i = \frac{z_i}{\left(1+V\left(K_i - 1\right)\right)} \qquad\qquad y_i = \frac{K_i z_i}{\left(1+V\left(K_i - 1\right)\right)}$$

This is about as far as algebra can carry us. We now have the equation reduced to only one unknown, V. But the equation is not explicit in V and must be solved by trial and error. Most spreadsheets have a SOLVER and a GOAL SEEK function which are intended to solve equations such as this. In our case, however, they cannot handle the circular references which we have. The values of x_i and y_i depend on values of z_i which come from the flowsheet. Once calculated, these values of x_i and y_i are fed back to the flowsheet to generate new values of z_i, which in turn cause different values of x_i and y_i. Another drawback of functions such as SOLVER is that their answers are dead. They do not update themselves automatically when the input to the equation changes. SOLVER must be invoked again for each iteration through the flowsheet. This can be achieved but it would require a very complicated macro.

There are many ways to solve this equation, but one which is very well suited to spreadsheets is the *Linear Inverse Interpolation* method. It may require more trials than some other methods to reach an answer, but this is not a major consideration. This method is stable, and it is live on the spreadsheet, meaning that the equation will automatically re-solve whenever any of its terms are changed.

For this solution we should open a new worksheet. We will name it *FLASH* by double-clicking on the tab at the bottom of the screen. In the upper left corner we will put the terms which will be in the equation. Below we will put a table, each of whose rows will be an iteration which guesses at the solution of the equation. If we set up the table properly, each guess will be closer to the solution than the previous guess. In the first row of the table we will solve the equation for three values of V. One will be the lowest possible value of V (called "Vlow" and equal to zero in this example), one will be the highest possible value (called "Vhi" and equal to 1.0 in this example) and the third (called "Vmid") will be interpolated between the first two, based on the values of the function calculated at Vhi and Vlow. The formula to find the interpolated value of Vmid is:

$$Vmid = \frac{Vlow * f(Vhi) - Vhi * f(Vlow)}{f(Vhi) - f(Vlow)}$$

For example, if the absolute value of f(Vlow) is small and the value of f(Vhi) is large, the guess at Vmid will be much closer to Vlow than to Vhi, because the value of the function at Vlow is much closer to zero. The layout of this table is shown in Fig. 3.

If any of our initial guesses is correct (not very likely) the value of the equation will be zero. If this is not the case, the correct value of V will either lie between the lowest V and the mid-point V or between the mid-point V and the highest V. We can tell which is the case by checking the signs of the solutions. If two solutions of the equation have opposite signs, then the correct solution (where the equation = zero) must lie between them. The second row of the table takes the two values of V which have been found to bracket the correct solution, interpolates between them to get a third value of V, and solves the equation three more times. Each succeeding row of the table repeats this procedure, with the range of the possible values of V being reduced each time. The total number of rows required depends on the necessary accuracy of the solution. Our example uses fourteen rows. Note in Fig. 3 how the value of f(Vmid) comes closer and closer to zero with each iteration (row). The value of Vmid (which is the final answer) homes in on the value 0.2112, which is the correct value of V.

	A	B	C	D	E	F	G	H	I	J
1										
2		z	k		SOLUTION OF FLASH EQUATION					
3	R1 =	0,1718	25		by LINEAR INVERSE INTERPOLATION					
4	R2 =	0,1002	2							
5	R3 =	0,0238	20							
6	P =	0,7041	0,04							
7	TOTZ	1,0000								
8				Vlow	f(Vlow)	Vhi	f(Vhi)	Vmid	f(Vmid)	
9				0	-4,000336	1	16,661889	0,1936063	-0,080647	
10				0,1936063	-0,080647	1	16,661889	0,1974906	-0,063191	
11				0,1974906	-0,063191	1	16,661889	0,2005226	-0,049838	
12				0,2005226	-0,049838	1	16,661889	0,2029069	-0,0395	
13				0,2029069	-0,0395	1	16,661889	0,204792	-0,031422	
14				0,204792	-0,031422	1	16,661889	0,2062889	-0,025068	
15				0,2062889	-0,025068	1	16,661889	0,2074812	-0,020044	
16				0,2074812	-0,020044	1	16,661889	0,2084335	-0,016054	
17				0,2084335	-0,016054	1	16,661889	0,2091954	-0,012876	
18				0,2091954	-0,012876	1	16,661889	0,2098061	-0,010339	
19				0,2098061	-0,010339	1	16,661889	0,2102961	-0,008308	
20				0,2102961	-0,008308	1	16,661889	0,2106897	-0,006681	
21				0,2106897	-0,006681	1	16,661889	0,2110061	-0,005376	
22				0,2110061	-0,005376	1	16,661889	0,2112606	-0,004328	
23										
24										
25										
26										

Fig. 3.

Table 4 gives the contents of the worksheet titled FLASH. All the cell addresses correspond to Fig. 3. The first eight cells list the inputs to the equation solution table.

Table 4.

Cell Name	Address	Contents	Formula	Final Result
ZR1	B3	Mol fraction of R1 in the flash feed	=EXITR1/SUM (FLOWSHEET!H15:H18)	0.1718
ZR2	B4	Mol fraction of R2 in the flash feed	=EXITR2/SUM (FLOWSHEET!H15:H18)	0.1002
ZR3	B5	Mol fraction of R3 in the flash feed	=EXITR3/SUM (FLOWSHEET!H15:H18)	0.0203
ZP	B6	Mol fraction of P in the flash feed	=EXITP/SUM (FLOWSHEET!H15:H18)	0.7042
KR1	C3	Relative volatility of R1	25	25
KR2	C4	Relative volatility of R2	2	2
KR3	C5	Relative volatility of R3	20	20
KP	C6	Relative volatility of P	.04	.04

The following six cells are the labels of the columns in the solution table.

	D8	Label of 1st Column of Table	Vlow	Vlow
	E8	Label of 2nd Column of Table	f(Vlow)	f(Vlow)
	F8	Label of 3rd Column of Table	Vhi	Vhi
	G8	Label of 4th Column of Table	f(Vhi)	f(Vhi)
	H8	Label of 5th Column of Table	Vmid	Vmid
	I8	Label of 6th Column of Table	f(Vmid)	f(Vmid)

The following six cells are the first row in the solution table.

	D9	Lowest possible value of V	0	0
	E9	Equation solved using V from D9	Equation for V (above)	– 4.00028
	F9	Highest possible value of V	1.0	1.0
	G9	Equation solved using V from F9	Equation for V (above)	16.66189
	H9	Interp. Between highest & lowest V's	=(D9*G9-F9*E9)/(G9-E9)	.193606
	I9	Equation solved using V from H9	Equation for V (above)	-.08065

Table 4. (continued)

Cell Name	Address	Contents	Formula	Final Result
		The following six cells are the second row of the solution table.		
	D10	New lower limit of V	=IF(E9*I9>0,H9,D9)	.193606
	E10	Equation solved using V from D10	Equation for V (above)	-.08065
	F10	New upper limit of V	=IF(E9*I9>0,F9,H9)	1.0
	G10	Equation solved using V from F10	Equation for V (above)	16.66189
	H10	Interp. between new highest & lowest V's	=(D10*G10-F10*E10)/ (G10-E10)	.197491
	I10	Equation solved using V from H10	Equation for V (above)	-.06319

The remaining twelve rows of the solution table are merely copies of the second row. The solution can be read from the value of Vmid in the bottom row (Cell H22).

The key cells of the solution table are D10 and F10 (and the corresponding cells in subsequent rows). The logic in these cells decides what the new limits of V will be for the next iteration. For instance, cell D10 checks the values of f(Vlow) and f(Vmid) to see if they have the same sign. If they have not, then the solution is somewhere between Vlow and Vmid, so cell D10 takes the value of Vlow, which becomes the new Vlow. If the signs are the same (as they are in this example), then the solution must lie between Vmid and Vhi, so D10 takes the value of Vmid, which becomes the new Vlow. Cell F10 behaves in a similar manner.

Having solved the main equation for V, we can use the result to calculate the vapor and liquid flows leaving the flash for each of the components (Table 5, see next page).

The final results (liquid and vapor flows of each component) are carried back to the flowsheet page and used for the feed to the Separator and recycle to the Reactor. Since we are not simulating the second step of the Separator, we will assume that 1% of the R1 fed to the Separator exits with the product, and the remainder exits as an overhead stream which joins the recycle to the Reactor. 2% of the R2 exits with the product, and the remainder exits in the side draw which continues to be 100% R2. The pertinent cells on the page FLOWSHEET are shown in Table 6, see page 183.

Table 5.

Cell Name	Address	Contents	Formula	Final Result
YR1	B31	Mol fraction of R1 in the vapor	=KR1*ZR1/(1+$H$22*(KR1—1))	.7077
YR2	B32	Mol fraction of R2 in the vapor	=KR2*ZR2/(1+$H$22*(KR2—1))	.1655
YR3	B33	Mol fraction of R3 in the vapor	=KR3*ZR3/(1+$H$22*(KR3—1))	.0949
YP	B34	Mol fraction of P in the vapor	=KP*ZP/(1+$H$22*(KP-1))	.0353
TOTY	B35	Sum of all vapor mol fractions	=SUM(YR1:YP)	1.0034
XR1	C31	Mol fraction of R1 in the liquid	=ZR1/(1+$H$22*(KR1—1))	.0283
XR2	C32	Mol fraction of R2 in the liquid	=ZR2/(1+$H$22*(KR2—1))	.0827
XR3	C33	Mol fraction of R3 in the liquid	=ZR3/(1+$H$22*(KR3—1))	.0047
XP	C34	Mol fraction of P in the liquid	=ZP/(1+$H$22*(KP-1))	.8833
TOTX	C35	Sum of all liquid mol fractions	=SUM(XR1:XP)	.9991
VAPR1	D31	Mol/hr of R1 in the vapor	=YR1*TOTV	38.95
VAPR2	D32	Mol/hr of R2 in the vapor	=YR2*TOTV	9.11
VAPR3	D33	Mol/hr of R3 in the vapor	=YR3*TOTV	5.22
VAPP	D34	Mol/hr of P in the vapor	=YP*TOTV	1.94
TOTV	D35	Total mols of vapor	=H22*TOTFED	55.04
LIQR1	E31	Mol/hr R1 in the liquid	=XR1*TOTL	5.82
LIQR2	E32	Mol/hr R2 in the liquid	=XR2*TOTL	17.00
LIQR3	E33	Mol/hr R3 in the liquid	=XR3*TOTL	.97
LIQP	E34	Mol/hr P in the liquid	=XP*TOTL	181.51
TOTL	E35	Total mols of liquid	=TOTFED-TOTV	205.49
TOTFED	F35	Total mols fed to flash	=SUM(FLOWSHEET!H15:H18)	260.53

Table 6.

Cell Name	Address	Contents	Formula	Final Result
	H24	Mols/hr R1 in liquid from flash	=LIQR1	5.8
	H25	Mols/hr R2 in liquid from flash	=LIQR2	17.0
	H26	Mols/hr R3 in liquid from flash	=LIQR3	1.0
	H27	Mols/hr P in liquid from flash	=LIQP	181.5
	F24	Mols/hr R1 in overhead from Sep.	=H24-H33	5.8
	F25	Mols/hr R3 in overhead from Sep.	=H26	1.0
	H33	Mols/hr R1 in product	=.01*H24	0.1
	H34	Mols/hr R2 in product	=.02*H25	0.3
	H35	Mols/hr P in product	=H27	181.5
	J29	Mols/hr R2 in side draw	=H25-H34	16.7
RECYR1	E15	Mols/hr R1 in the recycle	=F24+VAPR1	44.7
RECYR2	E16	Mols/hr R2 in the recycle	=VAPR2	9.1
RECYR3	E17	Mols/hr R3 in the recycle	=F25+VAPR3	6.2
RECYP	E18	Mols/hr P in the recycle	=VAPP	1.9

Data Tables

The next bombshell dropped by the Research Department is the news that conversion of R1 in the reactor is not fixed at 70% but is an equilibrium reaction, where the exit gas composition is dependent on temperature and pressure. They have provided the following table of equilibrium constants (expressed as partial pressures) vs. Temperature.

Temperature (°C)	K_{eq} (Atm^{-1})
100	250.0
200	100.0
250	34.0
300	25.0
350	17.0
400	13.0
450	10.0
500	8.0

where $K_{eq} = \dfrac{pP^2}{(pR1)(pR2)^2}$

First we will look at several ways to extract the equilibrium constant at a given temperature for use in the flowsheet. A good beginning is to copy the above table to a new sheet on our spreadsheet, starting in cell B1. We will name this new sheet INTERP (for reasons which will become evident later) by double-clicking on the tab at the bottom of the screen. Then give the name KTABLE to the two columns of figures (not including the titles). The address of KTABLE will be B3:C11. Then back on the flowsheet, find a blank cell such as J6 and name it TEMP. There we will enter the reaction temperature we want to consider. Under it, name cell J7 KEQ. Here we will read the equilibrium constant corresponding to TEMP.

In the cell KEQ we will enter a VLOOKUP function to find the equilibrium constant from the table on the sheet INTERP. A VLOOKUP function requires three bits of information (function arguments). First it must know the independent variable we will use to enter the data table. In our case, this variable is TEMP. Second it must know which data table to use. In our case this is KTABLE. Third it must know in which column of the data table to find the answer. In our case this is column 2. So the entry in cell KEQ will look like this:

=VLOOKUP(TEMP,KTABLE,2)

Try entering the value 150 in cell TEMP. Cell KEQ will return the value 100.0. The spreadsheet has gone down the left hand column of the table until it found a value equal to the value in TEMP. Then it looked in the second column of the table for the answer. But what if the value of TEMP is not one of the temperatures for which we have a value of KEQ in our table? Try entering 160 in TEMP. Try 170. You will see that the answer does not change, although we know that the actual value of Keq will be decreasing as the temperature rises. If the spreadsheet cannot find an exact match for the value of TEMP, it will use the next smaller number in the table. That is, if we enter 160, the spreadsheet will use 150. If we enter 140, it will use 100. Also investigate what happens if we use temperatures above or below the range given in the table.

If we are interested in readings of KEQ closer than every 50° we will need to interpolate between the values given in the table. Now on either side of KTABLE (in columns A and D) number the rows of data 1–9. This will give two columns of index numbers, one on either side of the table. If we knew the table would always have temperatures in even increments of 50°, we could simplify this procedure, but the method given here will apply to any table– even if the independent variable is not always given in even increments.

Now we need to name two ranges. TABLE1 will include the first three columns of the table (A3:C11). TABLE2 will include the last three columns of the table (B3:D11). In column G we will have a number of named cells in which we do the actual interpolation calculations. Each of these cells should be labeled in the corresponding row of Column F for ready identification. The column named Final Result in Table 7 gives values for a temperature of 170°.

Table 7.

Cell Name	Address	Contents	Formula	Final Result
TABLE1	A3:C11	KTABLE + 1st Index Col.		
TABLE2	B3:D11	KTABLE + 2nd Index Col.		
TTEMP	G3	Temperature to enter table	=IF(TEMP<100,100,IF (TEMP>499,499,TEMP))	170
LOWINDEX	G5	Row above TTEMP	=VLOOKUP(TTEMP, TABLE2,3)	2
HIGHINDEX	G6	Row below TTEMP	=LOWINDEX +1	3
LOWTEMP	G8	Listed temp. at LOWINDEX	=VLOOKUP(LOWINDEX, TABLE1,2)	150
HIGHTEMP	G9	Listed temp. at HIGHINDEX	=VLOOKUP(HIGHINDEX, TABLE1,2)	200
LOWK	G11	Value of K at LOWTEMP	=VLOOKUP(LOWTEMP, TABLE2,2)	92.0
HIGHK	G12	Value of K at HIGHTEMP	=VLOOKUP(HIGHTEMP, TABLE2,2)	55.0
LININTERP	G14	Linear interpolation between LOWK and HIGHK	=LOWK+(HIGHK-LOWK) *(TTEMP-LOWTEMP) /MAX (.01,HIGHTEMP-LOWTEMP)	77.2
LOGINTERP	G15	Logarithmic interpolation between LOWK and HIGHK	=EXP((LN(HIGHK)-LN(LOWK)) *(LN(TTEMP)-LN(LOWTEMP))/ MAX(.01, (LN(HIGHTEMP)-LN (LOWTEMP)))+LN (LOWK))	73.55

Some explanatory notes on the above table:

- The formula for TTEMP puts a lower limit of 100° and an upper limit of 499° on the temperature entered into the sheet. Otherwise the interpolation procedure will give an error message.
- The formula for LININTERP is the same one which we would use on a pocket calculator. It assumes a straight line relationship between temperature and equilibrium constant. The MAX function is present to prevent the possibility of having the denominator equal zero.
- If we graph temperature vs. equilibrium constant we will find it gives a straight line on log-log paper. This means that a logarithmic interpolation is probably more accurate than a linear interpolation– especially at low temperatures. This formula is the same as for linear interpolation except that each term is replaced by its natural logarithm and finally the anti-logarithm of the entire expression is taken to provide the answer.
- Note the value of LOWK is actually higher than the value of HIGHK. This is because LOWK corresponds to LOWTEMP and HIGHK corresponds to HIGHTEMP.

If we expect to be faced with frequent interpolations of this data table, a better solution might be the user-defined function written in Visual Basic for Applications and shown as "KeqInterp(DegC)" in the Appendix to this chapter. The cell addresses used in the code of this macro can easily be changed to fit the location of any two-column data table. To use this function, we need only enter the formula =KEQINTERP(AnyTemp) into any cell. AnyTemp represents a temperature or the address of a cell containing a temperature. The function will return the value of K_{eq} for the temperature AnyTemp.

Data Regression

An alternative to using interpolation of the data table is to express the equilibrium constant in equation form as a function of temperature. This is the method we would have to use if our information on K_{eq} was in the form of raw experimental data with many measurements made at each temperature. In our case we could easily use the interpolation techniques just described. If, however, we can successfully represent the data by an equation, we can avoid having the table in place on the spreadsheet, and we can just use the equation in cell KEQ to find the value of the equilibrium constant for any temperature. Some sets of data do not lend themselves to equations, and in these cases interpolation may be the only accurate technique. One way to see if we have a well-behaved set of data is to graph them. In Excel, as we've already seen, this is very easy to do.

To graph Keq vs. Temperature, we first select the two columns of data (KTABLE in our example). Note that the titles of the data columns are not included in the selection. Then click on INSERT, Chart, and As New Sheet from the menu bar. This will invoke the ChartWizard, which will start by asking us if the selected data are the ones we want to graph. If so, click on the Next button. The next step asks us to choose the type graph we want from among fifteen different types shown. We want the XY Scatter Plot shown in the center of the page. (Note: Do not succumb to the temptation to choose a Line Graph. This is mainly for business applications where the X-Axis is evenly spaced and expressed by labels such as JAN, FEB, MAR, APR, etc.)

Having selected the graph type, we click again on the Next button and are asked to choose which type of XY graph. The choices include graphs with lines or without lines, semi-log and log-log plots. For now we will choose selection 2 which has both lines and points shown in rectangular coordinates. When we click the Next button we immediately get a small preview of the graph, as shown in Fig. 4. The shape of the curve indicates an exponential rather than a straight-line relationship between temperature and equilibrium constant. To explore this, we can click the Back button and change our selection of the type of XY plot. This time we pick one with a semi-log scale on the Y axis (selection 4).

Returning to Step 4, the preview of the graph (as shown in Fig. 5) is much closer to a straight line, but still shows some curvature. Click the Back button again, and we will try a log-log plot (selection 5). The preview of this graph shows a fair-

Fig. 4.

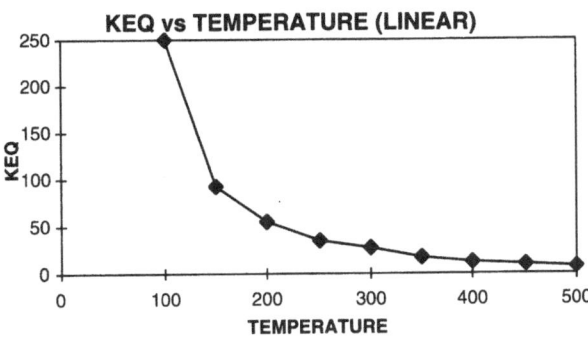

Fig. 5.

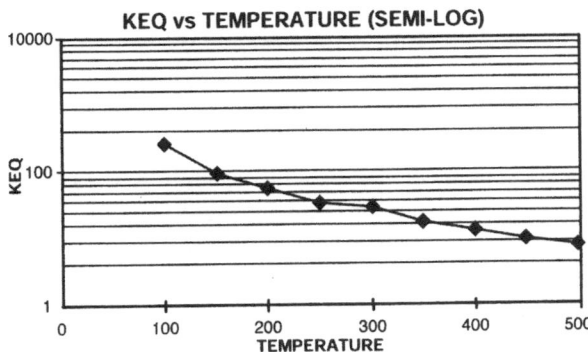

ly straight line. This is at least worth completing the graph so we can save it. The other choices we have in this step are all right as shown, so we can click the Next button and proceed to step 5. Here the Wizard asks if we want a legend (the sample graph shows one on the right hand side). Since there is only one dependent variable, we have no need for a legend– so we click No. We then fill in anything we want as titles for the chart and the two axes. Then we hit Finish, and the graph appears on a new worksheet titled Chart1. By double-clicking on the new sheet tab, we can change its name to something meaningful such as "Keq vs. Temp". (Note that spaces are allowed in sheet titles, but subscripts are not.)

The ChartWizard has selected our axes for us, but by looking at the chart we can see that all our points are crowded into the right-hand third of the chart. There is no need to have an X-axis going all the way from 1 to 1000. We can correct this by double-clicking on the X-axis to bring up the Format Axis box. There, on the page titled Scale we can delete the number 1 in the Minimum box and replace it with the number 100. When we return to the graph, it should look like Fig. 6.

Now that we know there is a log-log relationship between temperature and K_{eq}, we can expect an equation such as shown below will allow us to calculate K_{eq} pretty accurately.

$$\mathrm{Ln}(K_{eq}) = a_0 + a_1{*}\mathrm{LnT} \quad \text{or} \quad K_{eq} = \exp(a_0 + a_1{*}\mathrm{LnT})$$

Fig. 6.

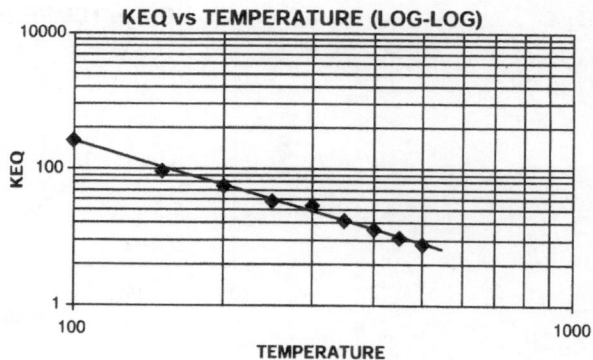

All that remains is to calculate the values of a_0 and a_1. This can be done by linear regression, a data analysis tool which is provided by all the major spreadsheets. The regression procedure finds the values of a_0 and a_1 which minimize the sum of the squares of the errors (measured Keq – calculated Keq). In Excel the data analysis functions are "add-ins" so if you click on the TOOLS menu and do not see a "Data Analysis" selection, you will have to load it. For this, click on the "Add-Ins", and select "Data Analysis ToolPak" from the list offered.

Since we will be regressing the log of T vs. the log of K, we need to get the figures out into a blank part of the sheet where we can work with them. In the sheet INTERP, tab once to the right and once down to find cell K23, and in cells K23:K33, copy the values of temperature from cells B1:B11. Similarly copy the values of K_{eq} from C1:C11 to L23:L33. In the next columns (M and N) we will compute the values of Ln(temperature) and Ln(K_{eq}) for each set of data. Give each column a title in row 24, such as LnT and LnK.

Now to actually do the regression analysis, we click on the TOOLS menu, select Data Analysis and select Regression from the list of statistical methods offered. The Regression dialog box asks for the location of the independent and dependent variables. Our dependent variables are in the range INTERP!N24:N33 (the values of Ln(K_{eq})), and our independent variables are in the range INTERP!M24: M33 (the values of Ln(temperature)). Note that the title rows are included. Excel will use these titles to identify the terms of the regression equation if we put a check in the box called Labels. Leave the other boxes in the input section blank. In the output section, choose the New Worksheet option so the results will be printed out onto a new sheet. In the lower section of the box, put a check in the box called Residuals. This will cause Excel to print out the calculated values of Ln(K_{eq}) for each set of data.

All that remains is to click OK and a new worksheet called Sheet1 will open. The important portion of this sheet are shown in Fig. 7. Many of the numbers on this sheet have quite a few decimal places, so (even if we may not believe all that accuracy) we should reformat the sheet so we can see them all. The entire sheet is already selected, so just choose FORMAT from the menu bar, select Column

	A	B	C	D	E	F	G	H
5	R Square	0,99599						
6	Adjusted R Square	0,99542						
7	Standard Error	0,07611						
8	Observations	9						
10	ANOVA							
11		df	SS	MS	F	Significance F		
12	Regression	1	10,07374	10,07374	1739,09137	0,00000		
13	Residual	7	0,04055	0,00579				
14	Total	8	10,11429					
16		Coefficients	Standard Error	t Stat	P-value	Lower 95%	Upper 95%	
17	Intercept	15,10687	0,28163	53,64065	0,00000	14,44092	15,77281969	
18	LnT	-2,09237	0,05017	-41,70241	0,00000	-2,21101	-1,97372531	
20	RESIDUAL OUTPUT							
21	Observation	Predicted LnK	Residuals	Observation	Predicted LnK	Residuals		
22	1	5,47116	0,05030	6	2,84992	-0,01671		
23	2	4,62278	-0,10099	7	2,57052	-0,00557		
24	3	4,02084	-0,01351	8	2,32408	-0,02149		
25	4	3,55394	-0,02758	9	2,10363	-0,02418		
26	5	3,17246	0,15974					

Fig. 7.

and then AutoFit Selection. This will give each column whatever width is necessary to show all its contents.

A lot of the data on this sheet are unimportant to us, but we need to check a few of them. In Cell B5 is the value of "R Squared". This value tells us that over 99.5% of the variability in the data has been explained by the effect of temperature. The rest is probably experimental error. This is such a good match that we need not worry about trying to find anything else that influences K_{eq}. Below in Cell B17 is the value of a_0 (called intercept), and in Cell B18 is the value of a_1 (the coefficient of LnT). These are really the answers we were looking for. Just to their right, in Cells C17 and C18 are values of the Standard Error of each coefficient. A rule of thumb is that the standard error must be less than half the value of the coefficient in order for the coefficient to be valid. A quick glance will show that we are OK here. In Cells B25:B33 we have the calculated values of LnK for each set of data, and in Cells C25:C33 the errors (measured $Ln(K_{eq})$ – calculated $Ln(K_{eq})$).

To see how well we have done, we should copy cells SHEET1!B24:B33 alongside the measured values of LnK in INTERP!O24:O33. Next to them in Column P we will take the anti-logarithm of each to get the calculated values of K_{eq}. (e.g. the formula in Cell P25 will be =EXP(O25)). For a visual picture of how well we did, we can again invoke the ChartWizard and create an XY Scatter Plot, without lines, of the measured vs. the calculated values of K_{eq}. It should look about like Fig. 8. A 45° line has been drawn on the chart using the drawing tools just to show where all the points would lie if we had done a perfect job.

Fig. 8.

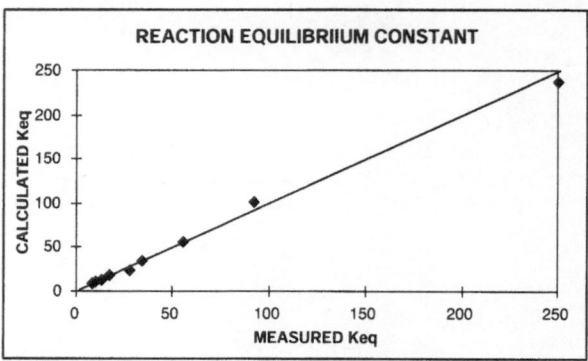

The regression procedure detailed above is a very general one which would apply even if K_{eq} were a function of several different independent variables. In this case, since we know K_{eq} to be a function only of temperature, there is another, easier way to get an equation for its relation to temperature. When we plotted K_{eq} vs. temperature on a linear plot we could have used Excel's trendline feature to find the line of best fit. We do this by clicking on the line between data points on the graph and choosing the command INSERT, Trendline.

The Trendline dialog box has two pages. On the page titled Options put check marks by the boxes titled "Display Equation on Chart" and "Display R-Squared Value on Chart". On the page titled Type, six types of curves are shown (Linear, Logarithmic, Polynomial, Power, Exponential and Running Average). We must choose one of them. We have already plotted our variables on a linear graph and seen it to be a poor fit, so we need not bother selecting the Linear curve. The Running Average curve does not really apply to this situation, so we can ignore it. If we select each of the others in turn, Excel will plot the variables according to that type curve on the same graph as our data, and will print the equation of best fit and the value of R^2 (goodness of fit).

The reader should try each of these possible curves to see how well they compare with the raw data. The best fit will turn out to be the Power curve, and the equation (after a little algebraic manipulation) will be the same one we found by regression in the example above.

The example in this chapter will proceed using the interpolated values of K_{eq} calculated as described in the previous section. If, however, we preferred to use the regression equation, we need only go to Cell KEQ on sheet EQN SOLVE and insert the formula =EXP(15.10687–2.09237*LN(TEMP)).

More Equation Solving

But so far we have only done half the job. We now know the equilibrium constant for any temperature, but we must calculate the flows of P, R1 and R2 leaving the reactor. This requires solving a non-linear equation for the number of mols of R1 reacted. First we must define some terms and expressions.

Mols R1 reacted = X   Mols R2 reacted = 2X   Mols P formed = 2X

PRES = Anywhere between 1 and 5 atmospheres. For this example, PRES = 4.0

INLETR1	=	FEEDR1 + RECYR1 – PURR1
INLETR2	=	FEEDR2 + RECYR2 – PURR2
INLETR3	=	FEEDR3 + RECYR3 – PURR3
INLETP	=	RECYP – PURP
TOTMOLS	=	(INLETR1-X)+(INLETR2–2X)+ (INLETP+2X) + INLETR3
	=	INLETR1 + INLETR2 + INLETR3 + INLETP – X

$$K_{eq} = \frac{(pP)^2}{(pR1)(pR2)^2} \qquad \text{(in terms of partial pressures)}$$

$$K_{eq} = \frac{[P]^2}{[R1]*[R2]^2 * PRES} \qquad \text{(in terms of mol fractions)}$$

$$K_{eq} = \frac{EXITP^2 * TOTMOLS}{EXITR1* EXITR2^2 * PRES} \qquad \text{(in terms of molar flows (mol/hr))}$$

$$K_{eq} = \frac{(INLETP+2X)^2 * TOTMOLS}{(INLETR1 - X)*(INLETR2 - 2X)^2 * PRES}$$

(in terms of inlet conditions and mols reacted)

$$\frac{(INLETP+2X)^2 * TOTMOLS}{(INLETR1 - X)*(INLETR2 - 2X)^2 * PRES} - K_{eq} = Zero$$

(Final equation referred to as "Equation for X" in table below)

This equation does not lend itself well to solution by the Linear Inverse Interpolation method which was used for the flash calculation. In this case, picking xhi at 99.99% reacted makes the denominator of equation for X above very near zero. This, in turn, makes the value of the expression at xhi very large ($>10^8$). This forces the Linear Inverse Interpolation method to pick a value of xmid almost identical to xlow, so it creeps up on the solution very slowly.

A similar method which works well in this case (indeed, almost any case) is the Bisection method. It differs only in its method of picking the value of xmid.

Instead of interpolating, the Bisection method simply averages the values of xhi and xlow. For comparison the formulas for xmid used by the two methods are:

Bisection: $$\frac{\left(xhi + xlow\right)}{2}$$

Linear Inverse Interpolation: $$\frac{\left(xlow * f(xhi) - xhi * f(xlow)\right)}{\left(f(xhi) - f(xlow)\right)}$$

Once again the second row of the solution table picks the new values of xlow and xhi by comparing the signs of the values of f(xlow), f(xhi) and f(xmid) calculated in the first row. The new values of xlow and xhi are picked so that the values of the equation calculated for these two values will have different signs, and therefore the solution will be somewhere between them. For this calculation we will create a new sheet titled EQN. SOLVE. Table 8 gives the contents of this sheet.

Table 8.

Cell Name	Address	Contents	Formula	Final Result
INLETR1	F2	Flow of R1 into Reactor	=FEEDR1+RECYR1-PURR1	136.5
INLETR2	F3	Flow of R2 into Reactor	=FEEDR2+RECYR2-PURR2	209.56
INLETR3	F4	Flow of R3 into Reactor	=FEEDR3+RECYR3-PURR3	6.2
INLETP	F5	Flow of P into Reactor	=RECYP-PURP	1.63
KEQ	F7	Equilibrium Constant	= LOGINTERP	73.55

The following six cells are the lables of the columns in the solution table.

	G8	Label of 1st Column of Table	xlow
	H8	Label of 2nd Column of Table	f(xlow)
	I8	Label of 3rd Column of Table	xhi
	J8	Label of 4th Column of Table	f(xhi)
	K8	Label of 5th Column of Table	xmid
	L8	Label of 6th Column of Table	f(xmid)

Table 8. (continued)

Cell Name	Address	Contents	Formula	Final Result

The following six cells are the first row in the solution table.

	Address	Contents	Formula	Final Result
	G9	Lowest possible value of X	0	0
	H9	Equation solved using X from G9	Equation for X (above)	−294.199
	I9	Highest possible value of X	=MIN(.9999*INLETR1, .9999*INLETR2/2)	104.77
	J9	Equation solved using X from I9	Equation for X (above)	7.97E+8
	K9	Avg. of highest & lowest X's	=(I9+G9)/2	52.3859
	L9	Equation solved using X from K9	Equation for X (above)	−290.503

The following six cells are the second row in the solution table.

	Address	Contents	Formula	Final Result
	G10	New lower limit of X	=IF(H9*L9>0,K9,G9)	52.3859
	H10	Equation solved using X from G10	Equation for X (above)	−290.503
	I10	New upper limit of X	=IF(H9*L9>0,I9,K9)	104.77
	J10	Equation solved using X from I10	Equation for X (above)	7.97E+8
	K10	Avg. of new highest & lowest X's	=(I10+G10)/2	78.5789
	L10	Equation solved using X from K10	Equation for X (above)	−250.56

The remaining eighteen rows of the solution table are merely copies of the second row.

The key cells of the solution table are G10 and I10 (and the corresponding cells in subsequent rows). The logic in these cells decides what the new limits of X will be for the next iteration. For instance, cell G10 checks the values of f(xlow) and f(xmid) to see if they have the same sign. If they do not, then the solution is somewhere between xlow and xmid, so cell G10 takes the value of xlow, which becomes the new xlow. If the signs are the same, then the solution must lie between xmid and xhi, so G10 takes the value of xmid, which becomes the new xlow. Cell I10 behaves in a similar manner.

The initial value of xhi depends on which of the reactants is limiting. At this stage we may not know whether R1 or R2 will be the limiting reactant. Therefore we use the MIN function to be sure that the value of xhi is really chemically possible. The value of X can never exactly equal the value of INLETR1 or INLETR2/2 because this would result in dividing by zero.

The final solution to the equation can be read from any of the values of X in the bottom row of the table. It is copied into a cell named SOLUTION. This is a very general method of solving any implicit equation with a single unknown. Now that we know how many mols of R1 react, we can divide that number by INLETR1 to get the equilibrium value of CONV.

In Fig. 9 we see the final version of the flowsheet as it would appear on the spreadsheet. The conversion of R1 is calculated from the equilibrium conversion at 170°C and 4 atm. The inert level has been set at 10% using the macro CON-VERGE listed at the end of this chapter.

To make our flowsheet a little more flexible, we need the ability to set the con-version to a specific value (as we have been doing up until now) or to use the equi-librium conversion calculated from the equation solution table. Accordingly, we will create a cell named SELECT in Cell E2 of the FLOWSHEET page, and we will change the formula of CONV to read =IF(SELECT=0,SOLUTION,SELECT). If we want to use the equilibrium conversion, we give SELECT the value zero so that CONV takes the value of SOLUTION. If we want to specify the conversion, we sim-ply enter the conversion we want into SELECT, and CONV takes that value.

We now have a fairly complicated flowsheet which takes values from other work-sheets to calculate the various flows. As we converge this flowsheet to any new set

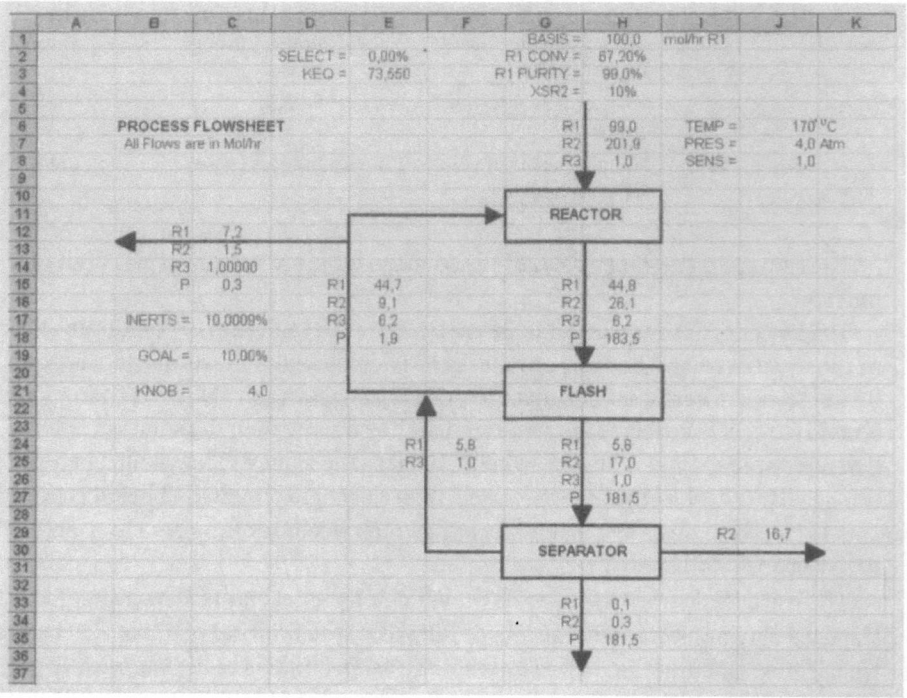

Fig. 9.

of conditions, we may encounter situations where the flows change so rapidly that they overshoot the correct value and may become unstable. On the other hand, some values may change so slowly that it takes forever to reach convergence. It is sometimes valuable to be able to adjust the rate at which the flowsheet values change. In our case, where the flowsheet and the equation solution interact, the ability to dampen or accelerate the changes can be very important.

To accomplish this, we will name a cell SENS (for sensitivity). It can be located anywhere, but we will put it in cell J8 on the FLOWSHEET page. If this cell has a value of 1.0, it will have no effect on the calculations. If its value is greater than 1.0, it will accelerate change, and if less than 1.0, it will dampen the changes by using only part of the difference in conversions from one iteration to another.

This requires that we also keep a record of the value of conversion from the previous iteration so we will know how much the value changed from one iteration to the next. First we will change the name of the cell SOLUTION on the sheet EQN.SOLVE to the name NEWCONV (in cell M37). Then we create a new cell which must be located above NEWCONV and name it OLDCONV (cell M35). In this cell we enter the formula =NEWCONV. As Excel recalculates the spreadsheet, it works generally downward and to the right. On this sheet it will first recalculate the equation solution table. Then it comes to OLDCONV and gives it the value which was NEWCONV in the last iteration. Then it proceeds to NEWCONV and gives it the new value from the equation solution table. OLDCONV, however, has already been recalculated, and will not be changed again until the next iteration.

Now we create a new cell named SOLUTION (the value of conversion which the flowsheet will use) and in it we put the formula = OLDCONV+SENS* (NEWCONV-OLDCONV). This way the change between one iteration and the next will be either magnified (if SENS>1.0) or reduced (if SENS<1.0). The final version of this example keeps the value of SENS at 1.0.

Damage Control

We have created a rather complex spreadsheet to represent the flowsheet of our plant. The actual flowsheet page draws inputs from three other sheets, all of which change with each iteration. It is easy to make a mistake or to generate an unstable flowsheet. We could easily find ourselves looking at four sheets full of nothing but dreaded error messages such as #####, #DIV/0!, or #NUM!.

A major source of error on the flowsheet is when the individual flows take on extreme values. These are computationally possible, even though they are physically impossible (fortunately). Some of this can be stopped by including limits in the formulas for the component flows. We might, for instance, limit all the recycle flows so they can never be less than zero. Yes, they might sometimes try to be negative if we don't prevent it. Here are some formulas for the recycle flows which prevent their taking on negative values and therefore dampen the wild swings which may sometimes occur during convergence.

RECYR1 = MAX(.01,F24+FLASH!D37)
RECYR2 = MAX(.01,FLASH!D38)
RECYR3 = MAX(.01,F25+FLASH!D39)
RECYP = MAX(.01,FLASH!D40)

Another error we can prevent is a purge flow larger than the recycle flow from which it came. Here are some formulas which will prevent that:

PURR1 = MIN(RECYR1,PURR3*RECYR1/MAX(.01,RECYR3))
PURR2 = MIN(RECYR2,PURR3*RECYR2/MAX(.01,RECYR3))
PURP = MIN(RECYP,PURR3*RECYP/MAX(.01,RECYR3))

On other sheets we have already seen other examples of this technique. In the flash calculation we limited the value of V to between zero and 1.0. In the equilibrium calculation we limited the value of X to between zero and 99.99% of the available reactants. In the interpolation procedures we limited the temperature to the range covered by the data table by using the formula:

=IF(TEMP<100,100,IF(TEMP>499,499,TEMP)).

Note that a temperature of 500°, although actually included in the table, will result in a #DIV/0! error in cells LININTERP and LOGINTERP.

In cases such as discussed above for the data table, we might want to go even a step farther. If the user enters a temperature outside the range of the table, the formula above will prevent an error message, but will not give a correct value of K_{eq} for the temperature actually entered. The user should be warned of this. We can do this by entering, in a cell near TEMP (such as cell FLOWSHEET!I10) the formula:

=IF(OR(TEMP<100,TEMP>499),"WARNING! TEMPERATURE IS OUTSIDE OF TABLE RANGE","")

So long as TEMP is within the working range of the table (100–499°C) the cell will remain blank (shown by the double quotes at the end of the formula). Whenever TEMP takes a value below 100 or above 499 however, the warning message will be displayed.

Despite our best efforts, errors will occur, but we can stop them from propagating and eventually correct them. The most common source of errors is on the flowsheet page, so we intervene whenever data from this page are transferred to another page (such as FLASH or EQN. SOLVE). If the cell in the flowsheet which is providing the data contains an error message, the cell receiving the data on the other sheet will take on some nominal value in the correct range. This value may not be correct, but it will not be an error; and it will stop the error from propagating to the receiving sheet. Some examples are shown on the sheet EQN. SOLVE.

INLETR1=IF(OR(ISERROR(RECYR1),ISERROR(PURR1)),100,
FEEDR1 +RECYR1-PURR1)

INLETR2 = IF(OR(ISERROR(RECYR2),ISERROR(PURR2)),200,
    FEEDR2 +RECYR2-PURR2)
INLETR3 = IF(OR(ISERROR(RECYR3),ISERROR(PURR3)),6,
    FEEDR3 +RECYR3-PURR3)

These formulas check the input cells to see if any of them contains an error message. If so, the formula gives a result somewhere close to a normal value for that variable, such 100 for INLETR1, 200 for INLETR2, or 6 for INLETR3. If there are no error messages, the receiving cell takes the value of its normal formula.

After a major catastrophe, even if it has not propagated beyond the flowsheet, we must still repair the damage. The best way to resurrect the flowsheet (or any other iterative calculation) is to work on one or two cells which are most heavily involved in the iteration process. In our flowsheet, these would probably be the recycle flows. In one or more of these cells, replace the formula with a number which represents a reasonable value for that cell. For instance, RECYR1 might be replaced by 40, RECYR2 by 10 and RECYR3 by 6. Then recalculate a few times to see if the other values come into line. If not, replace some other cells with values that will not change from one iteration to another until the remaining cells stabilize. Then, one at a time, replace the artificial values with the correct formulas in the cells which we have changed. Another stabilizing influence would be to set the conversion at a constant value (such as 70%) so it will not change until we are sure the flowsheet is stable.

Documentation

One difficulty with any large and complicated spreadsheet is that it is difficult for others to follow (or for ourselves to follow when we have been away from it for a few months). There are several things we can do, however, to make our spreadsheets easier to understand.

- Add some explanatory text to each sheet, explaining the purpose of that sheet, where its input comes from and where its output goes.
- Add explanatory text to each macro, stating the purpose of the macro, the shortcut key attached to it, and the rationale behind any significant lines of code.
- Use cell names instead of addresses wherever possible. Well-chosen cell names add a great deal of clarity to formulas. Each named cell should have a label in an adjacent cell showing what is contained in the named cell.
- Provide a table of cell names. In this example it is on the sheet named INDEX, and included in this chapter as Fig. 10. All spreadsheets have a function which will create such a table giving the name and address of each named cell. In Excel, the command is INSERT, Name, Paste, Paste List. The program will insert a list of cell and macro names at the cursor. The spreadsheet provides the list of names and addresses (in Columns A & B) but we should add a short comment about what the cell represents. Note that this is a dead table. It will not update

	A	B	C
1	NAME OF CELL OR RANGE	ADDRESS OF CELL OR RANGE	FUNCTION OF CELL OR RANGE
2			
3	BASIS	=Flowsheet!$H$1	Value of R1 feed rate on which flowsheet is based
4	CONV	=Flowsheet!$H$2	Conversion of R1 per pass through the Reactor
5	CONVERGE	=#REF!$A$1:$A$21	A macro which adjusts the inerts in the flowsheet to a desired value
6	EXITP	=Flowsheet!$H$18	Flow rate of the product "P" in the Reactor exit stream
7	EXITR1	=Flowsheet!$H$15	Flow rate of unreacted R1 in the Reactor exit stream
8	EXITR2	=Flowsheet!$H$16	Flow rate of unreacted R2 in the Reactor exit stream
9	EXITR3	=Flowsheet!$H$17	Flow rate of R3 in the Reactor exit stream
10	FEEDR1	=Flowsheet!$H$6	Flow rate of makeup R1 fed to the Reactor
11	FEEDR2	=Flowsheet!$H$7	Flow rate of makeup R2 fed to the Reactor
12	FEEDR3	=Flowsheet!$H$8	Flow rate of makeup R3 fed to the Reactor
13	GOAL	=Flowsheet!$C$19	Desired percent inerts (R3) in the flowsheet recycle
14	HIGHINDEX	=Interp!$G$6	Number of row in KTABLE with temperature above the temperature for which value of K_{eq} is desired
15	HIGHK	=Interp!$G$12	Value of K_{eq} from KTABLE in row HIGHINDEX
16	HIGHTEMP	=Interp!$G$9	Value of temperature from KTABLE in row HIGHINDEX
17	INERTS	=Flowsheet!$C$17	Percent inerts (R3) in flowsheet recycle stream
18	INLETP	='Eqn. Solve'!$F$5	Sum of all flows of P into the Reactor
19	INLETR1	='Eqn. Solve'!$F$2	Sum of all flows of R1 into the Reactor
20	INLETR2	='Eqn. Solve'!$F$3	Sum of all flows of R2 into the Reactor
21	INLETR3	='Eqn. Solve'!$F$4	Sum of all flows of R3 into the Reactor
22	KEQ	=Flowsheet!$E$3	Calculated value of reaction equilibrium constant
23	KNOB	=Flowsheet!$C$21	Sensitivity of adjustment of PURR3 to control inerts
24	KP	=Flash!$C$6	K value (vapor/liquid ratio) of P at Flash conditions
25	KR1	=Flash!$C$3	K value (vapor/liquid ratio) of R1 at Flash conditions
26	KR2	=Flash!$C$4	K value (vapor/liquid ratio) of R2 at Flash conditions
27	KR3	=Flash!$C$5	K value (vapor/liquid ratio) of R3 at Flash conditions
28	KTABLE	='Eqn. Solve'!$B$3:$C$24	Table of K_{eq} at various temperatures
29	LININTERP	=Interp!$G$14	Value of Keq from KTABLE by linear interpolation
30	LIQP	=Flash!$E$34	Flow rate of P in liquid from Flash
31	LIQR1	=Flash!$E$31	Flow rate of R1 in liquid from Flash
32	LIQR2	=Flash!$E$32	Flow rate of R2 in liquid from Flash
33	LIQR3	=Flash!$E$33	Flow rate of R3 in liquid from Flash
34	LOGINTERP	=Interp!$G$15	Value of Keq from KTABLE by logarithmic interpolation
35	LOWINDEX	=Interp!$G$5	Number of row in KTABLE with temperature below the temperature for which value of K_{eq} is desired
36	LOWK	=Interp!$G$11	Value of K_{eq} from KTABLE in row LOWINDEX
37	LOWTEMP	=Interp!$G$8	Value of temperature from KTABLE in row LOWINDEX
38	NEWCONV	='Eqn. Solve'!$M$37	Value of conversion of R1 calculated from most recent solution of equation for K_{eq}
39	OLDCONV	='Eqn. Solve'!$M$35	Value of conversion of R1 calculated from previous solution of equation for K_{eq}
40	PRES	=Flowsheet!$J$7	Reactor pressure in absolute atmospheres
41	PURITY	=Flowsheet!$H$3	Purity of makeup R1 fed to the Reactor
42	PURP	=Flowsheet!$C$15	Flow rate of P in the purge
43	PURR1	=Flowsheet!$C$12	Flow rate of R1 in the purge
44	PURR2	=Flowsheet!$C$13	Flow rate of R2 in the purge
45	PURR3	=Flowsheet!$C$14	Flow rate of R3 in the purge
46	RECYP	=Flowsheet!$E$18	Flow rate of P in the recycle
47	RECYR1	=Flowsheet!$E$15	Flow rate of R1 in the recycle
48	RECYR2	=Flowsheet!$E$16	Flow rate of R2 in the recycle
49	RECYR3	=Flowsheet!$E$17	Flow rate of R3 in the recycle
50	SELECT	=Flowsheet!$E$2	Selected conversion of R1 per pass in the Reactor
51	SENS	=Flowsheet!$J$8	Relative weighting of OLDCONV and NEWCONV used for final value of conversion of R1
52	SOLUTION	='Eqn. Solve'!$M$38	Final value of conversion of R1 after damping with SENS
53	TABLE1	=Interp!$A$3:$C$11	KTABLE plus first index column

Fig. 10.

	A	B	C
	NAME OF CELL OR RANGE	ADDRESS OF CELL OR RANGE	FUNCTION OF CELL OR RANGE
54	TABLE2	=Interp!$B$3:$D$11	KTABLE plus second index column
55	TTEMP	=Interp!$G$3	Temperature used to enter KTABLE
56	TEMP	=Flowsheet!$J$6	Temperature at which K_{eq} is to be calculated
57	TOTFED	=Flash!$F$35	Total mols/hr fed to Flash
58	TOTL	=Flash!$E$35	Total mols/hr of liquid from Flash
59	TOTV	=Flash!$D$35	Total mols/hr of vapor from Flash
60	TOTX	=Flash!$C$35	Total of mol fractions in liquid from Flash
61	TOTY	=Flash!$B$35	Total of mol fractions in vapor from Flash
62	VAPP	=Flash!$D$34	Flow rate of P in vapor from Flash
63	VAPR1	=Flash!$D$31	Flow rate of R1 in vapor from Flash
64	VAPR2	=Flash!$D$32	Flow rate of R2 in vapor from Flash
65	VAPR3	=Flash!$D$33	Flow rate of R3 in vapor from Flash
66	XP	=Flash!$C$34	Mol fraction of P in liquid from Flash
67	XR1	=Flash!$C$31	Mol fraction of R1 in liquid from Flash
68	XR2	=Flash!$C$32	Mol fraction of R2 in liquid from Flash
69	XR3	=Flash!$C$33	Mol fraction of R3 in liquid from Flash
70	XSR2	=Flowsheet!$H$4	Excess R2 (above stoichiometric) fed to Reactor
71	YP	=Flash!$B$34	Mol fraction of P in vapor from Flash
72	YR1	=Flash!$B$31	Mol fraction of R1 in vapor from Flash
73	YR2	=Flash!$B$32	Mol fraction of R2 in vapor from Flash
74	YR3	=Flash!$B$33	Mol fraction of R3 in vapor from Flash
75	ZP	=Flash!$B$6	Mol fraction of P in feed to Flash
76	ZR1	=Flash!$B$3	Mol fraction of R1 in feed to Flash
77	ZR2	=Flash!$B$4	Mol fraction of R2 in feed to Flash
78	ZR3	=Flash!$B$5	Mol fraction of R3 in feed to Flash
79			

Fig. 10. (continued)

itself automatically to reflect changes in cell names or locations or named cells added after its own creation. To update these, the table creation sequence must be run again.

- Avoid mental arithmetic. If we wanted to change the R1 feed from lb-mol/hr to SCFM, we could use the formula =FEEDR1*6.32. It would be much better to use the formula =FEEDR1* 379.4/60. Though the answer is the same, this formula contains two conversion factors familiar to many engineers: 379.4 SCF/lb-mol and 60 minutes/hour. The reader has a reasonable chance of being able to figure out what the author has done.

- Use the INSERT Note command to attach notes to important cells. These cells are then designated by a small red dot in the upper right hand corner. Examples are on the *FLOWSHEET* sheet, cells E2 and J6, and on *FLASH*, cell B2. By selecting the TOOLS, Options, View page we can choose whether or not to show the little red dots and the corresponding notes. If we choose to show them, the note will appear whenever the cursor is on that cell, even if the cell is not selected. The shortcut key Shift-F2 will also open the window for inserting a note in the cell where the cursor is located.

Summary

We have seen, using the medium of a process flow diagram, how spreadsheets can be used to perform iterative calculations on a recycle system, look up data from a table, interpolate between table values, apply linear regression to data, solve implicit equations and display graphs. It is to be hoped that chemical engineers, even if they may not be involved in flowsheeting, will recognize the wider application of spreadsheets to their everyday problems. Spreadsheets are too useful to be left to the accountants.

References

Rachford HH, Rice JD (1952) Procedure for Use of Electronic Digital Computers in Calculating Flash Vaporization Hydrocarbon Equilibrium. J Pet Technology 4(10), sec 1, p 19 & sec 2, p3, October, 1952

Appendix: Some Typical VBA Macro Code

Macros are computer programs which instruct the spreadsheet to perform a series of actions in sequence. In one way or another they have been features of electonic spreadsheets almost since their first appearance. They may contain loops, branches and calls to subroutines just like programs in FORTRAN or other high-level languages.

On the Excel module pages where they are created, VBA code normally appears with different types of statements in different colors. In these examples, lines which start with an apostrophe (') and are in normal type font are comments – they are ignored by the computer. Lines without an apostrophe and in bold face type are the actual code to be executed.

The Converge Macro

The Subroutine CONVERGE is used to set the inert (R3) level in the flowsheet shown on the sheet FLOWSHEET.

```
Sub CONVERGE()
      'This Macro, written in Microsoft's Visual Basic for Applications,
      'adjusts purge to give desired inert (R3) concentration in
      'the recycle stream of the flowsheet.
      'Be sure Calculation is set for "Manual", "Iteration", "Max Itera-
            tions"=5
      'Calculate the magnitude of the error function. The range INERTS cal-
            culates the percent R3 in the recycle stream. The range GOAL is
            where we enter what we want the percent R3 to be.
Calculate
1 CORR = Range ("INERTS") – Range ("GOAL")
If Abs (CORR) > 0.00001 Then
```

'If inert level is too far from goal, adjust inerts in the purge to
' equal inerts in the feed plus error.
'Knob is a measure of sensitivity which can be set to give a fast
' but stable solution

Range("PURR3") = Range ("FEEDR3") + CORR*Range ("KNOB")
Calculate
GoTo 1
Else

'In the final flowsheet, R3 in the purge must equal R3 in the feed.

Range ("PURR3") = Range ("FEEDR3")
Calculate
Beep
End If
End Sub

The Function KeqInterp

This is a user-defined function, also written in Microsoft's Visual Basic for Applications. It interpolates values of K for temperatures between those listed in the table on sheet INTERP. It can be used for any other table by adjusting the cell addresses. It includes provisions for handling temperatures outside the range of the table without causing an error message.

User Defined Functions differ from subroutines (such as CONVERGE, above) in that they cannot make any changes to the spreadsheet. They can only receive data from the spreadsheet, process it and return their output back to the cell on the spreadsheet from which the function was originally called. They are used in the same way as the spreadsheet's internal functions such as SUM or SQRT.

This function was originally written by Prof. David E. Clough of the Department of Chemical Engineering of the University of Colorado.

Function KeqInterp (DegC)

'If the temperature is below the range of the table, the function will
'return the value of K_{eq} corresponding to the lowest temperature of
'the table.

If DegC < Sheets("INTERP").Cells (3, "B").Value Then
KeqInterp = Sheets("INTERP").Cells (3, "C").Value

'If the temperature is above the range of the table, the function will
'return the value of K_{eq} corresponding to the highest temperature of
'the table.

```
ElseIf DegC >= Sheets("INTERP").Cells (11, "B").Value Then
    KeqInterp = Sheets("INTERP").Cells (11, "C").Value
```

'If the temperature is within the range of the table, the function
'performs a linear interpolation between the values given in the table.

```
Else
    For RowIndex = 3 To 11
    If Sheets("INTERP").Cells (RowIndex, "B").Value > DegC Then
        LowRow = RowIndex - 1
        HighRow = RowIndex
        Exit For
End If
    Next RowIndex
    TempLow = Sheets("INTERP").Cells (LowRow, "B").Value
    TempHigh = Sheets("INTERP").Cells (HighRow, "B").Value
    KeqLow = Sheets("INTERP").Cells (LowRow, "C").Value
    KeqHigh = Sheets("INTERP").Cells (HighRow, "C").Value
    KeqInterp = (DegC - TempLow) / (TempHigh - TempLow) *_
        (KeqHigh - KeqLow) + KeqLow
End If
End Function
```

Spreadsheets in Molecular Biology

G. Shaw

Introduction

The central dogma of biology is that deoxyribonucleic acid (DNA) makes ribonu-cleic acid (RNA) makes protein (Alberts et al. 1994; Lodish et al. 1995). The DNA molecule is a simple linear chain each link of which contains one of four differ-ent chemicals called bases. These bases are adenine, thymine, guanine and cyto-sine and are usually abbreviated as A, T, G and C respectively. Since any one of these bases can be found at each link in the chain the potential informational content of even short lengths of DNA is enormous. For example, a modest twen-ty base DNA segment can be constructed in 4^{20} different ways, corresponding to 1.099×10^{12} combinations. In living cells segments of DNA generate comple-mentary RNA molecules with the same information content. Certain of these RNA molecules are used to direct the production of proteins. Appropriate parts of the RNA sequence are read in sets of three bases, called codons. The 64 (4*4*4) possible codons each have a specific meaning, resulting in the addition of one specific amino acid to a growing protein or the ending of that chain. Like nucle-ic acids, proteins are also linear chains, with each link in the chain being repre-sented by one of twenty different types of amino acid. The potential information-al content of protein sequences is even greater than nucleic acid sequences. For example there are $20^{10} = 1.024 * 10^{13}$ possible 10-amino acid peptides. The specif-ic nucleic acid sequence is therefore decoded into a specific protein sequence, which in turn defines the folded three-dimensional structure and positioning of chemically reactive amino acid side chains on this structure, both of which deter-mine the function of the protein. Even the simplest free living organism charac-terized to date, the bacterium *Mycoplasma genitalium,* consists of several hun-dred different proteins (Fraser et al. 1995). Since the average sized protein is sev-eral hundred amino acids long, the combinatorial complexity of life can only be described as astronomical.

Recently there has been an enormous increase in the number of nucleic acid and protein sequences available, primarily as a result of improvements in the effi-ciency of DNA sequencing. For example, at the time of writing (March 1997) the entire genomes of three prokaryotic species (*Haemophilus influenzae, Mycoplas-ma genitalium* and *Methanococcus jannaschii*) have been determined, and the entire genome of one eukaryote, the yeast *Saccharomyces cerevisiae* has also been completed (Fraser et al. 1995; Fleischmann et al. 1995; Bult et al. 1996; Williams 1996). Data on the first three is available on the World Wide Web at URL *http://www.tigr.org*, while the yeast genome can be found at URL *http://speedy.mips.biochem.mpg.de/mips/yeast/*. These projects alone have generated a total of over 16 million bases of nucleic acid sequence and over 11,500 deduced protein sequences. More sequence information is continuously being generated and the nucleic acid sequence of the entire human genome is expected to be determined around the turn of the century. The entire genomes of the nematode worm (*Caenorhabditis elegans*), the fruit fly (*Drosophila melanogaster*) and several other species are also expected to be completed in the near future.

This vast amount of sequence information is interpreted primarily by a wide variety of computer programs. Sometimes the individual researcher wishes to answer a specific question which may not be addressed by the particular programs to which he or she has access or possibly in any available program. It is therefore advantageous to be able to program oneself or to have access to specialists for this purpose. However, for many kinds of sequence interpretation a modern spreadsheet program provides a surprisingly attractive alternative. Spreadsheets are extremely versatile, widely available, quite easy to use and include a wide variety of graphical and statistical procedures. Many of the commercial sequence interpretation packages cost hundreds or even thousands of dollars, compared to inexpensive spreadsheets which may already have been purchased for other purposes. Spreadsheets are also particularly useful for teaching the essentials of sequence analysis and have been successfully employed for this purpose. Students can appreciate exactly how these methods work and perform simple exercises aimed at further increasing their understanding. This approach is also efficient for developing new sequence analysis algorithms, since the effects of alterations in the scoring or weighting parameters can be seen on a graphical output essentially immediately; it is not necessary to recompile the program after each modification as would be required during the development of, for example, a C++ program. Finally, knowing how spreadsheets work and being able to use them effectively is a useful skill applicable to a variety of other purposes both scientific and non-scientific. The present author has previously published a report pointing out the utility of spreadsheets for protein sequence analysis (Shaw 1995). This chapter provides many more sophisticated examples of this approach including applications involving nucleic acid sequences.

Analysis of Protein Properties

Key attributes of any protein are the molecular weight, amino acid composition and charge. All of these attributes can be accurately calculated from the amino acid sequence, and all of these calculations can be conveniently performed using a spreadsheet program. In the examples provided in this chapter, I have made use of Microsoft Excel version 5.0 for the IBM-PC and compatibles. There are functionally identical versions of this program for the Apple Macintosh and other computers and other widely available spreadsheet programs can be made to function in a similar way. In all of the methods described here the protein or nucleic acid sequence to be analyzed is inserted into the first column of the first sheet of an Excel spreadsheet, which is always labeled Calculations. Different ways to insert protein or nucleic acid sequences into this column are discussed in the section headed Some Practical Considerations at the end of this chapter. Various manipulations of the sequence data are performed on the first and subsequent sheets, and one of the sheets also contain plots of different kinds of data. The last sheet of each workbook, always labeled Sequences, contains several other example sequences which can be copied into the first column of the Calculations sheet to replace the original sequence. This is easily done by selecting the relevant column label (where the letter address is visible), which selects and allows the copying of the entire contents of that column. Having copied a column of sequence data from the Sequences sheet, select the label for the "A" column on the Calculations and paste. This will replace of the entire contents of that column with the data copied from the Sequences worksheet. The selected column is highlighted during this process and should look like the "A" column in Fig. 1. Replacing the original sequence causes all the data in the spreadsheet to be recalculated, so that the effect on the plots and tabular data can be immediately assessed.

For a simple example examine the first sheet of CHARGED.XLS, as illustrated in Fig. 1. This spreadsheet plots out an estimate of the local charge down a pro-

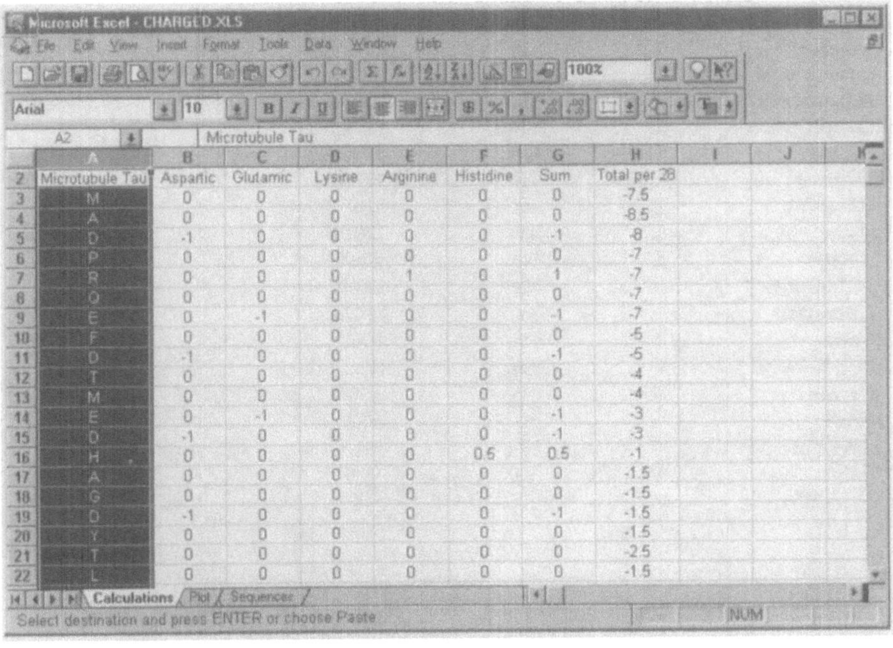

Fig. 1. View of the CHARGED.XLS Calculations sheet

tein molecule. Under physiological conditions certain amino acids are positively or negatively charged. The amount and location of these charges are important characteristics of the protein, and regions of positive or negative charge concentration are also associated with particular types of binding properties; for example basic regions are often found in DNA binding proteins. The protein sequence of the microtubule associated protein tau has been loaded into the first column of the Calculations sheet. The sequence is expressed in the single letter amino acid code, a concise method of showing sequence data used by most protein sequence databases and described in Table 1. As will be discussed in more detail below under slightly acidic conditions only five amino acids contribute to the overall charge of a protein. The basic amino acids lysine and arginine have charge of close to +1, the mildly basic amino acid histidine has a charge of about +0.5 and the acidic amino acids glutamic acid and aspartic acid have a charge close to –1. The Calculations sheet of the spreadsheet will be used to assign values of –1, 0.5 and 1 to the appropriate amino acids and 0 to all the others. This is performed in a very simple manner in the second and subsequent columns of the Calculations sheet. The column labeled "Aspartic" contains a –1 every time the first column contains a D, the single letter code for aspartic acid (see Table 1). To

Table 1. Amino Acids and their properties

Full Name	Single letter code	Three letter code	Molecular Weight in Daltons*	pKa	Ka
Alanine	A	Ala	71.09	–	–
Cysteine	C	Cys	103.15	10.46	3.47×10^{-11}
Aspartic acid	D	Asp	115.1	3.9	1.26×10^{-4}
Glutamic acid	E	Glu	129.13	4.07	8.51×10^{-5}
Phenylalanine	F	Phe	147.19	–	–
Glycine	G	Gly	57.07	–	–
Histidine	H	His	137.16	6.04	9.12×10^{-7}
Isoleucine	I	Iso	113.17	–	–
Lysine	K	Lys	128.19	10.79	1.622×10^{-11}
Leucine	L	Leu	113.17	–	–
Methionine	M	Met	131.31	–	–
Asparagine	N	Asn	114.12	–	–
Proline	P	Pro	97.13	–	–
Glutamine	Q	Gln	128.15	–	–
Arginine	R	Arg	156.2	12.48	3.311×10^{-13}
Serine	S	Ser	87.07	–	–
Threonine	T	Thr	101.12	–	–
Valine	V	Val	99.15	–	–
Tryptophan	W	Trp	186.23	–	–
Tyrosine	Y	Tyr	163.19	10.13	7.41×10^{-11}
(N-term)	–	–	–	2.33	4.46×10^{-3}
(C-term)	–	–	–	9.75	1.66×10^{-10}

*Molecular weight is for a single amino acid within a protein molecule. Since each link of the protein chain is built up by removing one water molecule between two amino acids, the molecular weight of each amino acid when free in solution is 18.02 Daltons more

accomplish this the B3 cell contains the equation =if($A3="D",-1,0). The meaning of this equation is simply IF(condition = TRUE, THEN assign some value, ELSE assign some other value). This equation returns a value of –1 if A3 contains a D, and 0 if it does not. The $ sign means that if this equation is copied into other columns the equation will still process data only from column A. However the lack of a $ sign in front of 3 in $A3 means that if this equation is copied down the column, this number (3) will be automatically incremented to match the position of the cell into which the equation is copied. In other words copying the contents of cell B3 to cell B4 will result in this cell containing =IF($A4="D",-1,0). This automatic re-indexing is a powerful feature of modern spreadsheets often met under the name of absolute and relative addressing. So, if the contents of B3 are copied down the rest of the protein sequence the result is to query every amino acid in column A for identity to D, and put a –1 in the appropriate position of the spreadsheet if an D is found. The same approach has been used in the Glutamic, Lysine, Arginine and Histidine columns as shown in Fig. 1, bearing in mind that the single letter code for these amino acids are E, K, R and H respectively (see Table 1). Cell C3 therefore contains the equation =IF($A3="E",-1,0). The equation in cell G3 is =SUM(B3:F3), which simply adds the results in each of the five preceding columns together. The G column therefore contains the charge of the amino acid in column A; 1 for K or R, 0.5 for H, -1 for D or E and 0 for any other amino acid. Plotting out such results is usually performed by counting the total charge over a "window", a selected length of amino acid sequence. In this case H3 contains the equation =SUM(G3:G30), counting over an arbitrarily selected window of twenty-eight amino acids. These seven equations can each be typed in the third row of the spreadsheet. When this has been done all seven can be selected and copied down the length of the sequence, and will be automatically re-indexed to refer to the correct cell in the A column, as was done in Fig. 1.

The data can be plotted out by selecting the H column from H2 to twenty eight entries from the end and pressing the ChartWizard button. In this case the "Line" plot was selected. In addition a white background was chosen, the data line was plotted in yellow, 50 was chosen as the number of units between tick marks and tick mark labels, and the scale for the X-axis was put at the bottom. These changes are performed simply by selected the desired item and choosing the required alteration from a menu. This plot was copied to the Plot sheet of the workbook. The data as plotted out using this method produces a rather jagged printout. Frequently such data is subjected to "smoothing" by one of a variety of methods. This can be performed in Excel by selecting the data-line in the plot then going to the Insert menu and selecting Trendline. A moving average over a fourteen amino acid window was selected in this case. This line shows graphically the acidic, negatively charged N-terminal region of the tau protein at the left and the even more acidic negatively charged C-terminal region at the right. In the middle the charge is more positively charged or basic. The graph plotted covers a greater range than is necessary for the tau sequence in order to accommodate some of the other sequences included on this spreadsheet.

Two other highly charged proteins, the neurofilament subunit NF-L and the synaptic protein synapsin are included in the Sequences sheet of this spreadsheet and can be copied into the first column of the Calculations sheet. The effect of this substitution is to recalculate the entire spreadsheet, the results of which can be determined by looking at the Plot sheet. The NF-L sequence has a basic N-terminal globular region at the left and a very acidic C-terminal extension, which are very obvious in the plot of this sequence. The synapsin protein is positive in overall charge but has regions neutral and negative charge.

The approach used in this example is, though simple to view and understand, rather inefficient. The first six calculations in the spreadsheet can all be performed in one column by nesting several IF commands. An IF command can be inserted into another IF command so that the equation now becomes IF(condition 1 = TRUE, THEN assign value 1, IF(condition 2 = TRUE, THEN assign value 2, ELSE assign default value)). Excel has an upper limit of seven such nested IF commands per cell. In the example given here the rather cumbersome equation;

=IF($A3="K",1,IF($A3="R",1,IF($A3="H",0.5,IF($A3="E", -1,IF($A3="R",-1,0)))))

could be placed in cell B3 of the Calculations sheet and copied down the rest of the column. This assigns the appropriate value to each amino acid as outlined above, or else 0 if the amino acid is not one of these five, and also renders the SUM command superfluous. As an exercise make a copy of the CHARGED.XLS spreadsheet then modify it to make use of this equation.

A more complicated example is shown in the spreadsheet AACOMP.XLS. In this case the sequence of the muscle protein tropomyosin was loaded into the "A" column of the Calculations sheet. In this example the number of occurrences of each amino acid in the sequence will be counted. Then, since each amino acid has a characteristically different molecular weight (see Table 1), the molecular weight will be multiplied by the number of occurrences of that amino acid, giving the total molecular weight fraction for each amino acids. The sum of these molecular weight fractions gives the total molecular weight of the protein and allows the calculation of the percentage content of each amino acid. The composition data will also be used to calculate the charge properties of the entire molecule as described below. A simple though inelegant means to count the number of each amino acid in a sequence is to use an IF loop to assign a value of 1 if the amino acid is present and 0 if not, similar to the approach used above. This would require twenty columns, and the sum of the "1" entries in each column obviously gives the number of the particular amino acid in the whole sequence. A much more efficient approach is to use Excel's COUNTIF function, which has the syntax COUNTIF(Range,Criteria). This function counts the number of non-blank cells within a range which meet certain user-defined criteria. In the Calculations sheet of AACOMP.XLS Cell C6 contains the equation =COUNTIF ($A$3:$A$500,"A"). This looks at the information in the cells from A3 to A500 and increments the value in C6 every time a cell in this range contains an "A". The result is therefore to count the number of occurrences of "A" in the first 500 cells of the A column. The number 500 was chosen arbitrarily and provides enough cells to accommodate the example proteins in the spreadsheet; this number could be increased or decreased as appropriate. The equation was copied over into the neighboring 19 cells of row 6. The $ signs in front of both the letter A and the row numbers 3 and 500 ensure that when this equation is copied into neighboring cells it will still refer only to cells in the range from A3 to A500. All that is required to make the equation count, for example cysteines rather than alanines, is to change the "A" to "C", the single letter code as shown in Table 1. Cell X6 contains the equation =SUM(C6:V6), which determines the total number of amino acids by adding up the numbers in each of these twenty cells. The molecular weight of each amino acid is included in the fifth row of the Calculations sheet in Dalton units. The total molecular weight of each particular amino acid in the protein is therefore the product of the molecular weight of that amino acid (in row 5) and the number of occurrences of that amino acid (in row 6), and is calculated in the seventh row, in the case of A or alanine by inserting =C5*C6 into cell C7. The sum of all the values in Dalton units is presented in cell X7 using a SUM command. The molecular weight of each amino acid in the spreadsheet assumes that it is within a protein, so that it lacks a water molecule. Since the two termini of the protein are not within the protein and share a single water molecule between them, this SUM command is incremented by the weight of one water molecule, 18.02 Dalton units. In general protein molecular weights are expressed in kilo-Dalton (kDa) units rather than Dalton units. The contents of cell X7 therefore correct for the water molecule and express the calculation in kDa using the equation =(SUM(C7:V7)+18.02)/1000.

The method used here not only generates the total number of amino acids and the molecular weight but also the numbers of each type of amino acid in the protein. This information, called the *amino acid composition*, is very useful since these twenty values are quite specific for different proteins, although classes of related proteins tend to have somewhat similar profiles. Amino acid composition data is also easy and inexpensively obtained experimentally and can be used to identify proteins (Eckerskorn et al. 1988; Shaw 1993). The AA Comp. sheet of the AACOMP. XLS workbook calculates the percentage composition of each amino acid. The sixth row of the spreadsheet contains the number of each amino counted in the Calculations sheet. To do this Cell C6 contains the equation =Calculations!C6, which therefore transposes over the number of amino acids calculated in the Calculations sheet. Cell X6 uses a SUM command to find the total number of amino acids, which is then used in the seventh row to calculate the percentage content of each amino acid using the equation, in the case of cell C7, =C6*100/$X6.

It is often useful to display the amino acid composition graphically, as has been done at the top of the sheet tagged AA Comp. Plot. The block of cells from C7 to V7 on the AA Comp. sheet were selected with the mouse and the ChartWizard button pressed. The ChartWizard allows a variety of different types of graph to be constructed and in this case a so-called *Radar* plot was produced and plotted out in the AA Comp. Plot sheet. In such plots the length of each vertex is linearly related to the percentage amount of the relevant amino acid. Such plots provide an easily appreciated overview of the amino acid composition data and are useful for comparing two different amino acid composition profiles. The order of amino acids in this plot is clockwise and alphabetical in terms of the single letter code, though any other desired order could easily be utilized.

A problem with this type of radar plot is that certain amino acids are much more abundant in proteins than others. For example Leucine constitutes 9–10% of the "average" protein while Tryptophan is only about 1–2%. This means that the "average" protein will produce a very asymmetric plot with this plotting method, and it is not immediately obvious except to an expert which amino acids are present in unusually high or unusually low amounts. It is therefore useful to compare the composition of a particular protein with that of the average protein. The composition of the average protein can be determined by simply counting the number of occurrences of each amino acid type in a large protein sequence database. In row 9 are placed the percentage content of the various amino acids in a current protein sequence database, in this case the latest available compendium of protein sequences released by GenBank. This GenBank database is release number 97 from November 1996 and contains almost 200,000 protein sequences and over 60 million amino acids from a wide variety of forms of life. The amino acid composition of this database is entered in rows 9 and 11; row 9 is used for the following calculations, and can be replaced with other figures in rows 10–15 when desired. Row 18 calculates the ratio of the values in row 9 with those in row 7 by simply dividing one by the other. The cells in row 20 contain values of 1 for plotting out the average profile. The block from B17 to V19 were selected and plotted out using the ChartWizard as above and the result is shown on the right of the AA Comp. Plot sheet. Comparison of this profile with the average, shown as a circle in the radar plot, now reveals that tropomyosin is unusually rich in glutamic acid, aspartic acid, alanine and lysine, but has an unusually low content of most of the other amino acids. The previous plot gives the impression that tropomyosin is also unusually rich in leucine, but examination of the new plot shows that this is not the case; there is more than in the average protein but not by very much.

As noted above four complete genomes have now been sequenced. The composition of the average protein in these genomes is given in rows 11 to 14 of the AA Comp. sheet. The content of each amino acid in these genomes is generally similar but not quite identical across these four species. For particular purposes it may be more appropriate to compare with the content of one of these species rather than from the Genbank database. Also is included an older protein sequence database, from the Protein Identification Resource (PIR) release number 46, which contains about 82,000 sequences and over 25 million amino acids. To compare a particular protein to any of these other databases simply copy the data from the appropriate row into row 9 of the AA Comp. sheet.

For specific purposes plots reflecting solubility or other chemical properties of the amino acids can easily be constructed. For example, altering the order of the data to be plotted on the AA Comp. sheet so that the most charged amino acids are at the right and the most hydrophobic are on the left would generate radar plots in which the relative amounts of these two types of amino acids would be reflected by the left/right asymmetry of the plot. This has been performed by copying the relevant cells in the desired order into the block C21 to V23 of the AA Comp. sheet, and the data are plotted out in the figure labeled Amino acid Composition ordered by Chemical Property in the AA Comp. Plot sheet. The replacement of the tropomyosin sequence with that for rhodopsin or synapsin, both of which are to be found in the Sequences sheet, produce recognizably different profiles with this kind of plot. For example rhodopsin contains far fewer of the charged amino acids and far more of the hydrophobic amino acids. Excel allows the easy construction of a wide variety of plots in this general way.

Another key property of proteins is their isoelectric point, defined as the degree of acidity or alkalinity at which the protein has no overall charge. Proteins at their isoelectric point can be precipitated out of solution, and will not bind well to charged resins used for protein purification. Knowledge of this property therefore has some practical significance. The isoelectric point can be calculated from the amino acid composition reasonably accurately, and these calculations can be readily performed in a spreadsheet. The acidity or alkalinity of a solution is a function of the concentration of hydrogen ions, usually referred to as H^+. For convenience H^+ concentration is expressed as pH units which are the negative Log_{10} of the concentration of H^+ in Moles/litre. Very acid solutions have pH values of 1–2, corresponding to high H^+ concentrations, while very basic solutions have pH values of 10–11, containing very low H^+ ion concentrations. Neutral pH, such as is found in many biological situations and in pure distilled water has a pH of 7, therefore containing 10^{-7} M/liter H^+. In acid solutions several types of amino acid side chain respond to the high H^+ concentration by binding H^+ ions. Conversely in basic solutions the low H^+ concentrations favors the loss of these ions from certain amino acid side chains. The addition or removal of these charged ions to individual amino acids obviously changes the charge properties of the entire protein.

Only aspartic acid, glutamic acid, lysine, arginine, histidine, cysteine and tyrosine can lose or gain H^+ ions. In addition the beginning and end of the protein chain, the N- and C-termini, each have one potential H^+ acceptor. The total charge properties of a protein are therefore dependent on the content of these seven amino acids and the N- and C-termini. For example glutamic acid and aspartic acid both contain a single reactive carboxylic acid side chain (-COOH). At pH = 7.0 virtually all of the -COOH groups have dissociated to produce a negatively charged COO^- attached to the protein, and a free H^+ ion. Increasing the H^+ ion concentration tends to favor the addition of H^+ to the COO^- and the production of the uncharged side chain COOH. At the much higher H^+ concentration of pH = 4.07 half of the glutamic acid molecules in a solution of distilled water will be neutralized while the other half will not. Similarly at a pH of 3.9, half of the aspartic acid molecules will be neutralized and half not. These pH values are termed the pKa for these particular amino acids (see Table 1). Like pH, the pKa is a negative Log_{10} of a concentration, and this concentration, the K_a (also known as the equilibrium constant), is by definition the H^+ concentration at which half of the amino acid is neutralized. Since these amino acids behave in approximately the same way when part of protein molecules, the charge properties of a whole protein can be estimated from the content of these and the other potentially charged amino acids (Ferscht 1977). For a protein in solution the amount of H^+ bound to one particular type of amino acid in a protein can be calculated from the equation:

$$Np = Nt[H^+]/([H^+] + Ka)$$

Where Np = number of amino acids of a particular type bound to H^+ ions, Nt = total number of that type of amino acid in the protein examined, $[H^+]$ = H^+ concentration in M/l and Ka = equilibrium constant = concentration of H^+ at which half of the amino acid has bound a H^+ ion. For example if we imagine glutamic acid at pH = 1, the H^+ concentration is 1 M/l, so that $Np = Nt*1/(1+Ka)$. For glutamic acid the pKa = 4.07 so that K_a is 0.0000851 M/l H^+. The value of 1+ K_a in the equation is therefore 1.0000851, which is approximately 1, so that Np is almost equal to Nt. Therefore virtually all glutamic acid molecules have bound H^+ ions and are therefore neutralized at pH=1, so that glutamic acid has almost no charge at such very high H^+ concentrations. However, at neutral pH, pH = 7.0 so the H^+ concentration is 10^{-7} = 0.0000001 M/liter. Under these conditions Np = $Nt*0.0000001/(0.0000001+0.0000851)$. The relatively small Ka value now has a much larger effect, so that $Np = Nt/852$, meaning that almost all glutamic acid side chains are not bound to H^+ ions and are hence not charged. Calculating the charge versus pH relationship for aspartic acid, cysteine, tyrosine and the C terminus can be performed in exactly the same way using the appropriate Ka values from Table 1.

The basic amino acids lysine, arginine and histidine and the N-terminus all contain nitrogen (N) atoms. These side chains take up H^+ to produce what can be loosely described as NH^+. Since lysine, arginine and the N-terminus have pK_a values of ~10 or more (see Table 1) these ions bind H^+ very strongly and are therefore positively charged except in very basic solutions. The pK_a for histidine is only 6.04 meaning that this amino acid becomes positively charged only at relatively high H^+ ion concentrations. At neutral pH this amino acid therefore has only a slight positive charge. The pK_a values for cysteine and tyrosine are >10.00 (see Table 1). These amino acids can therefore lose a H^+ ion, becoming negatively charged at very low pH. However, since such very basic conditions are outside the range of most biological phenomena, they have little effect on the charge of most proteins.

A spreadsheet can be used to compute the charge at differing pH for each amino acid in a protein, as has been performed in the I.E.P. sheet of AACOMP. XLS. The amino acid composition values from the Calculations sheet have been transposed to row 5 of the I.E.P. sheet. These values are then copied into the table in this spreadsheet, which contains in the first row pH values from 1 to 14, incremented by half a pH unit; the corresponding H^+ concentrations are given in column D. Columns E-S contain the equations with the relevant pKa value for each of the potential H^+ accepting amino acids (see Table 1). These equations are exactly as given above for the amino acids lysine, arginine, histidine and the N-terminus for which addition of a H^+ ion produces a positive charge. These positive values are added together in column Q of the spreadsheet. In the case of glutamic acid, aspartic acid, tyrosine, cysteine and the C-terminus the loss of the H^+ ion produces a negative charge. The equation as given above can be used but must be modified to take into consideration that in these cases addition of the H^+ neutralizes the amino acids while loss of the H^+ ion produces a negative charge. The total negative charge due to a particular amino acids (Nneg) equals the amount with a H^+ ion added (and hence neutralized) minus the total number of residues which can potentially lose a H^+ using the equation.

$$Nneg = Nt[H^+]/([H^+] + Ka)-Nt$$

Note that this equation is actually reversed so that it produces a negative value which corresponds to the negative nature of the charge produced by these amino acids. These values are in columns I-N and the sum of negative charges is in column R. Adding the positive and negative charges gives the total charge at each pH, and is shown in column R. The total positive, negative and net charge as a function of pH can be plotted out as on the top the I.E.P. Plot sheet by selecting the block of cells from O7 to R36 and using the ChartWizard button as before. In this case a "Line" plot was selected. This graph shows the total charge of the protein at different pH, and also the separate positive and negative components of this total. The plot also allows the estimation of the iso-electric point of the protein in question, the pH at which it has no net charge. This simply corresponds to the pH at which the "total" curve intercepts the 0 value on the Y axis, which in the case of tropomyosin is about 4.2.

The selection of the block from E7 to N36 plots out the behavior of each amino acid at differing pH as also shown in the I.E.P. Plot sheet. This graph makes clear how each type of amino acid contributes to the total charge properties of a molecule at different pH conditions, and shows graphically the meaning of the pKa values. As before different sequences can be copied from the Sequences sheet into the first row of the Calculations sheet to view different charge profiles.

Predicting Protein Structure

A major determinant of protein structure and stability is the interactions of so-called hydrophobic, or water-fearing amino acids. These amino acids tend to associate together on the interior of proteins and are also found in regions of protein which traverse cellular membranes. Many important proteins, including most receptor molecules for hormones and growth factors, contain one or more membrane-spanning domains, and the identification of such regions has implications for the topology and function of the protein in question. About twenty amino acids are required to span a typical cellular membrane, and sequences of amino acids this long containing predominantly hydrophobic amino acids are frequently membrane spanning regions (Van Heijne 1987). Such regions may be relatively easy to spot with computer programs since they are rich in hydrophobic amino acids and because the neighboring sequences tend to be much richer in charged and neutral amino acids. A wide variety of numerical values have been used to measure the degree of hydrophobicity of amino acids (summarized in Van Heijne 1987). These are typically based on a compilation of amino acid properties typically including their solubility properties in water compared with organic solvents and the frequency at which they are found inside membranes or in contact with other hydrophobic amino acids in known proteins. Since there is no standardized way of performing these calculations, different authors have provided different values, and the choice of which to use is the decision of the individual investigator. One or other of these data sets is selected and the appropriate values assigned to each amino acid. When these values are averaged out over a window of twenty amino acids membrane-spanning domains appear as distinct peaks or trough, depending on the particular numerical values employed. Frequently the data is then statistically smoothed as shown in the CHARGED. XLS spreadsheet example above.

Calculations of this sort can be readily performed in a spreadsheet program as shown in the MEMB.XLS spreadsheet. The sequence of bacterial rhodopsin has been inserted into the first column of the Calculations sheet of this worksheet. This is a well studied protein now known to contain seven trans-membrane regions. We could use a set of IF commands to assign the appropriate values to each amino acid similar to the method used in the CHARGED.XLS example. However this would require at least three columns since Excel can only perform a maximum of seven IF commands per cell. The B column illustrates a much more versatile and efficient method of assigning numerical values to each of the twenty amino acids using a lookup table, in this case located in block F5 to G24 of the Calculations sheet. The equation in cell B3 is =LOOKUP(A3,$F$5:$F$24,$G$5:$G$24). This takes the contents of cell A4 and compares them with the contents of the cells in the table from F5 to F24. If a match is found the function returns the value found in the same row of the table but in the range G5 to G24. In the case of cell B4, A4 is an "M" so the LOOKUP function moves down the table from cell F4 until it finds the M in cell F14. LOOKUP then reads the corresponding numerical value in G14. Although this approach is an elegant one, one must be careful since the LOOKUP function is unreliable unless both sets of data are arranged downwards in ascending order. It is for this reason that the amino acids are inserted in alphabetical order (which corresponds to ascending ASCII character order), and the assigned numbers are in ascending numerical order.

Columns H to M contain numerical values given by various authors to each of the amino acids to reflect specific properties relevant to different types of analysis. Those in row H were published by Argos and coworkers and are based on a compilation of solubility and structural properties of the different amino acids and were developed specifically to identify membrane spanning regions in proteins (Argos et al. 1982). Charged amino acids have negative values and strongly hydrophobic amino acids have positive values. As a result regions rich in hydrophobic residues which might represent membrane spanning regions should have positive average values, and should appear as peaks when this data is plotted out. These values are loaded into column C of the Calculations sheet. This cannot be done with the LOOKUP function since these values are not organized in ascending order. Instead we will use another efficient and versatile function, the OFFSET function. This has the syntax OFFSET(Reference,Rows,Columns, Height,Width). This command looks at a reference cell (Reference) in the spreadsheet, moves to some other cell as specified by the number of rows down and the number of columns to the right from the reference point (Rows and Columns), and returns the value of the cell or group of cells at that position. Cell C4 contains the equation =OFFSET($G$5,B4,$F$3,1,1). The $G$5 defines this cell as the reference point from which the function should start. The B4 cell contains the numerical value corresponding to the amino acid in cell A4 as determined by the LOOKUP function. The OFFSET function looks this number of cells down from the reference point, which results in OFFSET now pointing to the row in the table appropriate for the amino acid in A3. The $F$3 value tells OFFSET how may cells to move to the right, which is the contents of cell F3, which in the unmodified spreadsheet contains 1. The final two OFFSET arguments tell the function how many cells, horizontally and vertically, to address. In this case both are 1, so data is extracted from just one cell. The result of this approach is to assign the correct values to each amino acid in column C by reading the values from column H. Note that changing the value in cell F3 from 1 to 2 causes the values in column I to be loaded, and so on. Any of the sets of values in columns H to M can be loaded by simply changing the number in cell F3, and further desired values can be put in column N and beyond and loaded in the same way.

The D column averages the values in column C over a window of twenty amino acids using the Excel's built-in AVERAGE() function. In this case the first equation is put in D12, corresponding to the middle of the twenty amino acid window. The Plot Sheet shows a plot of the data from column D using the line plot and the curve was smoothed using the moving average set to a window of ten, essentially as performed in the original Argos paper (1982). To do this simply select the data line with the mouse, then go to the Insert menu and select "Trendlines". Several different options can be selected from the Trendlines menu. Figure 2 shows a diagram generated by copying the graph from Excel to Microsoft Powerpoint and adding a block diagram of the seven putative transmembrane domains. This figure is virtually identical to that published in the original Argos paper (1982). The first two hydrophobic trans-membrane segments are clearly seen as two sharp peaks. The next three and the following two trans-membrane segments are somewhat less well defined, but can be seen with a little imagination. More recent

Fig. 2. Diagram of hydrophobicity profile as plotted out
from MEMB.XLS using the
parameters of Argos and
co-workers (1982). The faint
jagged line is the hydrophobicity values over a 20 amino
acid window. The denser line
is a "Trendline" made using
a 10 amino acid moving
average. The diagram at the
top indicates the location of
the seven trans-membrane
segments

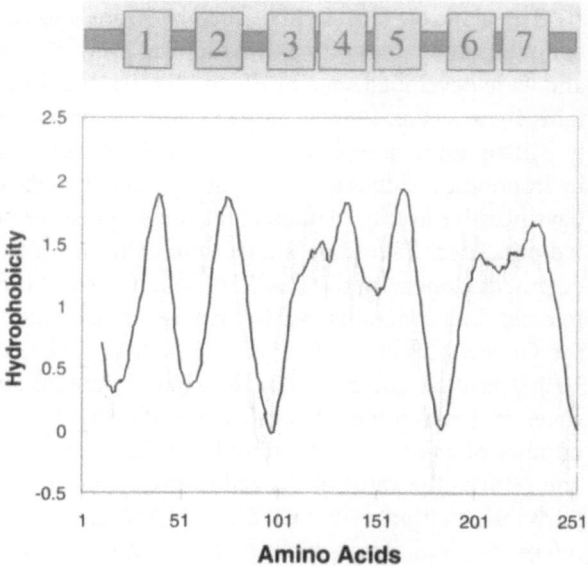

structural evidence shows that rhodopsin does indeed have seven membrane
spanning regions as predicted by this method. As another example the Sequences
sheet contains the sequence of the β2-adrenergic receptor (β2-AR) and the
angiotensin type 1a receptor (At1a), both members of a large family of proteins
which contain seven trans-membrane domains. Various other proteins without
transmembrane regions are tropomyosin, an antibody light chain (AB light) and
the β subunit of a trimeric G protein (G-β).

The other scales included in Sheet 1 of MEMB.XLS are those of Kyte and Doolit-
tle (K/D), Engelman, Goldman and Steitz (EGS), Von Heijne (v. Heijne), Hopp and
Woods (H/W) and Welling and coworkers (Welling, all these values are presented in ref. 11). The K/D, EGS and Von Heijne parameters are fundamentally rather
similar to those of Argos and coworkers, designed primarily to find hydrophobic membrane spanning regions of molecules, although the EGS and Von Heijne
values produce troughs instead of peaks for strongly hydrophobic regions. The
Von Heijne parameters are optimized to find hydrophobic leader sequences, which
are segments of about fifteen amino acids at the N-termini of proteins which
allow the newly synthesized protein to pass through the cell membrane and be
expressed outside the cell. The antibody molecule contains such a sequence. The
Hopp and Woods and Welling and coworkers values are based on attempts to predict regions likely to be on the surface of proteins and hence to produce sites where
antibodies and other proteins might bind. In both cases peaks represent such predicted regions. Since such binding sites are usually relatively small a suitable window size for such parameters would be about seven amino acids.

A major problem in molecular biology is the prediction of the three-dimensional structure of proteins from sequence data. Presently it is much easier to obtain

a protein sequence than a protein structure, meaning that many researchers are forced to deduce the most likely structure from the amino acid sequence. Proteins contain only a few fundamental types of structure; these are the α-helix, the β-sheet and the turn (Alberts et al. 1994; Lodish et al. 1995). Each has a characteristic content and arrangement of amino acids, and there are several computer methods aimed at identifying these types of sequences. One popular method of predicting protein structure was published by Chou and Fasman (1978) and a workable version of this can be implemented in Excel as shown in the CHOFAS. XLS spreadsheet. This version is one of a wide variety which could be constructed based on the Chou and Fasman algorithm, which is not unambiguously defined in the literature.

The two functions LOOKUP and OFFSET are used to assign appropriate values to each amino acid in the same way as in MEMB.XLS above. In this case three sets of values are entered in columns C-E of the Calculations sheet from data in the table in blocks N4 to R23. These three values reflect the frequency at which the particular amino acid is found in α-helices, β-sheets or turns in known proteins. For α-helices and β-sheets these values are averaged over windows of seven amino acids as in columns F and G of the Calculations sheet of the CHOFAS. XLS spreadsheet. Two turn probability factors, called Pt and pt are calculated. Pt is the turn probability averaged over a four amino acid window using the factors shown in column E, while pt is the product of the turn probabilities calculated using the factors in columns S to V over the same four amino acid window determined as described below.

A region is predicted to form an α-helix if the helix probability over a seven amino acid window is > 1.05 and also greater than the β-sheet probability over the same range. This determination is performed by the equation =IF(AND(F6>G6,F6>1.05)=TRUE,2,1) in cell J6. The AND function returns a value of TRUE only if the α-score is greater than the β-score (F6>G6) and the α-score is greater that 1.05 (F6>1.05). If AND is TRUE a value of 2 is returned to the cell, and if not 1. The numbers 2 and 1 are chosen to produce a readable plot and have no intrinsic meaning beyond this. The determination of likelihood of β-sheet is similar, in that the β-sheet probability must be greater than the α-helix probability and also greater than 1.05. The equation =IF(AND(G6>F6,G6>1.05)=TRUE,4,3) in cell K6 therefore returns a value of 4 if the region meets these criteria and 3 if it does not; again the numbers 4 and 3 are chosen simply for plotting purposes. The turn predictions are more complex, and a turn is predicted if the Pt value > 1.00 and also greater than both the α-helix and β-sheet probabilities and if the pt value $> 7.5 * 10^{-5}$. The Pt values are calculated in the same general way as the helix and sheet calculations as shown below. The pt values are also used and are based on a table which enumerates how frequently particular amino acids are found in turns in known proteins, these figures being included in columns S to V of the Calculations sheet. These values refer to how frequently an amino acid occurs in the first, second, third and fourth positions in a turn (f i, f i+1, f i+2 and f i+3 respectively). The pt value is calculated as the product of the appropriate four values and is determined for each set of four amino acids using the rather cumbersome equation in cell I6:

=PRODUCT(OFFSET($S$4,B4,0,1,1),OFFSET($S$4,B5,1,1,1),
OFFSET($S$4,B6,2,1,1),OFFSET($S$4,B7,3,1,1))

This multiplies the appropriate values of the sequence A4-A7 by using the OFF-SET function to extract the appropriate values from S, T, U and V columns. The L6 cell contains the equation:

=IF(AND(H6>F6,H6>G6,H6>1,I6>0.000075)=TRUE,6,5)

to see if these four conditions are met, returning 6 if they are, and 5 if they are not. The data are plotted out on the Plot sheet of the CHOFAS.XLS spreadsheet. Peaks indicate a prediction for the relevant structure, while no peaks predict absence of the same structure. The Sequences sheet contains the sequence of the neurofilament subunit NF-L, tropomyosin, an antibody molecule and ubiquitin. NF-L has a N-terminal region believed to be non-helical and is probably a β-sheet structure, followed by a classical α-helical segment. Tropomyosin is know to be an α-helical rod with no β-sheet and few turns. Ubiquitin is mostly α-helical with several turns while the antibody molecule is mostly β-sheet interspersed with turns. As can be seen this version the Chou and Fasman algorithm predicts these features quite accurately.

An even more complex example is the use of Excel to predict regions likely to form α-helical coiled coils as shown in COILCOIL.XLS. The algorithm was described by Lupas and coworkers and is quite complex (1991). The α-helical coiled coil is a periodic "heptad" arrangement of hydrophobic and charged amino acids (Alberts et al. 1994; Lodish et al. 1995). The amino acids in a coiled-coil segment are labeled a, b, c, d, e, f and g. The a and d positions usually contain large hydrophobic amino acids, frequently leucine, while the remaining amino acids are usually charged. The a and d positions of two α-helical coiled-coils interact and are therefore on the inside of the structure. Lupas and coworkers determined how frequently each of the amino acids were found in the seven possible positions in known coiled coil proteins. This data, which is reproduced in Table 2 here, gives the relative frequency at which each amino acid is found in an α-helical coiled-coil compared to that expected in an average protein by chance. There are seven different scores for each amino acid depending on which position it may be found in a coiled-coil. The next amino acid also gets seven scores, but the scores must be shifted by one position relative to the previous amino acid, since the next amino acid is one amino acid further down any possible coiled-coil. The score for position "a" for amino acid N must therefore be linked to that for the "b" position for amino acid N+1 and so on. A window of twenty eight amino acids is used since this appears to be the minimum length for coiled coils in nature. The seven sets of scores for each possible 28 amino acid peptide are multiplied together and the 28th root is determined. The highest of the seven possible scores for each twenty eight amino acid peptide is converted into a probability based on

Table 2. Frequency at which each amino acid is found in each of the seven possible positions in known coiled coil proteins (data from reference 14)

	a	b	c	d	e	f	g
L	3.167	0.297	0.398	3.902	0.585	0.501	0.483
I	2.597	0.098	0.345	0.894	0.514	0.471	0.431
V	1.665	0.403	0.386	0.949	0.211	0.342	0.360
M	2.240	0.370	0.480	1.409	0.541	0.772	0.663
F	0.531	0.076	0.403	0.662	0.189	0.106	0.013
Y	1.417	0.090	0.122	1.659	0.190	0.130	0.155
G	0.045	0.275	0.578	0.216	0.211	0.426	0.156
A	1.297	1.551	1.084	2.612	0.377	1.284	0.877
K	1.375	2.639	1.763	0.191	1.815	1.961	2.795
R	0.659	1.163	1.210	0.031	1.358	1.937	1.798
H	0.347	0.275	0.679	0.395	0.294	0.579	0.213
E	0.262	3.496	3.108	0.998	5.685	2.494	3.048
D	0.030	2.352	2.268	0.237	0.663	1.620	1.448
Q	0.179	2.114	1.778	0.631	2.550	1.578	2.526
N	0.835	1.475	1.534	0.039	1.722	2.456	2.280
S	0.382	0.583	1.052	0.419	0.525	0.916	0.628
T	0.169	0.702	0.955	0.654	0.791	0.843	0.647
C	0.824	0.022	0.308	0.152	0.180	0.156	0.044
W	0.240	0.000	0.000	0.456	0.019	0.000	0.000
P	0.000	0.008	0.000	0.013	0.000	0.000	0.000

the scores obtained with known coiled-coil molecules, and this probability is plotted out. This program has been remarkably effective and predicted α-helical coiled coils in several molecules, and in several cases these predictions have been verified.

The Calculations sheet of COILCOIL.XLS uses the LOOKUP function to assign appropriate numerical values to each amino acid in column B. Column C labeled "Index" contains the value 1 in cell C3. The cell below this contains C3+1, and when this copied down the spreadsheet the result is to sequentially number the rows. This number is used to shift the reading frame of the scoring method as we go down the sequence. Columns D to J, labeled RF1 to RF7, assign the appropriate values in each of the possible reading frames to each amino acid using the OFFSET function and the data in cells V5 to AD24. D3 contains the equation =OFFSET(\$X\$4,\$B3,MOD(\$C3,7),1,1) and the neighboring cell E3 contains =OFFSET(\$X\$4,\$B3,MOD(\$C3+1,7),1,1). The MOD function returns the modulus, that is the remainder, when the value in cell C3, the index number, is divided by 7. This value must always be in the range 0 to 6 and is therefore ideal for telling OFFSET which column to address for the amino acid scores. As can be seen the equation in E4 is the same as in D4 except that the MOD function now contains (\$C3+1,7) instead of (\$C3,7). This means that the equations in this column all address the next set of values in the table, and in the subsequent RF columns this value is further incremented. As we go down the columns the MOD values are also incremented since the value in the C column is also incremented. The result of all this is that each twenty-eight amino acid segment is scored in seven different ways, the spreadsheet searching for the amino acid characteristics of α-helical coiled coils in each of the seven possible reading frames. Columns K to Q determine the 28th root of the product of scores for a 28 amino acid segment of each RF column using the equation, in cell K3 for example, =POWER(PRODUCT(D3:D30), 1/28). The R column, marked "Largest", returns the largest of these scores using the MAX function, in this case simply =MAX(K3:Q3). Finally the probability of a coiled coil is calculated in the pCC column using the following nested IF expression, which is a simple description of the probability curve described by Lupas et al., assigning probabilities based on the highest score obtained by each twenty-eight amino acid segment.

=IF(R3>1.55,1,IF(R3>1.48,0.9,IF(R3>1.4,0.8,IF(R3>1.36666,0.6,IF(R3> 1.33333,0.4,IF(R3>1.22,0.2,IF(R3>1,0.1,0)))))))

This equation is placed in cell S30 and below. The T column is labeled pCC plot and cell T3 contains the equation =MAX(S3:S30). This ensures that any twenty-eight amino acid segment predicted to contain an α-helical coiled coil is plotted out as a line twenty-eight amino acids long, rather than as a point. The data in this column provides the output of the program which is plotted out in the pCC Plot sheet.

The Sequences sheet contains several different sequences which can be copied into the first column of the Calculations sheet. These proteins include a β-subunit of trimeric G proteins which was predicted and is now known to contain an α-helical coiled coil at the N-terminus. Another example is the protein coronin, predicted to contain a similar region this time at the opposite end, the extreme C-terminus. Tropomyosin is an almost entirely α-helical coiled coil, and NF-L has three regions of coiled coil interspersed with non-helical linkers. All of these features are readily apparent when these sequences are copied into the Calculations sheet.

Numerous other protein properties can be calculated from sequence information data. As one further example of practical significance is the molar extinction coefficient. Proteins absorb ultraviolet light at a wavelength around 280 μm and the amount of absorption at this wavelength is determined almost exclusively by the content of only two amino acids, tryptophan and tyrosine. Under standardized conditions in water 1 M of tyrosine has an optical density of 1.1×10^3 units at a wavelength of 278 μm. 1 M of tryptophan has an optical density of 5.2×10^3 units at a wavelength of 279 μm (Mahler and Cordes 1966). For practical purposes the wavelength of 280 μm is used experimentally which is close enough to both 278 μm and 279 μm to make no significant difference. The optical density of pure protein can be very easily determined under standard conditions and gives a value, known as the molar extinction coefficient ε, which can then be used to determine accurately the protein content if the mole content of tryptophan and tyrosine (nTrp and nTyr respectively) are known. 1 M of the pure protein has an absorption of nTrp x 5.2×10^3 plus nTyr x 1.1×10^3. In practice researchers tend to measure protein concentrations in mg/ml, so that a useful quantity is the optical density at 280 μm of a 1 mg/ml solution of the protein. 1 mg/ml of protein is equivalent to 1 g/l and a 1 M solution is defined as the molecular weight of the protein in g/l. So a 1 mg/ml solution is equivalent to a (1/molecular weight)M solution. It follows that to the following equation will calculate exactly the absorption of a 1 mg/ml solution of pure protein at 280 μm (A_{280}):

$$A_{280} = (nTrp \times 5.2 \times 10^3 + nTyr \times 1.1 \times 10^3)/\text{molecular weight}.$$

This equation included in a single cell of a spreadsheet as described above and could be performed as an exercise. Knowing that a specific amount of absorption corresponds to 1 mg/ml of protein allows the easy calculation of the amount of protein corresponding to an experimentally determined absorption value (Mahler and Cordes 1966).

Protein Sequence Analysis with Dot Plots

A popular method for graphically demonstrating relationships between two protein or nucleic acid sequences is the dot plot (Maizel and Lenk 1981). Two sequences to be compared are arranged at 90° to one another and a table is constructed. In this table a dot, cross or the relevant amino acid is inserted if the amino acid at the ordinate is the same as the amino acid at the abscissa. In the simplest case nothing is inserted if the amino acids are not identical. If the two sequences to be compared are identical or very closely related, a diagonal line of dots, cross or amino acids will stretch through the middle of the matrix. If the two sequences contain regions of similarity these are also manifested as diagonal lines, although not necessarily through the middle of the matrix. If a protein containing repeated sequences is compared with itself, several other diagonal lines besides the central one are seen. This approach is useful for comparing sequences either with themselves or with others, and is particular good at finding regions of similarity or segments of repeated sequence. Such plots can be performed in Excel fairly simply. Although one sequence could be inserted vertically and the other horizontally as described above, in the DOTPLOT.XLS spreadsheet the two sequences to be compared inserted into the A and B columns of the Calculations. This is to simplify the entry of sequences, which can be produced in the same way as the others described in this chapter. Two indices are needed to increment in the appropriate directions when the equations in the table are copied, and these are inserted in column 3 and row C, both starting at cell C3 which contains the value 0. Cells below this one are progressively incremented by 1, as are cells to the right, so that D1 has a value of 1 as does C4. Since the first index in either direction should be 0, cell C3 contains –1, and both C4 and D3 contain C3+1. The first cell within the matrix contains the equation:

IF(OFFSET($A$3,$C3,0,1,1)=OFFSET($A$3,D$2,1,1,1),
OFFSET($A$3,D$2,0,1,1),"")

The IF function uses the OFFSET function to compares the contents of the first row with that of the second row, and returns the contents of the first row (i.e. the contents of OFFSET($A$3,$C3,0,1,1)) if both rows contain the same amino acid, or nothing (" ") if they are not the same. The $C3 and D$2 values in the first two OFFSET functions use the index values in the C column and 3 row to increment the values by the required amount, allowing the equation to copied anywhere within the matrix and address cells appropriate for the new location.

The width of the columns is set to 2 units to give a tabular view similar to that of the more usual dot plots, in which the cells are square. This was done by selecting Format, then Column, then Width and entering 2.0. The example given provides the sequence of part of the rabbit neurofilament molecule NF-H which is compared with itself. Obviously a clear diagonal line is seen as expected (see Fig. 3). However numerous other diagonal lines are seen, and inspection of these diagonals reveals the presence of repeated sequences. The example sequence is extraordinarily rich in repeated sequences based on the consensus AKSPAE and AKSPVKEE (reviewed in Shaw 1991). The presence of these repeated sequences is very obvious in the dot plot, and the distance between these diagonals shows

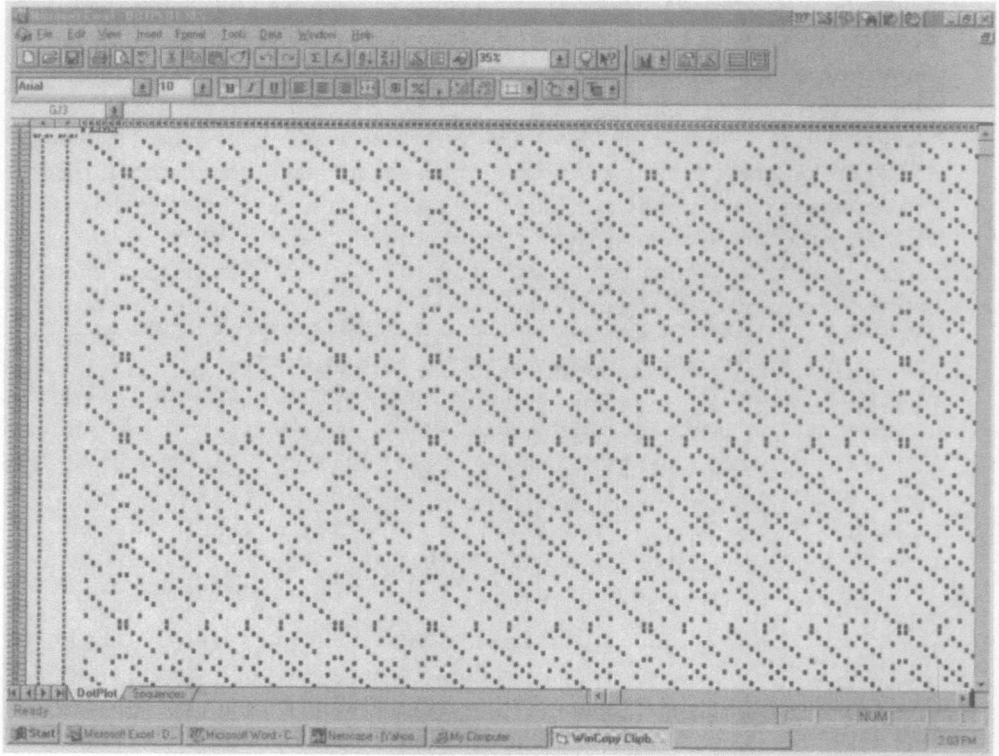

Fig. 3. View of the DOTPLOT.XLS Calculations sheet showing a highly repetitive neurofilament sequence

that the basic repeat unit is 6 and 8 amino acids. Another region of the rabbit NF-H molecule is included in the Sequences sheet as are the sequences of calmodulin and a region of the neurofilament molecule NF-M. The calmodulin and NF-M sequences both contain more subtle repeated sequences; what and where are these repeats?

The scale of the dot plot can be reduced to 35% to view diagonal lines more easily. This is done by replacing the scale factor, which is most likely set to a default value of 100%, in the "standard" tool bar with the number 25. If the standard toolbar is not visible it can be displayed by selecting the Toolbar option of the View menu. Under these conditions the letters cannot be viewed or printed well, and it is sensible to insert another more visible and printable character. The MS Linedraw font for the IBM PC and compatibles contains a dense black rectangle ideal for this purpose. If the third OFFSET function in the above equation is replaced by "Û" this will form ■ in the appropriate cell when the MS Linedraw font is selected. This approach has been taken with the spreadsheet shown in Fig. 3. Fonts can be changed using the Formatting toolbar, which is displayed in the same way as the standard toolbar described above. The equation in the first cell is therefore altered to;

$$IF(OFFSET(\$A\$7,\$C4,0,1,1)=OFFSET(\$A\$7,D\$3,1,1,1),"Û","\ ")$$

Since selecting such special characters is rather inconvenient in Excel, the Û character has been placed in the C3 cell, so that \$C\$3 can be substituted "Û" in the above equation. This simply displays the contents of the C3 cell, the ■ character, if the two OFFSET functions contain the same amino acid.

Analysis of Nucleic Acid Sequences

Nucleic acid sequences are also easily analyzed using spreadsheets. For example, DNA sequences rich in the bases G and C are often involved in the regulation of expression of genes (Alberts et al. 1994; Lodish et al. 1995). In particular the sequence CG is found unusually frequently in such regulatory regions in mammalian genes, usually known as promoters. In fact the presence of unusually large numbers of CG sequences is a widely used method for finding such promoter regions. Since the backbone of the DNA molecule is a phosphate group, the C and G residues are linked by a phosphate, and the regions rich in CG sequences are therefore often called CpG islands. Regions rich in G and C are easily displayed in spreadsheet programs, simply by converting G and C residues in 1, while making A and T residues 0. As shown in the Calculations sheet of DNA.XLS, the promoter sequence for hypoxanthine phosphoribosyltransferase (HGPRT) has been inserted in column A. This sequence is processed by inserting the equation =(IF A3="G",1,IF A3="C",1,0) into the B3 cell and copying this down the full length of the sequence. CG sequences are also found very easily in Excel. The equation =IF(AND(A3="C",A4="G")=TRUE,1,0) in column C uses the AND function to see if A3 contains a C and A4 contains a G. If this condition is TRUE, a 1 is returned.

Columns D and E simply calculate the percentage GC content over a window of 50 bases and the number of CG sequences in a window of 50 bases. The results are plotted out on the Plot sheet. The Sequences sheet contains the promoter of the superoxide dismutase (SOD) gene which also shows a prominent peak of GC content and particularly high CG sequence content immediately preceding the start of the gene, which is at the end of the spreadsheet. A region of the calmodulin gene not including the CpG island is included for comparison purposes. Obviously the approach used to find CG sequences shown here could be easily modified to find other desired short DNA sequences. Restriction endonucleases, useful for genetic engineering, cleave DNA only at specific sequences, usually 4–8 bases in length. Furthermore regulatory proteins also recognize and bind to short and specific DNA sequences. A spreadsheet can be easily set up to search for and plot out the location of such sequences. As an exercise modify the DNA.XLS spreadsheet to find and plot out the location of all occurrences of the sequence GATC, which is recognized by the widely used restriction endonuclease *SauIIIa*.

Nucleic acid sequences of proteins are often analyzed with dot plots as described above (Maizel and Lenk 1981). Dot plots are particularly informative

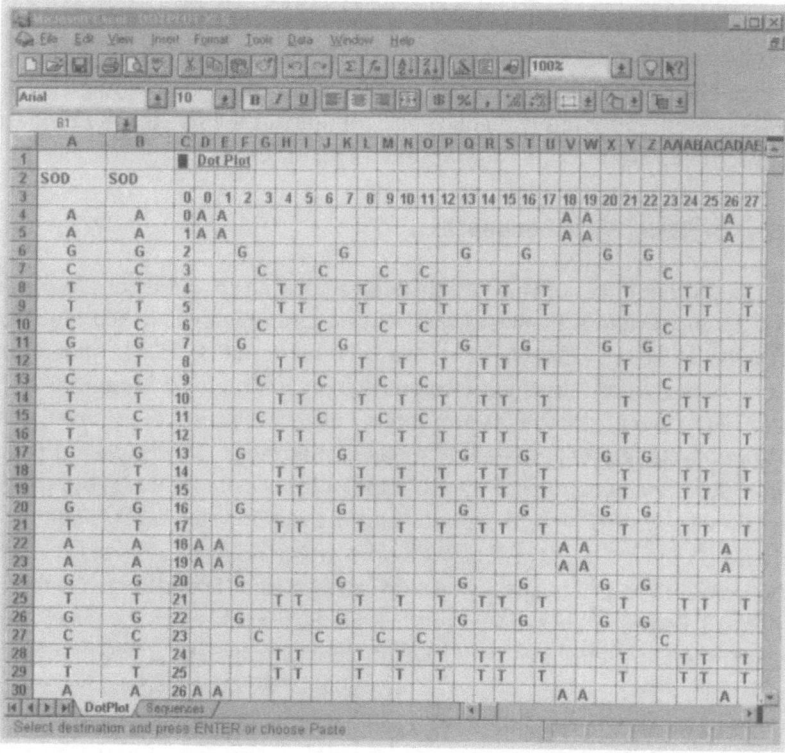

Fig. 4. View of the DOTPLOT.XLS Calculations sheet showing direct and inverted repeat sequences

	Name	Molecular Weight	Extinction Coefficient ε in A_{260} Units/μm
Table 3. Properties of bases in DNA	Adenine	313.21	15.4
	Cytosine	289.19	7.4
	Guanine	329.21	11.5
	Thymine	304.2	8.7

for nucleic acids since their sequences frequently contain inverted and palindromic sequences which are readily highlighted by this approach. For example refer again to the DOTPLOT.XLS spreadsheet. The Sequences sheet contains three nucleic acid sequences in columns G, H and I. If the SOD entry is copied into the A and B columns of the Calculations sheet, you should see something resembling Fig. 4. The central diagonal is expected since the sequence is being compared with itself. Note also the second diagonal from F26 to I29, also Z6 to AC9. This indicates that the short sequence GCTT is repeated. Note also the several diagonals perpendicular to these, for instance K17 to Q11. These indicate the presence of sequences which are identical when read in the reverse orientation, in the example given GTCTCTG. These are referred to as palindromic sequences and are of interest since they often act as binding sites for important regulatory proteins (Alberts et al. 1994; Lodish et al. 1995).

Other worksheets similar to those constructed for the analysis of protein sequences can be easily constructed. For example calculating the molecular weight of a nucleic acid sequence can be done using the approach shown in AACOMP. XLS above using molecular weight values as shown in Table 3 below.

Calculating the extinction coefficient as described for protein can also be performed. All four bases contribute to the absorption properties of DNA and the preferred wavelength for measurement is 260 μm rather than 280 μm, but the principle is the same. There are also numerous commercially available and in general very expensive programs designed to calculate the optimum temperatures at which short lengths of DNA will hybridize with longer DNA sequences. Such information is required for the efficient performance of a variety of powerful molecular biological techniques including various types of hybridization and the now ubiquitous polymerase chain reaction. These programs make use of published and relatively simple algorithms (Breslauer et al. 1986; Freier et al. 1986) and there is no doubt that they could also be implemented in a spreadsheet program.

Some Practical Considerations

The first requirement for using a spreadsheet program to analyze a protein or DNA sequence other than those given as examples here is to get the sequence into the first column of the spreadsheet, with one character per cell. Protein and nucleic acid sequences can be obtained from a variety of Internet sources, such as the comprehensive NIH databases at URL *http://atlas.nlm.nih.gov/Entrez/ index.html*. There are several possible ways to proceed from a Web page containing the desired sequence information, which is invariably in combination with reference material and other information. One simple approach is select and directly copy the sequence information using the mouse. Alternately the "Save Source" or other appropriate command on the Web browser can be used to save all the information as a HTML document. Sequence data, either as a HTML document or copied directly from a web page can then be loaded into Microsoft Word or a similar word processor for further manipulation. The sequence data contains not only amino acids or bases but also spaces, carriage return/line feed (CR/LF) characters, numbers and sometimes tabs. To get the information in a format suitable for loading into a column of Excel, all of these characters must be removed and a newline must be added after each amino acid or base. This could be performed manually but is obviously a tedious and error prone procedure. To make the process more straightforward the disc contains two automated methods for generating sequences in columns which can be pasted directly into Excel.

The MSDOS file included on the disk, called SPACER.EXE, will remove all characters except uppercase ACDEFGHIKLMNPQRSTUVW from text files and add a line feed after each character remaining. Note that a few databases give sequences only in lowercase letter. These can be changed to uppercase using the Word "Change Case" command found in the Format menu. Most other word processors contain a similar command. The sequence from the database is first selected with a mouse, copied into a new file which is then saved as a MSDOS text file. This procedure can be performed with any modern Windows based word processor. The MSDOS SPACER.EXE program is then run and the name of the appropriate file selected. This simple program asks for a name for a new file which will contain the processed data, and then generates this file. The newly generated file can then be loaded directly into Excel.

The second approach makes use of the considerable capabilities built into Microsoft Word versions 6.0 and beyond. These contain a sophisticated macro language called WordBasic which has been used here to write a program similar to but more convenient than the SPACER.EXE program. This program is included on the disc as a text file called SPACER.TXT. This text file contains a WordBasic procedure which removes line feeds, numbers, tabs, spaces etc. from a mouse selected segment of text, and it then adds newline characters after each remaining character. The sequence is therefore turned into a column ready for copying into Excel. The simplest method of getting the text file to function as a macro in Word version 6.0 is to select the "Tools" menu, then "Macro", then "Record", at which point a window asks for the name of the Macro to be recorded, which can be called Spacer or any other convenient name. (If you want the Spacer macro to appear as an option on the "Tools" pull down menu, or on a toolbar or define a hotkey for this macro this is most easily done at this stage). On pressing the Enter key a small box appears which indicates that any further key strokes will be recorded. Simply select the button with a square on it, saving the Spacer macro.

Then select "Tools", then "Macro" as before, but now select "Spacer" from the list of macros. Now select "Edit", and the contents of the macro should be:

- Sub MAIN
- End Sub

Which amounts to a macro which occupies memory and disk space but has no useful function. It can be made to function by opening the SPACER.TXT file in Word, copying the entire contents and replacing the two lines of the macro with this text file. The modified macro can be saved using the "Save Template" command, and can now be run by selecting a sequence to process, then selecting "Tools", then "Macro", selecting "Spacer" from the Menu and finally selecting "Run". Alternately, if you added this macro to a menu or a toolbar, or defined an appropriate hotkey you can conveniently run this macro at any time. The selected text is now processed to remove irrelevant characters and made into a column, which can then be selected and pasted into a column of Excel.

Conclusion

The potential applications of spreadsheets in molecular biology are by no means exhausted by the examples given here. Indeed it is certain that an entire book could be written just on the topic of this chapter and making use only of features built into current versions of Excel. The inclusion of Visual Basic for Applications (VBA) in Excel suggests that even a sequence analysis program which is difficult to implement in Excel could be adapted by custom programming to run in one of these spreadsheets. Presumably newer versions of Excel will be even more powerful and versatile. The numerous advantages to this approach were outlined above, to which must be added that an individual who has analyzed proteins and nucleic acids in the way suggested here will know exactly how the analysis was done and will thoroughly understand the strengths and limitations of the method being used. It is to be hoped that this chapter will encourage others to make use of and expand on these methods.

References

Alberts B, Bray D, Lewis J, Raff M, Roberts K, Watson J (1994) Molecular Biology of the Cell, 3rd edn. Garland, New York
Argos JK, Rao M, Hargrave PA (1982) Structural Prediction of membrane-bound protein. Eur J Biochem 128:565–575
Breslauer KJ, Frank R, Blocker H, Marky LA (1986) Predicting DNA duplex stability from the base sequence. Proc Natl Acad Sci USA 83:3746–3750
Bult CJ, White O, Olsen GJ, et al (1996) Complete genome sequence of the methananogenic Archaeon, Methanococcus jannaschii. Science 273:1058–1078
Chou PY, Fasman GD (1978) Empirical predictions of protein conformation. Ann Rev Biochem 47:251–276
Eckerskorn C, Jungblut P, Mewes W, Klose J, Lottspeich F (1988) Identification of mouse brain proteins after two-dimensional electrophoresis and electroblotting by microsequence analysis and amino acid composition analysis. Electrophoresis 9:830–838

Ferscht A (1977) Enzyme Structure and Mechanism. Freeman, Reading San Francisco

Fleischmann RD, Adamis MD, White O, et al (1995) Whole-genome random sequencing and assembly of Haemophilus influenzae. Science 269:496–512

Fraser CM, Gocayne JD, White O, et al. (1995) The minimal gene complement of Mycoplasma genitalium. Science 270:397–403

Freier SM, Kierzek R, Jaeger JA, Sugimoto N, Carruthers MH, Neilson T, Turner DH (1986) Proc Natl Acad Sci USA 83:9373–9377

Lodish H, Baltimore D, Berk A, Zipursky SL, Matsudaira P, Darnell J (1995) Molecular Cell Biology, 3rd edn.Scientific American books, New York

Lupas A, Dyke MV, Stock J (1991) Predicting coiled coils from protein sequences. Science 252:1162–1164

Mahler HR, Cordes EH (1966) Biological Chemistry. Harper and Row, New York

Maizel JV Jr, Lenk RP (1981) Enhanced graphic matrix analysis of nucleic acid and protein sequences. Proc Natl Acad Sci USA 78:7665–7669

Shaw G (1991) Neurofilament proteins. In: Burgoyne RD (ed) The Neuronal Cytoskeleton. Alan Liss, New York, pp 185–214

Shaw G (1993) Rapid identification of proteins. Proc Natl Acad Sci USA 90:5138–5142

Shaw G (1995) Protein sequence interpretation using a spreadsheet program. Biotechniques 19:978–983

Van Heijne G (1987) Sequence analysis in molecular biology. Academic Press, San Diego

Williams N (1996) Yeast genome sequence ferments new research. Science 272:481

Spreadsheet Applications in Materials Science

A. A. Gorni

Introduction

Until about 20 years ago, computer applications in Materials Science were frequently discouraged by the high cost of computer processing and by the huge efforts necessary to develop and debug a program in the harsh mainframe environment. Only in very specific cases – crystallographic texture analysis, for example (Bunge 1982) – was the use of computers justified. However, it is interesting to note that a book on Materials Science edited more than twenty years ago (Guy 1976) already included some simple Fortran programs related to applications in this field.

As computer power became affordable and widely available, new kinds of software arose. One example of these were the spreadsheet programs, which provided a friendly and intuitive interface for simple calculations, instead of the mainframe's painful environment. This kind of program soon proved to be of value for general scientific (Arganbright 1983) and metallurgical engineering applications (Dobson 1984).

The relatively recent development of new powerful microprocessors, like the 80486 and Pentium, and the continuous decrease in hardware prices made available computers with dozens of megabytes of RAM and hard disks with commonly more then a Gbyte of capacity. This was enough computing power to allow the implementation of graphical user interfaces in IBM microcomputers.

In the field of Materials Science, work on the use of modern spreadsheets included applications as diverse as a data base management system for the steel industry (Whipp 1993), for the design of exercises in a polymer laboratory course (McGee and Mattson 1993) and a review on their use in problems of ceramic materials (Skaar 1994).

In this chapter we will present some classical computer applications in Materials Science: quantitative metallography (determination of the fraction of second phase and grain size in microstructures), material toughness calculations, viscometric determination of the molecular weight of polymers, identification of phases using X-ray diffraction and the solubilization of microalloying elements in austenite.

All examples were developed using Microsoft Excel version 7.0. Unfortunately this means, insofar as some of them use the more advanced features of this later version, that incompatibilities with earlier Excel versions can not be excluded.

Quantitative Metallography

The properties of a material are clearly influenced by its microstructure. For example, its mechanical resistance is generally proportional to the fraction of hard phase present in the microstructure, as well as inversely proportional to its grain size. Therefore, a quantitative correlation between properties and microstructure can be a powerful tool for quality control and research. The mathematical foundations for the determination of several quantitative parameters of the microstructure – like fraction of the phases present, grain size, mean free path, interfacial area and so on – were developed some years ago (Underwood 1970). However, as materials belong to an imperfect and real world, microstructure is not exactly identical throughout the volume of a given material. This means that the quantitative parameters that describe a microstructure unavoidably have some associated level of statistical dispersion. For this reason, the determination of quantitative parameters of a microstructure requires that a minimum number of fields or portions of the microstructure must be analyzed in the microscope. In this way, enough data points can be obtained, assuring sufficient precision in the parameter being determined. The number of fields necessary to assure an adequate precision in the mean confidence – say, 1% – can eventually be unbearably high. Some heterogeneous microstructures can require dozens or even hundreds of fields to be analyzed. As the manual quantitative analysis of microstructure can be somewhat cumbersome and time consuming, this is a perfect task for a computer. In fact, this application only became widely feasible about ten years ago, when the price of digital hardware started to become affordable. Nowadays, digital image analyzers can perform this task in a quick, accurate and economical way.

However, even now not all laboratories have such equipment. Also, in some cases, the complexity or the image quality of the microstructures to be characterized does not allow a reliable automatic analysis. In these cases, the application of manual methods of quantitative analysis of microstructures is still required.

The main example to be developed here is the determination of the volumetric fraction of second phase in a microstructure using the point-count procedure. This method is standardized according the ASTM Standard E-562. Additional details can be found in (Underwood 1970) or (Subbarao et al. 1976). The fraction of second phase in each field examined in the microscope is determined as follows. A grid is superposed over the image of the microstructure, as shown in the left side of Fig. 1. This can be done by projecting the image generated by the microscope onto a translucent glass screen. On this screen there must be an attached film of transparent plastic with the selected grid drawn on it. Some microscopes have oculars with the grid already drawn on an additional lens.

The microstructure in Fig. 1 is only a schematic example, with the clear grains being the matrix, that is, the dominant phase, and the darker grains being the second phase. We can see that the distribution of fraction of second phase and the grain sizes are not uniform within the microstructure, so these values will vary according to the position of the field being analyzed.

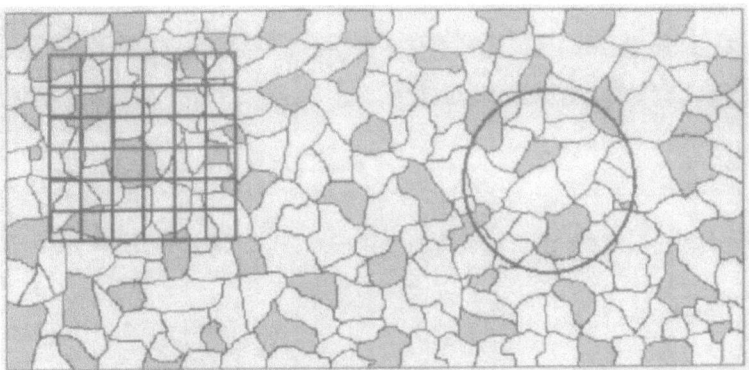

Fig. 1. Schematic representation of the manual measurement of the fraction of second phase and grain size in a microstructure

In order to minimize the number of fields to be examined, the size and spacing between the lines of the grid, as well as the magnification of the microscope, are selected according to standard rules, which can be seen in (Underwood 1970). For example, the grid and magnification used must be selected so that a given grain of second phase coincides with not more than one point of the grid.

Once the grid and the microscope magnification to be used are selected, the procedure for the determination of the fraction of second phase can effectively begin. For each field of microstructure being observed in the microscope, one must count the number of line intersections that coincide with the area determined by the darker grains of second phase. After that, the fraction of second phase can be calculated by dividing this number by the total number of line intersections available in the grid. In our case, as we have a grid with 7 × 7 lines, the total number of line intersections is 49. Generally this fraction of second phase is expressed as a percentage. This procedure must be repeated until a sufficient number of fields have been counted to generate an adequate level of confidence for the mean value of fraction of second phase. Of course, the location of the fields must be chosen randomly across the total surface of the sample being analyzed.

Let's examine the Excel worksheet VOLFREX.XLS. This was developed for the calculation of the mean second phase fraction in a microstructure with two phases. As can be seen in Fig. 2, this spreadsheet contains some data as an example of its use. The worksheet VOLFRAC.XLS, also provided with this book, is a "virgin" version, with no data included. In the cells from C5 to G5 there is a field for the identification of the sample being analyzed. In this case, a sample of dual phase steel is being studied. This kind of material shows a microstructure with 80 to 90% of soft ferrite matrix with 20 to 10% of harder martensite as second phase. This peculiar microstructure results in good forming properties and high mechanical resistance to sheets produced with this steel.

In the cell C7 we must input the total number of line interceptions observed with the grid being used. In our example this value is 49, as mentioned above. This

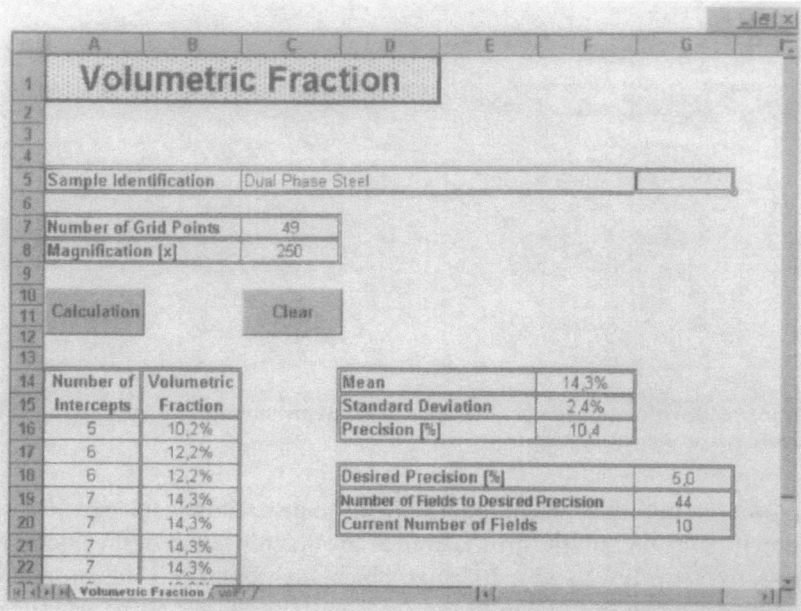

Fig. 2. An example of the use of the worksheet VOLFREX.XLS developed for the calculation of the second phase fraction in a microstructure using the point-count procedure

data will be used later in the calculations. The microscope magnification used during the observation of the microstructure may be introduced in the cell C8. As this parameter is not directly used in the calculations, its input is optional. However, we recommend including it to document the conditions under which the measurement was made. The magnification used in the present example was 250.

Now we can enter the number of line intersections which coincide with the second phase of the microstructure for each field observed. These data must be input from the cell A16 downwards, one data field per cell. That is, the value corresponding to the first observed field must be introduced in the A16 cell, the second in the A17 cell, and so on. In our example, we have data from 10 fields in the sequence A16:A25 randomly chosen in the microscope.

Now we must calculate the percentage of second fraction for each field observed. Firstly, we must select the cells B16:B25, and to attribute the percentage number format to them. It is advisable to limit precision to one decimal place. Then, we must enter the equation =A16/$C$7 into the cell B16. The percentage of second phase in the first field observed will appear in this cell. Now we must copy and paste this formula from the B17 to the B25 cell. This will generate results of percentage of second phase in the microstructure for all available data.

After the calculation of this parameter, we want to carry out a simple statistical analysis. The first statistical parameter to be calculated is the mean value of second phase fraction, **x**, which will be placed in the cell F14. The mean is calculated using the Excel function AVERAGE. So, we must enter the formula =AVERAGE(B16:B25) in this cell. Now we calculate the corresponding standard deviation s. This can be done using the Excel function STDEV. Now we enter the formula =STDEV(B16:B25) into the F15 cell. Of course, the format of the mean and the standard deviation must be identical to that used for the input data, that is, percentage format with one decimal place.

Next, we must determine a fundamental statistical parameter: the precision of the mean x, that is, its confidence interval. It is defined by the formula:

$$\frac{200\, s}{\sqrt{n}\; x} \tag{1}$$

where **n** is the number of analyzed fields. In our worksheet, this formula is introduced in the F16 cell, as =200*[F15]/(SQR(COUNT(B16:B25))*[F14]).

Well, considering the ten fields analyzed up to this moment, the mean value of the percentage of second phase has a precision of only 10.4%. A value normally chosen for the required precision **p** is 5%, as we can see in the cell G18. Now we can calculate the minimum number of fields needed to achieve this desired precision. This is given by:

$$\frac{200\, s}{(p\, x)^2} \tag{2}$$

This formula is introduced as =(200*[F15]/([G18] * [F14]))^2 in the cell G19. Now we can see that, for the current value of the standard deviation, we must analyze a total of 44 fields in the microstructure to get a mean with 5% of confidence. However, we must note that the standard deviation will certainly change as additional measurements are carried out. As the minimum number of measurements necessary to a given precision is directly proportional to the standard deviation, it also can change as more data is input to the worksheet. However, for a sufficient number of analyzed fields, this parameter tends to converge to a constant value. Thus, it is advisable to periodically calculate the number of minimum fields to be analyzed until it becomes relatively constant. For every twenty fields examined usually suffices. If we do not periodically check this trend, we can eventually waste time and work analyzing extra fields and generating a result more precise than we really need.

Finally, we can introduce into the cell G20 the formula =COUNT(B16:B25), which calculates the total number of fields analyzed. Optionally, for documentation purposes, we can also place the date and hour of calculation on the worksheet. This can be done introducing the formulas =TODAY() and =NOW() into the cells G7 and G8, respectively. Now our analysis is complete.

Now, one can ask, "Why 5% precision? Why not, for example, 10%?". Well, as we can see from the formula for the minimum number of fields necessary for a given precision (cell G19), this parameter is inversely proportional to the square of the given precision. Therefore the selection of the required precision is a balance between the accuracy desired in the mean value of percentage of second phase on one hand and the work necessary to get it on the other. Sometimes an improvement in the mean precision from 5% to 1% implies in an increase from hours to weeks of work! In practice, the normally accepted precision of 5% is a compromise value which can usually be achieved without exaggerated effort.

Now let us have a look at the effect of varying the required precision on the minimum number of fields to be analyzed. We can use the Scenario Manager option of Excel. It can be opened by clicking **Tools** and then **Scenario Manager**. We are presented with three options. In the first – **Best Precision** – the required precision is 1%. The accuracy of the results is very great; however, the number of required measurements is very high: 1088. The second option is the **Optimized Case**, that is, our default case, the required precision of 5%, condition which asks for 44 fields to be analyzed. In the **Non-recommended Case** the required precision is set to 10%. If we apply this scenario to the worksheet, we can see that the number of minimum fields is only 11. That is, when the required precision changes from 10 to 5%, the accuracy of the results doubled and the number of minimum fields to be analyzed shows a 4-fold increase. However, when the required precision changed from 5 to 1%, the accuracy improved 5 times, but the number of fields to analyze suffered an increase of almost 25 fold! Considering these three options, a precision of 1% is the ideal case, but the workload makes this goal impractical. A required precision of 10% requires minimum work, but generally the result is not accurate enough. Therefore, considering the practical limitations of metallurgical research, a precision of 5% in the results is generally enough and the workload is bearable.

Now our example is complete. However, suppose that you have to determine the fraction of second phase in some dozens, or even hundreds of metallographic samples. This is a common situation if you are working in research. Repeating again and again this worksheet calculation procedure can be very boring – copying and pasting the formulas to calculate the percentage of second phase in the B column, and then adjusting the cell sequences in the mean, standard deviation and precision formulas in the cells F14, F15 and F16 for each time you want to calculate the results of your measurements... Well, metallographic analysis can be boring, but your calculation procedure needn't be! Here we've automated it by developing an Excel VBA macro to do the job.

This macro is called **volFr** and can be activated by the buttons present in the worksheet. The right button – **Clear** – as its name says, clears the worksheet of all previously processed data, preparing it for a new calculation. It is the first step before any new calculation. Now, you can introduce your data. Although sample identification and magnification are optional, it is recommended to include them for documentation purposes. The total number of line intersections, in the C7 cell, is obligatory. Then you must input the data obtained from the metallographic analysis of the fields, that is, the number of line intersections that coincided with the grains of second phase, from the A16 cell downwards. After that, you are ready to execute the macro, pressing the left button, called (of course...)

Calculation. This macro initially will check your data, executing the VBA code listed below:

```
Do While Cells(counter, "A") <> ""
        Cells(counter, "A").Font.Color = RGB(0, 0, 0)
If Not IsNumeric(Cells(counter, "A")) Then
            flag = True
            Cells(counter, "A").Font.Color = RGB(255, 0, 0)
            Cells(counter, "A").Value = "Error!"
        End If
        counter = counter + 1
    Loop
If flag Then
        Beep
        bf1 = "Incorrect or missing data were detected."
        bf2 = "Correct the cells written in red!"
        MsgBox bf1 & Chr(13) & Chr(13) & bf2 & Chr(13),, "Warning"
        Exit Sub
    End If
```

Note that all data is examined until a cell with a null character is found. If a non-numeric cell is found, the variable **flag** is set to "true", while the message **Error!** is written in red in this cell. Then the macro will stop the calculation procedure, issue a beep and display a warning box.

If your data is OK, then the macro will format it, including the use of borders. This is done by the following VBA code fragment:

```
' All data is OK; program continues, generating
' a border around the grid intercepts data.
'
    str = "B16:B" & counter
    Range(str).Name = "Results"
    Range("A16:B" & counter).Select
    With Selection.Borders(xlLeft)
        .LineStyle = xlDouble
        .ColorIndex = xlAutomatic
    End With
    With Selection.Borders(xlRight)
        .LineStyle = xlDouble
        .ColorIndex = xlAutomatic
    End With
    Selection.Borders(xlTop).LineStyle = xlNone
    With Selection.Borders(xlBottom)
        .Weight = xlThin
        .ColorIndex = xlAutomatic
    End With
```

```
        Selection.BorderAround LineStyle:=xlNone
        Range("A" & counter & ":B" & counter).Select
        With Selection.Borders(xlBottom)
            .LineStyle = xlDouble
            .ColorIndex = xlAutomatic
        End With
        Selection.BorderAround LineStyle:=xlNone
    '
    ' Grid intercepts data is centered...
    '
        Range("A16:B" & counter).Select
        With Selection
            .HorizontalAlignment = xlCenter
            .VerticalAlignment = xlBottom
            .WrapText = False
            .Orientation = xlHorizontal
        End With
    '
    ' ...sorted...
    '
        Range("A16:A" & counter).Select
        Selection.Sort Key1:=Range("A16"), Order1:=xlAscending, Header:= _
            xlGuess, OrderCustom:=1, MatchCase:=False, Orientation:= _
            xlTopToBottom
    '
    ' ...and formatted.
    '
        Range("Results").Select
        Selection.NumberFormat = "0.0%"
```

The variable **counter** contains the number of the line which contains the last data set introduced in the worksheet.

Note that all worksheets in this chapter, except the last one, include similar tests for the validity of the input data format. Now the macro will perform all the necessary calculations. After that, you can save the results, if you wish. If you need to add more data, you can simply add it below the last input value in the A column. During the next execution, the macro will automatically detect the last data elements added, re-format the worksheet and perform a new calculation including the additional information. To begin a completely new calculation, press **Clear**, as said before. Note that the use of Clear, erasing all data previously introduced in the worksheet, automatically transforms an example worksheet into a "virgin" worksheet, identical, in this case, to VOLFRAC.XLS.

Before we turn to another topic, a couple of general words on the VBA macros presented here:

- Readers familiar with earlier spreadsheet macros like those in Lotus 1-2-3 may be missing the {WINDOWSOFF} command which switches off screen updating and thus disturbing flickering during macro execution. The VBA equivalent of this is Application.ScreenUpdating = False. It has intentionally **not** been used in all our examples. We leave it to the reader to experience for himself the reason for the command. You'll know it when you see it!!
- In those worksheets involving charts too large or impractical to fit on the screen we have included a View Graph button for the readers convenience. In most cases however we do not bring him back to the initial menu for a new calculation. We leave it as an exercise to write and include the code for this. Clue: find the macro code attached to the View graph button and edit that.

Another example taken from quantitative metallography, very similar to this one, is the determination of the grain size of a microstructure. In this case, we must count the number of intersections (P_L) between a test line of a given length L_T and the grain boundaries of the microstructure. For example, in the right side of the schematic microstructure of Fig. 1 we can see a circle. We can count approximately twenty intersections between the grain boundaries (disregarding the phase to which the grain belongs) and that circle. From that information we can calculate the mean diameter of the grains, that is:

$$\frac{L_T}{P_L \, M} \tag{3}$$

where M is the microscope magnification used in the observation of all fields. More information about this kind of measurement can also be found in [8,9]; the ASTM E-112 standard is a good guide to such procedures. This example can be seen in the worksheet GRSIZEX.XLS, included on the diskette The corresponding "virgin" version is called GRSIZE.XLS. It will be not discussed here, as it is very similar to the previous worksheet VOLFRAC.XLS. For practice, check the differences between them, including the VBA macros.

Material Toughness Calculation

Toughness is defined as the total energy applied to a material to cause its rupture. As usually measured it is taken to be the area under the stress-strain curve determined by, for example, an impact test. A common example is the Charpy impact test. A heavy pendulum of weight W is raised up to a point corresponding to a height h_1. The pendulum is then released, colliding with a notched sample of the material being tested, at the point at which the pendulum has maximum velocity. Generally the sample breaks due to the impact. After the collision, the pendulum still has enough kinetic energy to maintain its trajectory up to a point corresponding to a given height h_2. Obviously, h_2 is lower than h_1, as some energy was spent to break the material sample. This energy can be calculated from the difference of potential energy of the pendulum between these heights. Some versions of the test use a height h_i instead of h_1; h_i is defined as the height that the pendulum reaches after its release with no sample in the machine, that is, without any impact. This way the inevitable friction losses in the machine can be automatically compensated for in the calculation.

A more sophisticated version of this test is the instrumented Charpy test (Kaiser et al. 1992). In this test, adequate instrumentation and fast data acquisition systems collect the evolution of the load absorbed from the pendulum along the displacement of the material sample. The curve so obtained is very useful in fracture mechanics studies, as it allows the characterization of the nucleation and propagation behavior of the crack that breaks the sample. This information is vital, for example, for the correct dimensioning of a submarine hull or a gas pipeline in the Arctic, both situations where toughness requirements are very stringent.

This curve also allows the calculation of the energy absorbed along the displacement of the material sample, as well as the total energy absorbed up to its fracture. This energy can be determined through the calculation of the area **A** below the curve load **L** versus displacement **x**, which is defined by:

$$A = \int L\, dx \tag{4}$$

The Excel worksheet TOUGH.XLS was designed to carry out this calculation. Let's consider an example version of this worksheet, called TOUGHEX.XLS, which can

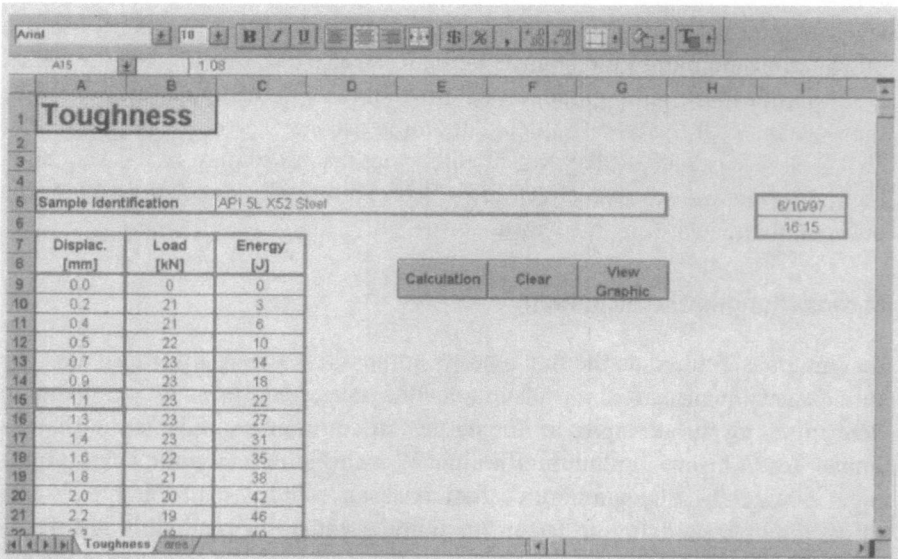

Fig. 3. Worksheet TOUGHEX.XLS, used for the determination of the toughness of materials from the curve load versus displacement

be seen in Fig. 3. Data from the curve load versus displacement, obtained experimentally from an instrumented Charpy impact test, must be supplied in the cells
A9 (displacement, in millimeters) and B9 (load, in kNewtons) downwards. Of
course, there is some space for the identification of the sample, in the cell interval C5:G5. In our case, a steel whose properties are defined by the standard API
5L-X52 is the material being tested; it is a microalloyed, fine grained steel suitable for pipelines and other applications that require good toughness.

Excel does not have a standard function for numerical integration. So we'll have
to implement one for ourselves! A procedure widely used for numerical integration uses the Simpson algorithm, which is described by the following formula
(Dorn and McCracken 1972):

$$A = \frac{h}{3}\left(L_0 + 4L_1 + 2L_2 + 4L_3 + 2L_4 + ... + 2L_{n-4} + 4L_{n-3} + 2L_{n-2} + 4L_{n-1} + L_n\right) \quad (5)$$

where h is the spacing between the abscissa values, which must be kept constant
along all the data sequence.

In our worksheet we created a macro called **SIMPSON**, which calculates Eq. (5).
Its arguments are Visual Basic variables. Of course, this routine could calculate
the value of the integral directly from the values of the worksheet cells. In that case,
however, its code would be somewhat complex, as direct indexing of the cells must
be done through the function RANGE. For example, the variable y(i) in the SIMP-
SON macro is equivalent to the cell reference **Range("B" & i+8)**. In this case it is
simpler to reference the Visual Basic variables, although this results in some waste
of memory. Besides that, the y() array was originally defined to contain up to 1000
elements of data; if the curve load versus displacement being studied results in
more than 1000 pairs of data, the arrays y() and z() would have to be redefined
to comply with this new condition.

When all data is introduced in the worksheet, we can calculate the toughness,
calling the macro by pressing the left button, called **Calculation**, which activates
a macro called **Area**. Unlike the last example, the execution of the macro in this
worksheet is mandatory. The macro initially will check the format of the data; if
non-numeric data is detected, then the computer will beep and a warning box
will appear on the screen. The program will abort. If all the data is OK, then the
worksheet will be formatted. Then, the absorbed energy in the material during
impact – that is, toughness – will be calculated by the SIMPSON macro. This energy will be calculated for every value of displacement available, so we can have the
evolution of the absorbed energy along the displacement. This value will be displayed from the cell C9 downwards.

Now, why not use Excel's graphical resources to display these results? We can
use the ChartWizard source code inside the macro to plot not only the evolution
of load, but also the evolution of the absorbed energy as a function of displacement. For this we use the code below, which is included in the macro called Area
cited above:

ActiveChart.ChartWizard Source:=Range("A7:C" & counter), Gallery:= _
xlXYScatter, Format:=6, PlotBy:=xlColumns, CategoryLabels:=1, _
SeriesLabels:=2, HasLegend:=1, Title:="", CategoryTitle:= _
"Displacement [mm]", ValueTitle:="Absorbed Energy [J]", _
ExtraTitle:=""

We must create a specific ordinate axis for each parameter, as their magnitudes
are somewhat different. This is done using additional commands in the macro.
The chart so generated can be seen on the right side of the worksheet. You can
move to it by pressing the View Graphic button. Finally, we can show the date and
time of calculation in the worksheet, for documentation purposes, in the cells I5
and I6 respectively .

As with the worksheet in the previous section, there is a macro to prepare the
worksheet for calculation. It can be activated by pressing the button titled **Clear**.
This macro, appropriately called Clear, not only clears all remaining values from
previous calculations in the worksheet, but also deletes the previous chart. This
is fundamental to avoid referencing problems during macro execution. A re-calculation using the same data will also delete the previous chart.

Viscosimetric Molecular Weight of Polymers

The molecular weight is a very important parameter in polymer science. Rough-
ly speaking, many polymer properties can be described by a linear relation with
the inverse of the molecular weight **MW** (Billmeyer 1971), that is,

$$PROPERTY = a - \frac{b}{MW} \qquad (6)$$

This is particularly valid for properties depending on the number of ends of polymer chains, like density, refractive index, mechanical strength, and so on.

There are several methods of determining the molecular weight of a polymer,
End group analysis, use of colligative properties or light scattering are only some
examples. However, these tests are generally complicated and expensive.

A relatively simple and low cost method for this purpose is based on the viscosity of a given polymer solution. Solution viscosity represents a measure of the
size of the spatial extension of polymer molecules. There is an empirical relationship between molecular weight and solution viscosity for linear polymers.

In this case, the measurement of the solution viscosity can be carried out by
comparing the efflux time required for a specified volume of polymer solution
to flow through a capillary viscosimeter and the corresponding efflux time for
the pure solvent. The latter time is also known as the standard efflux time. This
data is used in the determination of some viscosity parameters, which, in their
turn, can be used in an empirical equation that allows the calculation of the viscosimetric molecular weight M_v of the polymer.

As we can see, this method does not permit a direct determination of the molecular weight of polymers, as we must have a previously developed equation relating solution viscosity to polymer molecular weight. However, this test is much simpler and more inexpensive than methods used for the direct determination of molecular weight.

Let's see an example of this procedure. Suppose we have six solutions with different concentrations of polymethyl methacrylate (PMMA) in ketone, kept at a constant temperature of $30 \pm 0.2\,°C$. This temperature precision is needed to avoid undesired fluctuations in the efflux time. The capillary tube used for the determination of the efflux time was designed in such a manner that the efflux time for the solvent is equal to 100 s. Note that efflux time is kept relatively long in order to minimize experimental errors. Obviously, the efflux time of the polymer solutions will be greater than for the solvent, and will increase with polymer concentration.

Now we have a set of data with seven concentrations and efflux times of polymer solutions, from 0 to 2.108 g/ml. It is time to introduce the Excel worksheet VISEX.XLS, which can be seen in the Fig. 4. Please note that VISEX.XLS is an example, as it contains data; the "virgin" version of this worksheet, free of any data, is called VISMW.XLS. This worksheet makes the calculations needed in the determination of the viscosimetry molecular weight, which will be described below.

First of all, we must include some basic information about the experiment, such as the name of polymer being analyzed (cells D5 to G5), the solvent used (cells D6

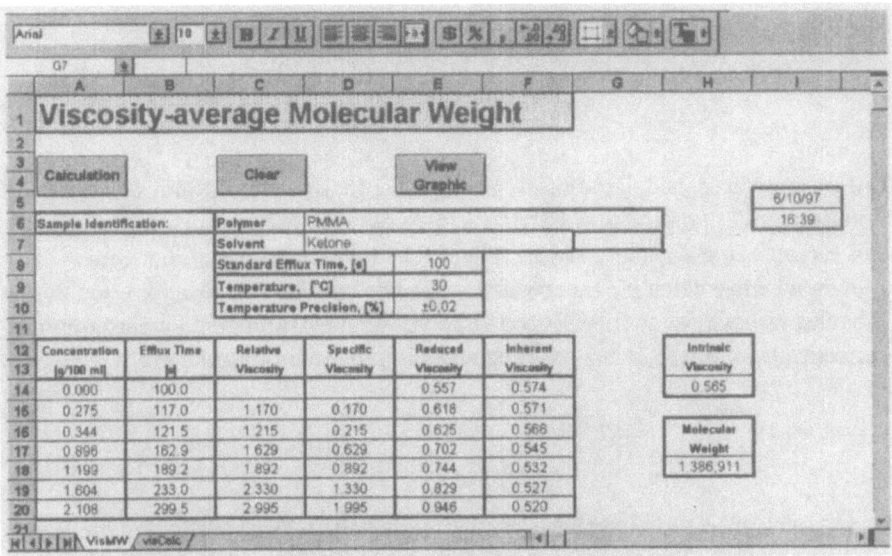

Fig. 4. Worksheet VISEX.XLS, used for the determination of the viscosimetric molecular weight of polymers

to G6), the standard efflux time t_0 (cell E7), the test temperature (cell E8) and the temperature precision (cell E9). This information is very important for documentation purposes.

Next we must enter the concentrations c vs. efflux times t data pairs in the cells A14 and B14 downwards. Note that we must start with zero concentration and standard efflux time.

Now we must calculate the relative viscosity η_r:

$$\eta_r = \frac{t}{t_0} \tag{7}$$

In our worksheet, we begin this calculation introducing the formula =B15/\$E\$7 in the C15 cell. Then, it can be copied and pasted in the cells from C16 to C20.

The next step is to calculate the specific viscosity, η_{sp}:

$$\eta_{sp} = \frac{(t - t_0)}{t_0} \tag{8}$$

Translating to spreadsheet language, this formula must be introduced in the D15 cell as =(B15-\$E\$7)/\$E\$7. Afterwards, it must be copied and pasted into the D16:D20 range.

The next parameter to calculate is the reduced viscosity η_{red}:

$$\eta_{red} = \frac{\eta_{sp}}{c} \tag{9}$$

that is, the cell E15 must contain the formula =D15/A15 and, obviously, it must be copied and pasted into the range E16:E20. At that time the cell E14 must still be empty.

Finally, the inherent viscosity η_{inh} must be calculated:

$$\eta_{inh} = \frac{\ln(\eta_r)}{c} \tag{10}$$

this parameter is calculated firstly in the cell F15; the corresponding worksheet notation for its equation is =LN(C15)/A15. As done before, we must copy and paste this formula in the range F16:F20. The cell F14 is also empty at that time.

Now we must calculate the intrinsic viscosity $[\eta]$. This viscosity is equal to the inherent viscosity or to the reduced viscosity, both extrapolated for zero polymer concentration; hence, it is a concentration independent parameter:

$$[\eta] = \left(\frac{\eta_{sp}}{c}\right)_{c=0} = \left[\frac{\ln(\eta_r)}{c}\right]_{c=0} \tag{11}$$

As the relationships between η_{red} or η_{inh} with polymer concentration are linear, we can calculate the corresponding equations of the straight lines by linear regression. After that, we can calculate $[\eta]$ extrapolating such equations to c = 0. Two values of $[\eta]$ are available, since it can be calculated by two different equations. They can be readily arrived at using the **INTERCEPT** function of Excel, as we can see in the E14 and F14 cells, which contain the intervals E15:E20;A15:A20 and F15:F20;A15:A20 respectively as arguments for this function. Of course, we can also use the average between them as a mean $[\eta]$ value, which is in the H14 cell.

The relationship between $[\eta]$ and the viscometric molecular weight generally assumes the form:

$$[\eta] = K\, M_v^a \tag{12}$$

This is the Mark-Houwink equation. The values of the factor **K** and the exponent **a** are tabulated for several polymers and solvents (Rodriguez 1970). In the specific case of PMMA diluted in ketone, **K** equals 4.8×10^{-4} and a is equal to 0.5. We have the following relationship between molecular weight and $[\eta]$:

$$M_v = \frac{[\eta]^2}{2.304 \cdot 10^{-7}} \tag{13}$$

This calculation is done in the cell H18, using the formula =(H14/4.8E-4)^2. Please note that 2.304×10^{-7} equals $(4.8 \times 10^{-4})^2$.

Additionally we can use the graphical resources of Excel to plot the viscosity values – η_{sp} and $\ln(\eta_r)$ – versus polymer concentration. We plot the straight lines that define the relationships between these parameters by selecting the sequence with the left button of the mouse, then pressing the right button, selecting the option **Insert Tendency Line**, then the **Linear** box and finally **OK**. This procedure must be repeated for both sequences. The corresponding chart for the example above can be seen in Fig. 5.

The procedure described above was programmed in the VBA macro called **Vis-Calc**. To use this macro, you only have to enter the standard efflux time t_0 in the E7 cell and the concentration/efflux time data set from the cells A14 and B14 respectively downwards. However, when you use this macro you *cannot* introduce the zero concentration/standard efflux time in the cells A14 and B14, as the macro will automatically do this for you. That is, the data introduced in the A and B columns must refer only to the solutions with non-zero polymer concentrations.

Fig. 5. Graphic of reduced and inherent viscosities versus polymer concentration generated by the worksheet VISEX.XLS

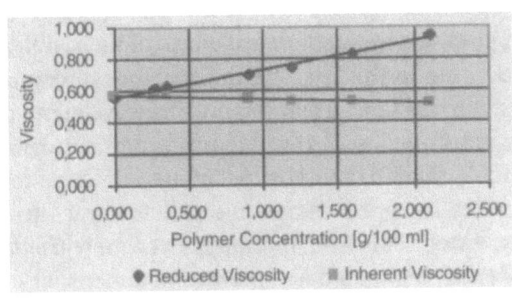

As usual it is advisable to include optional information in the worksheet: the names of the polymer and solvent used, test temperature and its precision. These data are not necessary for the calculations, but they can be of great value if the results of a test turn out to be strange.

Now you can press the button **Calculation**, in the left side of the worksheet. All the calculations will be carried out and the chart will be automatically generated. In a similar way to earlier examples, this macro first checks the validity of the data entered into the worksheet, automatically identifies the end of the data set, formats the report and calculates and plots the results, including the tendency lines. The basic graph – plotting of the results – was done using commands of the ChartWizard; the tracing of the tendency lines was carried out with additional commands. Finally, the macro inserts the date and time of calculation into the worksheet.

If you have to re-calculate the worksheet, it will automatically recognize the zero concentration point in the A14/B14 cells that was added by the macro in the previous calculations. This way you do not have to delete this point before the re-calculation.

To start a brand new calculation, press the button **Clear** on the right side of the worksheet. All previous data and calculations will be erased, and you will be ready to enter a new set of data.

A possible further development for this worksheet could be the inclusion of a small database of the constants **K** and **a** of the Mark-Houwink equation. Using these the macro could automatically identify the correct values of these constants from the names of the polymer and the corresponding solvent.

Identification of Phases using X-Ray Diffraction

The diffraction of visible light is a well known physical phenomenon; in fact, it is an experiment included in almost all basic science courses. Normally a diffraction grid is used in this experiment; its spacing must be nearly equal to the visible light wavelength, that is, 5500 Å. This is necessary to satisfy the physical requirements of this phenomenon.

The crystalline structure of materials is also able to diffract electromagnetic radiation, in a similar way as the diffraction grid used in the visible light diffraction experiments. However, as the spacing between the crystallographic planes is much shorter than the wavelength of the visible light, this kind of radiation can not be used in this case. As the spacings of crystallographic planes are about 1 Angstrom, we must use radiation of a wavelength close to this value. This requirement is fulfilled by X-rays, electrons and neutrons (Guy 1976).

The X-ray diffraction experiment allows us to measure the spacing between the crystallographic planes and to determine the crystal structure of a substance being tested. This information is extremely useful for materials studies, mainly for identifying the phases present in a material.

In one of the most popular variants of X-ray diffraction experiments a sample is placed in the path of a monochromatic beam of X-rays. The sample must be finely powdered, so it contains a multitude of very small, randomly oriented crystals. This is equivalent to having planes of atoms at all possible angles to the incident radiation. The diffracted X-rays will be scattered only according to specific angles, depending on the material being analyzed. This occurs because when the X-ray beam strikes two specific (**hkl**) planes of a crystal (**h, k** and **l** being the Miller indexes of these planes), the wave reflected by the second plane has a slightly longer path. The additional length that this beam must travel is determined by the expression

$$2\,d\sin(\theta) \tag{14}$$

where **d** is the spacing between the atomic planes and θ is the angle between the X-ray beam and the atomic plane. For the larger part of θ angle values, the waves from the two adjacent plans will be out of phase and will be annulled by destructive interference. However, if the additional length of the second wave is equal to the wavelength λ or a multiple of it (2λ, 3λ, etc.), then the waves will be in phase. Mathematically speaking, in those cases we have

$$n\lambda = 2\,d\sin(\theta) \tag{15}$$

This equation is known as the Bragg law; **n** is an integer value known as the diffraction order.

As a crystalline structure contains different (**hkl**) planes, there are several X-ray beams diffracted at different values of the angle θ, one beam for each plane. These values of θ can be determined through the use of an X-ray detector. This moves along a semi-circle whose center coincides with the sample location. The angle defined by the original X-ray beam and the diffracted beam is equal to 2θ.

After a typical X-ray diffraction experiment, we will have a list of values of the angle θ (or 2θ) at which a peak in the intensity of the diffracted X-ray beam was detected and the intensity of this diffracted beam.

In the case of cubic crystals we have the following relation between crystallographic plane spacing **d** and lattice parameter **a**:

$$d^2 = \frac{a^2}{(h^2 + k^2 + l^2)} = \frac{a^2}{Q^2} \tag{16}$$

so, for the particular case of cubic crystal, the Bragg equation becomes, with some rearranging,

$$\sin^2(\theta) = \frac{\lambda^2}{4a^2}(h^2 + k^2 + l^2) = C\,Q^2 \tag{17}$$

Well, as we can verify in Table 1, which is specific for cubic crystals, each type of crystal can contain only some specific types of crystallographic planes, that is, each crystal type has a specific sequence of $(h^2+k^2+l^2)$ values. The sequence relative to the simple cubic crystal structure has all possible values of $(h^2+k^2+l^2)$, while the diamond cubic, for example, has the following sequence: 3, 8, 11, 16, 19 and so on.

The value of C in Eq. (17) must be constant for the various values of θ detected in the diffraction experiment. In this way, we can identify the cubic crystalline system of a unknown material submitted to X-ray diffraction by calculating the value of C for each angle θ and the corresponding Q^2 values, for each system listed in Table 1. The cubic crystallographic system of the material being analyzed will be that one where the values of C are approximately constant.

Now it is time to introduce to the stage the worksheet DIFFREX.XLS, which can be seen in Fig. 6. Once again, note that this worksheet contains data to illustrate its use. Its "virgin" version, with no data, is called DIFFRAC.XLS. Now let's see how it works.

Of course, firstly this worksheet allows us to store some information about the material being tested. In the C5 to G5 cell range there is some room for sample identification. In the E6 cell we must enter the wavelength λ of the X-ray being used in the test; as we saw before, this parameter is vital for the calculations. In the E7 cell must be entered the maximum intensity of the original X-ray beam,

Table 1. Diffraction lines allowed for cubic crystals (Subbarao et al. 1976)

(hkl)	Q^2 $(h^2+k^2+l^2)$	Simple Cubic	Body Centerd Cubic	Face Centerd Cubic	Diamond Cubic
100	1				
110	2				
111	3				
200	4				
210	5				
211	6				
220	8				
300	9				
221	9				
310	10				
311	11				
222	12				
321	14				
400	16				
322	17				
410	17				
330	18				
411	18				
331	19				
420	20				

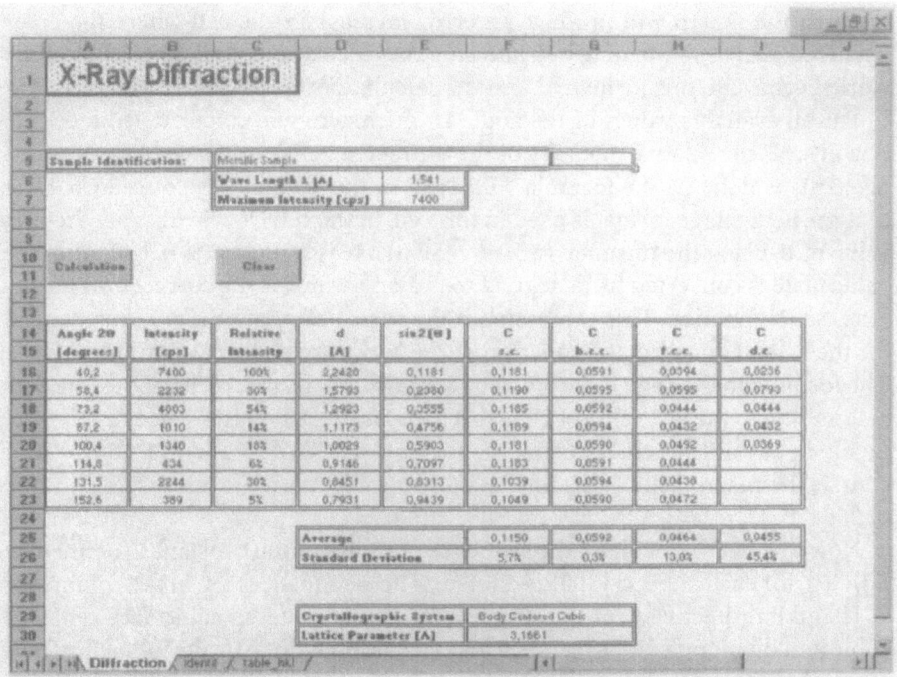

Fig. 6. Worksheet used for the identification of crystalline cubic systems using X-ray diffraction, DIFFREX.XLS

in cps units. This information is not directly used in the calculations, but should be present to characterize the conditions under which the experiment was carried out.

Now we must enter the data provided by the test itself. From the cell A6 downwards we must enter the values of the angle 2θ where diffracted X-rays with significant intensity were detected. These intensities are entered in the cell B6 downwards.

Now we can call a macro called **Identif**, which will perform all the calculations necessary. This can be done by pressing the left button on the worksheet, called **Calculation** (this is becoming somewhat monotonous, is it not?).

First of all, the macro will create a matrix called **hkl** with the data listed in Table 1. For the sake of clarity, this data was stored in a separate worksheet, called **table_hkl**, in the form of sequences of the values of $(h^2+k^2+l^2)$ for each cubic system. For each column, lines 1 and 2 are used to identify the crystalline system; line 3 stores the total number of diffraction lines available for that crystalline system in this worksheet; in the remaining lines below there are the specific Q^2 values for that crystalline system.

After that, the macro checks the validity of the values of X-ray wavelength, maximum intensity of beam, angle 2θ and relative intensity. If any errors are detected, then the macro will produce an error message in the cell where the error occurred, display a warning box and stop the calculation. If all data input in the worksheet is OK, it is formatted and the calculations begin.

For all available values of the angle 2θ the macro will calculate from cell C6 downwards the relative intensity of the diffracted beam, expressed as a percentage. This is done by the formula =B6/$E$7 in the specific case of cell C6. Note that the percentage format is used in this cell. In the cell D6 we calculate the first value of d, using the formula =$E$6/(2*SIN((A16*PI())/(2*180))). Note that the value of 2θ is converted from degrees to radians, which is the correct unit for the Excel's SIN function. This expression is the basic Bragg law, Eq. (14), rearranged for the calculation of **d** as a function of the X-ray wavelength λ and the angle θ. The formula in the cell D6 is copied downwards by the macro, until the line corresponding to the last value of 2θ is reached. The next step is to calculate the value of $\sin^2(\theta)$ in the cell E6, using the formula =SIN((A16*PI())/(2*180))^2. Of course, the macro copies this formula downwards, until the last available value of 2θ.

Now the macro calculates the sequence of values of the constant C, defined by Eq. (17), for each cubic crystalline system. These sequences are in the columns F, G, H and I, respectively for the simple cubic, body centered cubic, face centered cubic and diamond cubic systems. This calculation is done by the following fragment of VBA code, included in the Identif macro, which assembles the corresponding equations:

```
' Loop for the four cubic crystalline systems

For i = 1 To 4

' Offset compensation between worksheet and matrix hkl
    iend = counter – 15
' If the number of Q² values in the array hkl is lower than the number of
' diffracted beams, then assume that the maximum number of cases is
' equal to the minimum value between the two.
    If iend > hkl(i, 1) Then iend = hkl(i, 1)
' The value of i determines the crystalline cubic system being analyzed,
' and also determines the correct location of the result of the constant C:
' 1 = simple cubic, column F; 2 = body centered cubic, column G;
' 3 = face centered cubic, column H; 4 = diamond cubic, column I
    Select Case f
        Case Is = 1
            bf1 = "F"
        Case Is = 2
            bf1 = "G"
        Case Is = 3
            bf1 = "H"
        Case Is = 4
```

bf1 = "I"
End Select
' Loop for the data (sequence of values of the 2θ angle)
    For j = 2 To iend + 1
' Another offset calculation between worksheet and matrix hkl
    pont = j + 14
' Effective calculation of the constant C
    Range(bf1 & pont).Value = Range("E" & pont).Value / hkl(i, j)
    Next j
Next i

After the calculation of the sequence of values of the C constant for each crystalline system, the macro calculates the mean value for each crystallographic system, as we can see in the F25:I25 cell range. This is done using the Excel function **AVERAGE**.

As we saw before, the crystallographic system of the unknown sample being analyzed must show an approximately constant value of C. We can identify this system through the calculation of the standard deviation, expressed in percentage, relative to each sequence of C values. The system with the minimum value of this standard deviation is the answer to our question. For example, in the case of the simple cubic system, this statistical parameter is calculated in the cell F26 using the equation =STDEV(F16:F23)/F25. Obviously, we use the percentage format in this cell. The sample procedure is also carried out for the other crystalline systems, in the cells G26, H26 and I26.

Finally the macro compares the values in the cells F26:I26 and chooses the minimum of this range. In our case, the minimum value corresponded to the G column, that is, the substance being analyzed has a cubic centered cubic crystallographic system. The name of this system is written in the cells F29 to G29. After that, the macro calculates the corresponding value of the lattice parameter a as a function of the X-ray wavelength λ and the mean value of the constant C, using a rearranged version of Eq. (17):

$$a = \frac{\lambda}{2\sqrt{C}} \tag{18}$$

Finally, the macro inserts date and hour of calculation in the cells I5 and I6, respectively, and ends.

If desired, this macro can be re-executed with additional data. For a completely new calculation, press the **Clear** button on the right side of the worksheet.

This worksheet can be modified or extended to include analysis of other crystalline systems, like hexagonal, orthorhombic, tetragonal, and so on. The necessary information can be found in other references, such as (Weiss 1963) and (Mayr 1981).

Solubilization of Microalloyed NbTi Steels

In the section on Material Toughness Calculations we studied the response of an A.P.I. 5L-X52 microalloyed steel to an impact test. These kinds of steels, normally produced in the form of hot rolled plates and strips, present high values of mechanical resistance and toughness, two characteristics hardly found at the same time in a single material. This fact makes them suitable for use in shipbuilding, offshore platforms, pipelines, bridges and other applications that have severe requirements on mechanical resistance and toughness, particularly at low temperatures.

These unusual properties are due to the presence of minute amounts of some alloy elements, like niobium, vanadium or titanium, in amounts up to 0,050% in weight. One or more of these elements can be added to the alloy. During the so-called controlled hot rolling of microalloyed steel plates or strips, these elements – particularly niobium – can hinder the austenite recrystallization. This action can be promoted by the element itself, since it is solubilized in the austenite, or by its carbonitrides that precipitate from austenite during rolling, as the temperature is continuously decreased during this operation. The absence of recrystallization refines the grain of the final product, and it is well known that as the grains of a material became finer, both mechanical resistance and toughness increase simultaneously. In fact, grain refining is the only hardening mechanism that increases these two contradicting properties.

In addition the remaining solubilized microalloying elements can precipitate in the ferrite phase during the slow cooling of the hot rolled plate or strip, increasing its mechanical resistance by precipitate hardening. In this case though the toughness of the material can be somewhat impaired.

However, in order to achieve these desirable effects, a significant amount of microalloyed elements must be solubilized during the steel reheating operation that precedes hot rolling. The effect of microalloying elements on austenite recrystallization can only occur if they are previously solubilized in this phase, that is, they must form a solid solution with iron at high temperature. Therefore, it is necessary to solubilize the former precipitates existing in the steel slab before its hot rolling to plate or strip. The determination of the relationship between steel austenitization temperature and the effective amount of solubilized microalloying elements is of greatest importance to the calculation of the correct parameters for the reheating process that precedes the controlled rolling.

As this thermodynamic model must deal with two kinds of precipitates (carbides and nitrides) and eventually with more than one microalloying element, it can become mathematically very complex. The example to be seen in this section will deal with microalloyed steels containing niobium and titanium. Later we will see how to extend this procedure for other steels with more than two microalloy elements.

First of all, we must define the scope of our model. It must determine the fraction and equilibrium composition of the carbonitride precipitates, as well as of the austenite, when considering an alloy composed of iron with small amounts of niobium, titanium, carbon and nitrogen.

This model will be based on solubility data for the primary carbides and nitrides, available in the literature. Dilute solution behavior will be assumed, that is, the solute activities in the austenite can be described by Henry's law. The interactions with other solute atoms eventually present in commercial steels, like manganese, for example, will be considered negligible. They will be disregarded.

As we will consider the presence of two microalloying elements – niobium and titanium – both of which react with carbon and nitrogen, we will assume that the resulting carbonitride will have the formula $Nb_xTi_{(1-x)}C_yN_{(1-y)}$. This is perfectly legitimate since the precipitates are stoichiometrically perfect (that is, they can not contain vacancies) and since the solubility of iron in these compounds is neglected. The atomic fractions of the carbonitride elements must be positive and lower than the unity. In other words the following conditions hold:

$$0 \leq x \leq 1 \tag{19}$$

$$0 \leq (1-x) \leq 1 \tag{20}$$

$$0 \leq y \leq 1 \tag{21}$$

$$0 \leq (1-y) \leq 1 \tag{22}$$

That is, we will assume that the atoms of niobium, titanium, carbon and nitrogen will be independently mixed in two sublattices which have the same crystallographic structure, similar to the sodium chloride type.

The deduction of the thermodynamic model adopted here can be seen in references (Speer et al. 1987; Adrian 1992). Firstly, we have a system of four non-linear simultaneous equations that describe the thermodynamic equilibrium of the system Fe-Nb-Ti-C-N:

$$x \ln \frac{x\,y\,K_{NbC}}{[Nb][C]} + (1-x)\ln \frac{y(1-x)\,K_{TiC}}{[Ti][C]} + (1-y)^2 \frac{L_{CN}}{RT} = 0 \tag{23}$$

$$y \ln \frac{x\,y\,K_{NbC}}{[Nb][C]} + (1-y)\ln \frac{x(1-y)\,K_{NbN}}{[Nb][N]} + y(1-y)\frac{L_{CN}}{RT} = 0 \tag{24}$$

$$x \ln \frac{x(1-y)\,K_{NbN}}{[Nb][N]} + (1-x)\ln \frac{(1-x)(1-y)\,K_{TiN}}{[Ti][N]} + y^2 \frac{L_{CN}}{RT} = 0 \tag{25}$$

$$y \ln \frac{y(1-x)\,K_{TiC}}{[Ti][C]} + (1-y)\ln \frac{(1-x)(1-y)\,K_{TiN}}{[Ti][N]} + y(1-y)\frac{L_{CN}}{RT} = 0 \tag{26}$$

Well, suddenly things got really big! The meaning of the variables are as follows:

- K_{NbC}, K_{NbN}, K_{TiC}, K_{TiN}: solubility products of the respective precipitates, expressed in terms of the atomic fractions of the solubilized elements;

- L_{CN}: regular solution parameter associated to the mixture of carbides and nitrides. In the case of the titanium compounds, it was experimentally determined to be –4260 J/mol. The same value was assumed for the other microalloying elements, as no experimental data is currently available:
- R: Gas constant;
- T: Temperature.

This system of non-linear equations has six unknown quantities, that define the austenite composition ([Nb], [V], [Ti], [C] and [N] contents) and that of the carbonitride (x and y). Additional equations can be developed using the mass balance; they will allow us to solve the system. They are listed below:

$$Nb = \frac{x}{2} f + (1-f)[Nb] \tag{27}$$

$$Ti = \frac{(1-x)}{2} f + (1-f)[Ti] \tag{28}$$

$$C = \frac{y}{2} f + (1-f)[C] \tag{29}$$

$$N = \frac{(1-y)}{2} f + (1-f)[N] \tag{30}$$

where f is the molar fraction of the carbonitride in the austenite.

Data available for solving this system are the nominal steel composition (**Nb, Ti, C** and **N**), the austenitizing (reheating) temperature and the solubility products K_{MX}, corresponding to the binary compounds MX (that is, NbC, NbN, TiC and TiN). The solubility products are related to the atomic fractions of the solubilized compounds, according to the general equation:

$$\log_{10}([M][X]) = B - \frac{A}{T} \tag{31}$$

and so:

$$K_{MX} = [M][X] = \frac{(Fe)^2}{10^4 (M)(X)} 10^{\left(B - \frac{A}{T}\right)} \tag{32}$$

where

- M: Nb or Ti amounts;
- X: C or N amounts;
- A and B: Constants of the solubility equation; its values, for several carbides and nitrides of microalloying elements can be seen in Table 2 (Adrian 1992);
- (Fe), (M), (X): atomic masses of the respective elements.

Table 2. Solubility product data for carbides and nitrides in austenite [17]

Compound	A	B
VC	9500	6.72
VN	7070	2.27
TiC	7000	2.75
TiN	8000	0.322
NbC	6770	2.26
NbN	10230	4.04

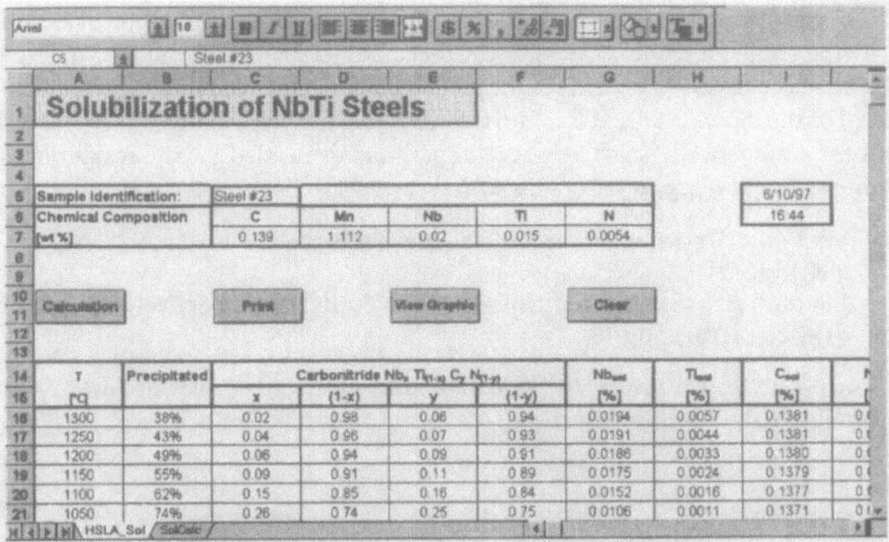

Fig. 7. Using NBTIEX.XLS to determine the microalloying solubility of elements in austenite as a function of temperature. View of the "input/output section"

Well, enough is enough! We have quite a problem on our hands. Let's see our tool to solve it – the worksheet NBTIEX.XLS. As with earlier examples, this worksheet contains some data used as an example; its "virgin" version, with no data included, is called NBTI.XLS. Fig. 7 shows the "input/output section" of NBTIEX.XLS. We can see in this figure that in the range C5:G5 we have our by now traditional space for the identification of the steel being studied. In row 7 we must enter its nominal chemical composition, in terms of the amount (weight percent) of C (cell C7), Mn (cell D7), Nb (cell E7), Ti (cell F7) and N (cell N7). The Mn content is not directly required in the calculation, as the model considers only a pure Fe-C-Nb-Ti-N system. However, as its content in real steels is significant, it is advisable to consider it in the calculation of the atomic fraction of the elements involved in the calculations.

Below the cell range intended for chemical composition, we can see a table where the final results of a successful run have been printed out. We will refer to this table later.

Now we are ready to see how this worksheet works, using the example of a specific steel composition, as we can see in Fig. 7. This is a low C-Mn-Nb-Ti steel, normally used for shipbuilding plates. We will use this worksheet to calculate the solute amounts of C, N, Nb and Ti as a function of temperature, more specifically, between 1300 and 800 °C, at intervals of 50 °C.

Now let's see the "work horse section" of the worksheet, where the calculations are performed. This section was isolated from the "input/output section" in order to simplify the lay-out of the output report. The calculation section is located in the right of the M column of the worksheet, and it can be seen in Fig. 8.

As the initial step to calculate the microalloy solubility as a function of austenitizing temperature, let's calculate the thermodynamic equilibrium for this specific steel at 1300 °C. We will see the motive for the choice of this temperature later. For the moment we enter it into the cell N15. Then we will note that the worksheet automatically calculates several parameters needed for the resolution of the system of non-linear equations, that is:

- the atomic fractions of C, N, Nb and Ti, respectively in the cells M12, N12, O12 and P12;
- the product solubilities of NbC, NbN, TiC and TiN, respectively in the cells M18, N18, O18 and P18;

Now we are ready to solve the system of non-linear equations. But how? As it is composed of non-linear equations, it can only be numerically solved. This gives

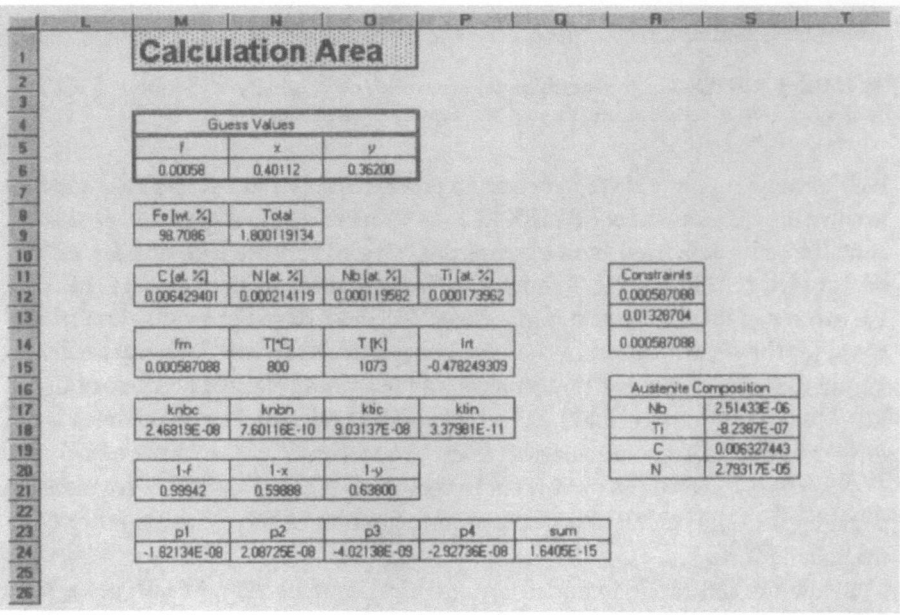

Fig. 8. Calculation section of the NBTIEX.XLS worksheet

us a good opportunity to use the **Solver** function of Excel. We can call it now, clicking on **Tools** and then **Solver**. The **Solver Parameters** window will appear in the screen, similar to that in Fig. 9. Now we can see how to use this powerful feature of Excel.

As you may know, the **Solver** function of Excel is traditionally used to calculate the value of variables to maximize or minimize a given function. We have a similar case here. Our system of equations is solved when all equations are equal to zero. We can define a general function, that is equal to the sum of Eqs. (23) to (26). The system of equations can be considered solved when the result of this general function is zero (or, at least, very near to that value!). This general function will be minimized by the **Solver** function. Translating this procedure to worksheet language, the Eqs. (23) to (26) were transposed to the cells M24 to P24. The cell Q24 contains the general function, that is, the sum of the squares of the Eqs. (23) to (26). The results of these equations were squared in order to keep them positive. This makes Q24 the **Target Cell** of **Solver**, as we can see in Fig. 9. Next, we must inform **Solver** that the required value of this cell is zero; this can be done setting Value of to zero. An alternative would be to use the **Min** option.

Now we must tell **Solver** which cells can be altered during the process of searching for a solution. Well, we can see that the unknowns in Eqs. (23) to (26) are f, x and y. These variables are located in the cells M6, N6 and O6, respectively. This information must be entered in the **By Changing Cells** box of the **Solver Parameters** window.

As our system was deduced from chemical laws, there are some constraints that must be respected in order to get a physically significant solution. Some of these constraints are quite obvious: f, x and y must be positive and less than one, as

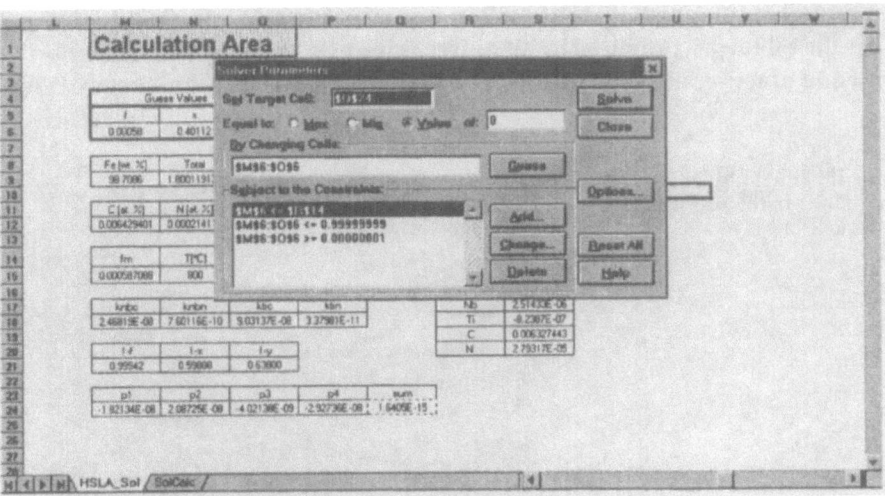

Fig. 9. Solver Parameters window in the NBTIEX.XLS worksheet

they are fractions that define chemical compounds. The last constraint is not so apparent, but it must be expressed to assure the continuity of the equations in our system. The value of f must be equal or less than a value defined by the minimum between 2*(Nb+Ti) and 2*(C+N). We can see these constraints in the box **Subject to the Constraints** of the **Solver Parameters** window, in the Fig. 9.

Now we must define the required precision for our solution. So, let's open the **Solver Options** window, this can be done pressing the **Options** button (obviously...) in the **Solver Parameters** window. The **Solver Options** window can be seen in Fig. 10. We will need a rather high precision in this problem, so we introduce 1E-20 into the **Precision** box. Now that the defining process of **Solver** is done, we can return to the **Solver Parameters** window, by clicking the **OK** button. However, before using **Solver**, we must return to the worksheet for a moment; this can be done by pressing the **Close** button.

Now we are ready to solve the solubilization model for the NbTi steel at a temperature of 1300 °C. However, we still have a serious problem: to define the initial values of f, x and y. Unfortunately, as the method used by **Solver** is iterative, it depends strongly on these values to converge to a satisfactory solution. This is particularly true for this case. Sometimes it is necessary to try several times until we get a set of initial values of f, x and y that can lead to an adequate solution, that is, a solution with the level of precision defined in the last paragraph. We select the highest temperature – 1300 °C – as the first case because, in this situation, the values of f, x and y are very small and with the same order of magnitude. This makes the process of selection of the initial values less arduous.

So – sigh – it is time to solve our system. We must introduce the initial values of f, x and y in their respective cells (repeating, M6, N6 and O6), call **Solver** and try to get a solution, pressing the button **Solve** in the **Solve Parameters** window. Then **Solver** starts to search for a solution to our problem. In most cases, we will not reach a solution with the required degree of precision. In this case, we can use the values just produced by the **Solver** as the next initial values. Generally this is good practice, and soon or later we will find an adequate solution for our first

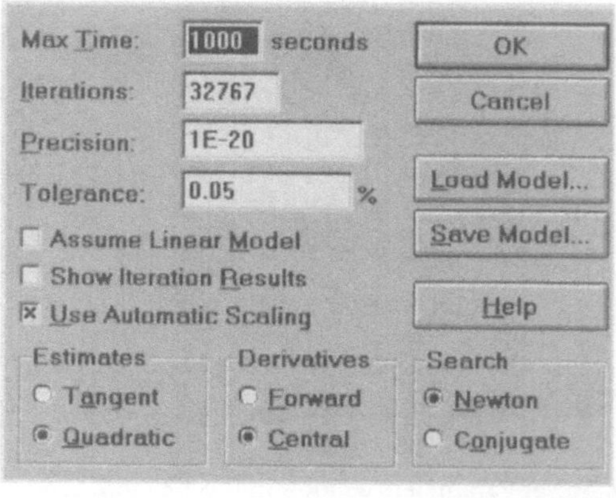

Fig. 10. Solver Options windows in the NBTIEX.XLS worksheet

Fig. 11. Solver Results window in the example of the NBTIEX.XLS worksheet

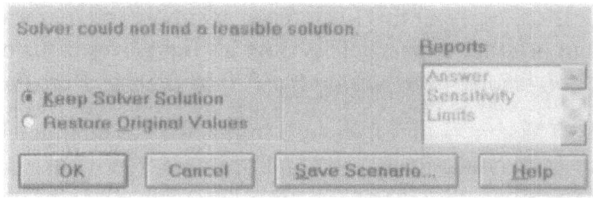

temperature studied. However, it is possible that the use of the values calculated by **Solver** as initial values leads to a deadlock, that is, they are simply repeated by **Solver** at the end of the calculation. In this case, it is necessary to propose new values for **f**, **x** and **y**. This can be cumbersome if you do not have some practice or mystical skills. However, after you've done this calculation for a large number of steels, you can use previous results for similar alloys as initial values – sometimes this saves a lot of time!

Well, finally we see a window similar to that of Fig. 11. Congratulations, we have finally found a good solution! Quick, press the **OK** button before it disappears! We just defined the solubilization equilibrium for the studied alloy at 1300 °C. The problem is, must we repeat these calculations and unlucky guesses for the next ten temperatures to be considered? Happily the answer to this question is no, because now we have (finally) two pieces of good news.

First, once an adequate solution is found for a case, its values can be efficiently used as initial values for the solution of the next lower temperature. That is, now we can use the values of **f**, **x** and **y** that solved the system of equations for the temperature of 1300 °C as initial values for the **Solver** call that will find a solution for the same system at the temperature of 1250 °C. These initial values will always lead to an adequate solution for the subsequent temperature. In the same way, the values of **f**, **x** and **y** that solve the system for the temperature of 1250 °C could be adopted as initial values for the calculation of the equilibrium at 1200 °C, and so on.

Fortunately, our trouble finding the "correct" initial values only occurs for the first temperature to be calculated. After that, we can use the solution of a temperature as initial values for the subsequent lower temperature, and **Solver** will always find the correct solution for the next temperature at the first try.

The second piece of good news is that now we have a macro called **SolCalc** that automatically makes the calls to the **Solver** function for each temperature and generates a complete report in the "input/output section" of the worksheet, as we can see in Fig. 7. In fact, only the first call to **Solver** must be manual, as it is hard to find the correct initial values at the first attempt. Once a good solution is found, this problem disappears, and the rest of the procedure can be done in automatic mode. Our macro solves the system for temperatures from 1250 to 800 °C, calculates the fraction of precipitated compounds as well the carbonitride and austenite compositions. This macro is activated by pressing the button **Calculation** (always the same name!), in the left side of the worksheet.

Note however that life is never as simple and that scarcely is one problem solved but another appears. This time it's the Solver call from inside the macro, this – "*SolverSolve userFinish:=True*" to be precise. Here there are two problems:

- The syntax of Solver calls in macros has changed in recent versions of Excel. This one is for Excel 7 and upwards. You'll have to consult the handbook for your version if you have problems running NbTiEx.XLS.
- For the worksheet to function correctly it is not sufficient only to have Solver installed (via *Tools Solver*). The library file Solver.XLA must also be made available by providing a path to it. You can do this directly from the error message which appears by browsing until you find it. It's usually in a directory with a name like Excel\Library\Solver.

Observe that we invoked the **Solver** from the macro using the command:

SolverSolve userFinish:=True

The **userFinish:=True** option is set to avoid the output of boxes after each **Solver** solution, informing whether the solution is adequate or not. The macro has a internal test for precision; if it is less than 0.001, a warning box is output and the macro stops.

Our worksheet also plots the amount of Nb, Ti and N in solution, as well the fraction of precipitated compounds, versus the austenitizing temperature; this chart can be seen in Fig. 12. It is not generated by the macro, but directly defined in the worksheet using the ChartWizard. You can move to it by pressing the View Graphic button on the worksheet. As we can see from this figure, the Ti solubilized content is low. This is because the most part of its precipitates will only dissolve when steel is in the liquid state. Please remember that the steel being studied is solid at the maximum temperature considered, 1300 °C. Nb is almost totally solubilized at 1300 °C; however, as temperature falls below 1150 °C it gradually precipitates, and about 800 °C it is totally in the combined form.

Fig. 12. Chart of the solubilized amounts of Nb, Ti and N, as well as the precipitated fraction as a function of austenitizing temperature, generated by the NBTIEX.XLS worksheet

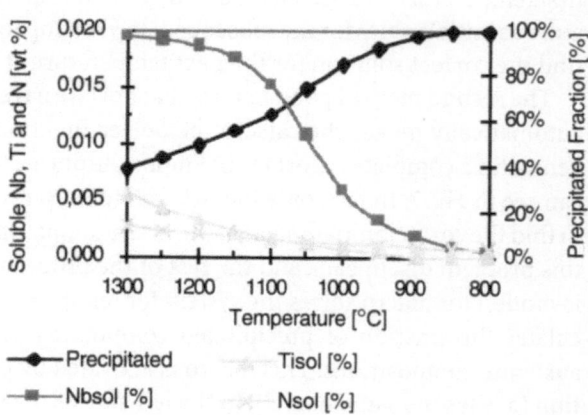

There are two other auxiliary macros: **Print,** which prints the table in the "input/ output section" and the graphical output, and **Clear,** which automatically clears the previous data and calculations, preparing the worksheet for a completely new calculation.

It is interesting to note that this worksheet normally allows the calculation of the solubilization profile as a function of the temperature of a steel to be completed in a few minutes, including the automatic plotting of charts. A previously developed program with this same objective, written using a statistical package, normally took some hours to perform the same task, as the intermediate results had to be re-introduced via keyboard for the different program modules. Besides that, the charts had to be manually done using separate plotting software.

The development of a similar worksheet for microalloyed steels with Nb and V or Ti and V would be simple: the mathematical development is analogous, except for the specific constants used in the solubility product equations. The data for the equations of solubility product of the precipitating compounds (NbC, NbN, VC and VN for the NbV steel; and TiC, TiN, VC and VN for the TiV steel) can be found in Table 2. Better still, you can develop a general procedure for steels with two microalloyed elements, which can identify the elements present from the chemical composition data, and then automatically define which equations of solubility product must be used.

A more ambitious task could be the development of a worksheet to solve the thermodynamic model for steels with three microalloying elements (Nb, Ti and V) or three microalloying elements plus aluminium. The necessary information for these models can be found in the paper by Adrian (Adrian 1992). However, it must be noted that the system of equations to be solved in these cases is more complicated, and new constraints must be applied. However, it will be a quite good challenge to check your newly-acquired **Solver** skills.

As an alternative for the reader's interest we provide another version of this worksheet (NbTiEx2.XLS) on the disk accompanying this book. This version avoids Solver calls from within the macro, making it more robust but infinitely more cumbersome. The Solver must be called manually for each set of new data. See what you think of it and how it can be improved.

Conclusion

In this chapter some basic examples of the use of spreadsheet software in Materials Science have been presented. With the exception of the examples on the determination of viscosimetric molecular weight of polymers and the solubilization profile of NbTi steels as a function of temperature, they can be applied to the three main fields of materials – metals, ceramics and polymers. These examples, however, have only scratched the surface of the possible applications of spreadsheet software in this field.

In fact, as modern laboratory equipment is normally controlled by microcomputers, data acquisition systems often present the collected information in the form of worksheets. It is almost obligatory that this kind of software supports data exportation in files written in Excel format, due to the extreme popularity of this software. This is an additional reason for the use of Excel as a first approach in data analysis – and, in most cases, this analysis will be the only one that is really needed!

Finally, compilations of scientific data now are being published ready for use in spreadsheet software. An example is the Handbook of Corrosion Data, recently edited by ASM International (Anon 1994), which includes a diskette with a corrosion materials performance data base using the Excel format. This is one further sign that commercial spreadsheets, in particular Excel, are really being accepted as a tool for general scientific data analysis.

References

Adrian H (1992) Thermodynamic model for precipitation of carbonitrides in high strength low alloy steels containing up to three microalloying elements with or without additions of aluminium. Materials Science and Technology, (8): 406–420

Anon (1994) Handbook of corrosion data. ASM International, Metals Park. 650 pp

Arganbright W (1983) Scientific applications for spreadsheet programs. Byte Books, New York, 285 pp

Billmeyer Jr FW (1971) Textbook of polymer science. John Wiley, New York, 598 pp

Bunge HG (1982) Texture analysis in materials science – Mathematical methods. Butterworths, London, 1982. 593 pp

Dobson WG, Wolff AK (1984) Engineering problem solving with spreadsheet programs. American Society for Metals, Metals Park.

Dorn WS, McCracken DD (1972) Numerical methods with Fortran IV case studies. John Wiley, New York, 568 pp

Guy AG (1976) Essentials of materials science. McGraw-Hill, New York, 435 pp

Kaiser HJ et al. (1992) Beurteilung der Zähigkeit von Rohrstählen mit instrumentierten Kerbschlagbiegeversuchen und Battelleversuchen. Thyssen Technische Berichte 24:107–112

Mayr M (1981) Bestimmung von Gitterparametern aus Röntgen-Pulverdiagrammen. Radex-Rundschau (4): 682–689

McGee WW, Mattson G (1993) Using an electronic spreadsheet in the design of exercises for a polymer laboratory course. Journal of Chemical Education 70: 756

Rodriguez F (1970) Principles of polymer science. McGraw-Hill, New York, 560 pp

Skaar EC (1994) CAD/CAM Review: Solving problems with spreadsheets. Ceramic Industry 143: 85–86

Speer JG et al. (1987) Carbonitride precipitation in niobium-vanadium microalloyed steels. Metallurgical Transactions 18A: 211–222

Subbarao EC et al. (1976) Experiments in materials science. McGraw-Hill, New York, 236 pp

Underwood EE (1970) Quantitative stereology. Addison-Wesley, Reading, 273 pp

Weiss RJ (1963) Solid state physics for metallurgists. Pergamon, Oxford, 410 pp

Whipp R (1993) Spreadsheet applications for steelmakers. Iron and Steelmaker 20: 35–37

Appendix A: The VBA Programming Language – a Conceptual Overview

J. Walkenbach

This appendix is intended to serve as a brief introduction to Visual Basic for Applications (VBA), the macro programming language that is included with:

- Excel 5
- Excel 95 (also known as Excel 7)
- Excel 97 (also known as Excel 8)

This appendix focuses on Excel 97, but also provides instructions for previous releases.

What is VBA?

Beginning with Lotus 1-2-3, all spreadsheets have included a macro language. In its broadest sense, a *macro* is a sequence of instructions that automates some aspect of working with a spreadsheet. You may create a macro, for example, to apply a specific type of formatting to a cell or range. Such a macro might incorporate several commands – change the font size, foreground color, numerical formatting, border style, and so on. Creating a macro would allow you to execute those commands with a single keystroke. Macros, of course, can be much more complex.

Macro languages have evolved significantly over the years, and VBA is arguably the most sophisticated macro language currently available. Another advantage of VBA is that the language itself is not limited to Excel. VBA is included in all of the Microsoft Office 97 applications, as well as in products from third-party vendors.

VBA was introduced in Excel 5. Prior to that version, Excel used an entirely different macro system known as XLM (that is, the Excel 4 macro language). VBA is far superior in terms of both power and ease of use. For compatibility reasons, however, the XLM language is still supported in later versions of Excel. This means that you can load an older Excel file and still execute the macros that are stored in it.

What You Can Do With VBA

VBA is an extremely rich programming language with thousands of uses. Listed below are a few typical uses for VBA macros:

- **Insert a text string or formula.** If you need to enter your company name into worksheets frequently, you can create a macro to do the typing for you.
- **Automate a procedure that you perform frequently.** For example, you may need to import a file and format it on a regular basis. If the task is straight-forward, you can develop a macro to do it for you.
- **Automate repetitive operations.** If you need to perform some action on twelve different workbooks, you can record a macro while you perform the task once – and then let the macro repeat your action on the other workbooks.
- **Create a custom command.** For example, you can combine several of Excel's menu commands so that they are executed from a single keystroke or from a single mouse click.
- **Create a custom toolbar button.** You can customize Excel's toolbars with your own buttons to execute macros that you write.
- **Create a simplified "front end" for users who don't know much about Excel.** For example, you can set up a foolproof data entry template.
- **Develop a new worksheet function.** Although Excel includes a wide assortment of built-in functions, you can create custom functions that greatly simplify your formulas.
- **Create complete, turnkey, macro-driven applications.** Excel macros can display custom dialog boxes and add new commands to the menu bar.
- **Create custom add-ins for Excel.** All of the add-ins that are shipped with Excel were created with Excel macros. You can create your own add-ins using only the tools supplied with Excel.

An Overview of VBA

VBA is probably the most complex feature in Excel, and it's easy to get over-whelmed. Following is a concise summary of how VBA works:

- You perform actions in VBA by writing (or recording) code in a VBA module and then executing the macro in a number of ways. VBA modules are stored in an Excel workbook, and a workbook can hold any number of VBA modules. To view or edit a VBA module in Excel 97, you must activate the Visual Basic Editor window (press Alt+F11 to toggle between Excel and the VBE window). In Excel 5 or Excel 95, modules appear directly in a workbook.
- A VBA module consists of subroutine procedures. A *subroutine procedure* is basically computer code that performs some action on or with objects. The following is an example of a simple subroutine called ShowSum (it adds 1 + 1 and displays the result in a so-called message box):

```
Sub ShowSum()
    Sum = 1 + 1
    MsgBox "The answer is " & Sum
End Sub
```

- A VBA module also can store function procedures. A *function procedure* returns a single value. A function can be called from another VBA procedure or even used in a worksheet formula. Here's an example of a function named AddTwo (it adds two values, which are supplied as arguments):

  Function AddTwo(arg1, arg2)
     AddTwo = arg1 + arg2
  End Function

- VBA manipulates objects. Excel provides you with well over 100 objects that you can manipulate. Examples of objects include a workbook, a worksheet, a range on a worksheet, a chart, and a drawn rectangle.
- Objects are arranged in a hierarchy, and objects can act as containers for other objects. For example, Excel itself is an object called Application, and it contains other objects such as Workbook objects. The Workbook object can contain other objects such as Worksheet objects and Chart objects. A Worksheet object can contain objects such as Range objects, PivotTable objects, and so on. The arrangement of these objects is referred to as an *object model.* Excel's object model is depicted in the online help system (see Fig. 1).

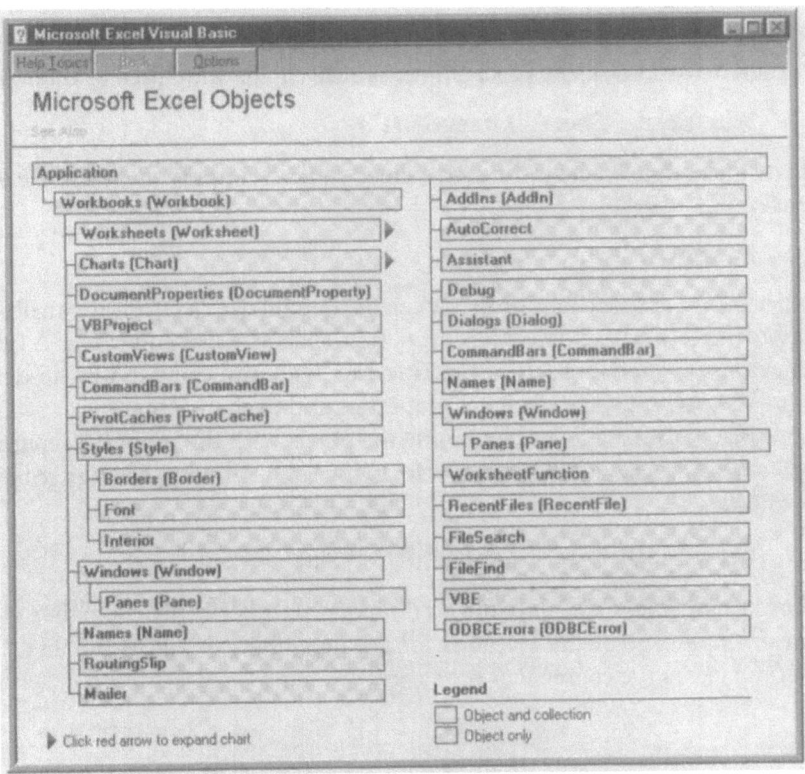

Fig. 1. A depiction of part of Excel's object model

- Like objects form a collection. For example, the Worksheets collection consists of all worksheets in a particular workbook. The CommandBars collection consists of all CommandBar objects (that is, menu bars and toolbars). Collections are objects in themselves.
- You refer to an object by specifying its position in the object hierarchy, using a period as a separator.

For example, you can refer to a workbook named analysis.xls as

        Application.Workbooks("analysis.xls")

- This refers to the analysis.xls workbook in the Workbooks collection. The Workbooks collection is contained in the Application object (that is, Excel). Extending this to another level, you can refer to Sheet1 in the analysis.xls workbook as follows:

        Application.Workbooks("analysis.xls").Worksheets("Sheet1")

- You can take it to still another level and refer to a specific cell (a Range object) as follows:

        Application.Workbooks("analysis.xls").Worksheets("Sheet1").Range("A1")

- If you omit specific references, Excel uses the *active* objects. If analysis.xls is the active workbook, the preceding reference can be simplified as follows:

        Worksheets("Sheet1").Range("A1")

- If you know that Sheet1 is the active sheet, you can simplify the reference even more:

        Range("A1")

- Objects have properties. A property can be thought of as a *setting* for an object. For example, a Range object has properties such as *Value* and *Name*. A Chart object has properties such as *HasTitle* and *Type*. You can use VBA to determine the current value of object properties and to change them.
- You refer to properties by combining the object with the property, separated by a period. For example, you can refer to the Value property of a Range object as follows:

        Worksheets("Sheet1").Range("A1").Value

- Objects have methods. A method is an action that is performed with the object. For example, one of the methods for a Range object is ClearContents. This method clears the contents of the range.

- You specify methods by combining the object with the method, separated by a period. For example, to clear the contents of the range A1:C12, use the following statement:

  Worksheets("Sheet1").Range("A1:C12").ClearContents

- You can assign values to variables. To assign the value in cell D4 on Sheet1 to a variable called *Interest*, use the following VBA statement:

  Interest = Worksheets("Sheet1").Range("D4").Value

- Variables, by default, are of the *variant* data type. This means that a variable can store any type of data. For increased efficiency, you should declare a variable as a particular data type by using a Dim statement. The statement below declares a variable as an Integer:

  Dim Counter as Integer

- A variable can also be an *array* – which is a collection of variables identified by an index number. The following statement declares a String array with 12 elements. You can use this array to store the names of the months: January, February, and so on.

  Dim MonthNames as String(12)

- Refer to elements within an array by using an index number. If you've created a 12-element array to hold the names of the months, you can refer to the fourth month as follows:

  MonthNames(4)

- VBA includes a wide variety of built-in functions that you can use in expressions. For example, the statement below uses VBA's Log function to calculate a natural logarithm:

  x = Log(y)

- A few built-in functions provide additional functionality. For example, the MsgBox function displays a small dialog box with text and buttons. The statement below displays a message to the user:

  MsgBox "Processing complete."

- The InputBox function is an easy way to solicit input from a user. The statement below asks the user to specify a value, and assigns the value to a variable named Value:

  Value = InputBox("Enter a value")

- VBA includes several types of conditional statements that determine program flow. The following subroutine use an If-Then structure to check the value in cell A1. If the value is positive, it is replaced by its square root.

```
Sub EnterSquareRoot()
    If Range("A1").Value > 0 Then
        Range("A1").Value = Sqr(Range("A1").Value)
    End If
End Sub
```

- In common with most other programming languages, VBA includes a variety of control structures that determine program flow. Following is a short subroutine that uses a For-Next looping structure to insert a series of values into a range:

```
Sub FillRange()
    x = 0
    For row = 1 to 10
        Cells(row, 1).Value = x
        x = x + 10
    Next row
End Sub
```

- VBA can also work with objects contained in custom dialog boxes. These objects include controls such as buttons, listboxes, and checkboxes. A custom dialog box is created on a UserForm in Excel 97's Visual Basic Editor. In Excel 5 and Excel 95, a custom dialog box is created on a dialog sheet in a workbook.
- VBA also includes features that allow you to customize user interface elements such as menus and toolbars.
- As a further indication of the power of VBA, the language also supports calls to the Windows API (Application Programming Interface). This feature allow advanced users to perform sophisticated feats that are normally impossible.
- VBA includes sophisticated debugging tools to help you identify problems in your code. For example, you can set a *breakpoint* in your code. When execution reaches the breakpoint, you can step through your code line-by-line and examine the current values of variables and properties.

Creating VBA Macros

Excel provides two ways to create macros:

- Turn on the macro recorder and record your actions. Excel's macro recorder is useful for creating simple macros.
- Enter the VBA code directly into a module. This requires some degree of knowledge of programming and Excel's object model.

Recording a Macro

In this section, I describe the basic steps that you take to record a VBA macro. In most cases, you can record your actions as a macro and then simply replay the macro; you needn't look at the code that's generated. If this is as far as you go with VBA, you don't need to be concerned with the language itself.

It's important to understand that you cannot record a macro that performs any type of repetitive action like looping, or uses conditional statements such as IF THEN. However, you can often record a basic macro and then modify it by manually adding such structures to make it more versatile.

Basic Steps

Excel's macro recorder translates your actions into VBA code. To start the macro recorder, choose *Tools Macro Record New Macro* (if you're using Excel 5 or Excel 95, choose *Tools Record Macro Record New Macro*). Excel displays the Record Macro dialog box that is shown in Fig. 2. This dialog box presents several options:

- **Macro name.** The name of the macro. By default, Excel proposes names such as Macro1, Macro2, and so on.
- **Shortcut key.** You can specify a key combination that executes the macro. You can also press Shift when you enter a letter. For example, pressing Shift while you enter the letter H makes the shortcut key combination Ctrl+Shift+H.
- **Store macro in.** The location for the macro. Your choices are the current workbook, your Personal Macro Workbook, or a new workbook. If you record it in your Personal Macro Workbook, the macro will always be available since Excel loads this workbook automatically at start-up.
- **Description.** A description of the macro. By default, Excel inserts the date and your name. You can add additional information if you like.

To begin recording your actions, click OK. When you're finished recording the macro, choose *Tools Macro Stop Recording* (in Excel 5 or Excel 95, choose *Tools Record Macro Stop Recording*).

Fig. 2. The Record Macro dialog box

Examining the Macro

If you're using Excel 97, activate the Visual Basic Editor (VBE) to view the macro.
To activate the VBE press Alt+F11. The Project window displays a list of all open
workbooks and add-ins. This list is displayed as a tree diagram, which can be
expanded or collapsed. To view a recorded macro you need to select the workbook and module for the appropriate workbook. Fig. 3 shows a sample recorded
macro, as displayed in the VBE. This macro inserts a new row at the active cell
position, and then changes the interior color of the new row.

If you're using Excel 5 or Excel 95, the recorded macro is stored in a Macro sheet
in the workbook you specified. Simply click the tab (probably labeled Module1)
to view the macro. If you specified your Personal Macro Workbook as the destination for the macro, you'll need to unhide the Personal Macro Workbook in order
to view the macro. Use the *Window Unhide* command to unhide the Personal
Macro Workbook.

Playing Back a Recorded Macro

To execute a macro, choose *Tools Macro Macros* (in Excel 5 or Excel 95, choose
Tools Macro) to display the Macro dialog box. Select the macro from the list and
click the Run button.

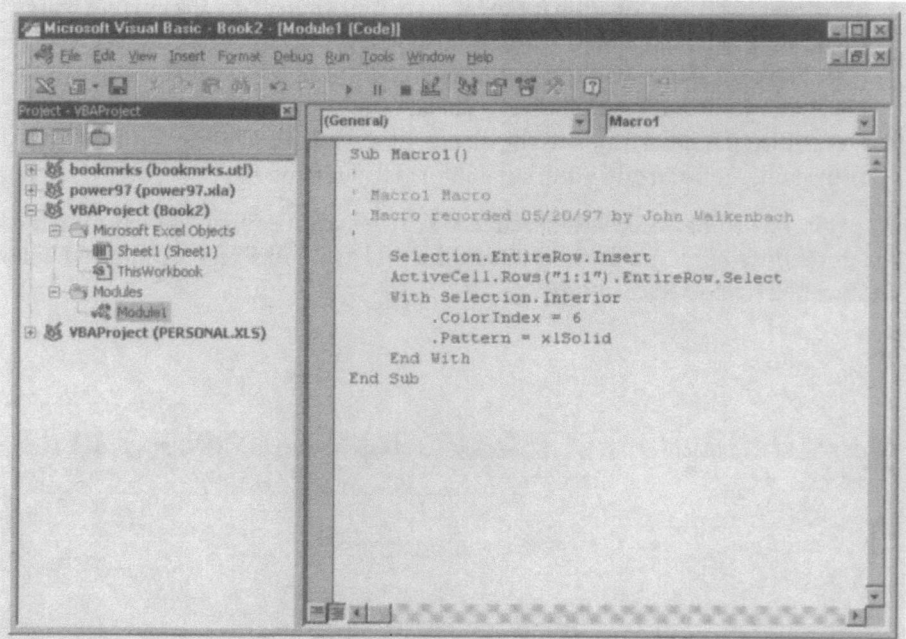

Fig. 3. This macro was generated by Excel's macro recorder

Absolute vs Relative Recording

When recording a macro, it's important that you understand the concept of *relative* versus *absolute recording*. Normally, when you record a macro, Excel stores exact references to the cells that you select (that is, it performs absolute recording). If you select the range B1:B10 while you're recording a macro, for example, Excel records this selection as

> Range("B1:B10").Select

This means exactly what it says: "Select the cells in the range B1:B10." When you invoke this macro, the same cells are always selected regardless of where the active cell is located.

If you want your macro to be more general in nature, you'll need to record the macro in relative recording mode. To set the recording mode to relative, click the Relative Reference button on the Stop Recording toolbar (this toolbar appears automatically while you are recording a macro). In Excel 5 or Excel 95, choose the *Tools Record Macro Use* Relative References command.

The recording mode – either absolute or relative – can make a *major* difference in how your macro performs. Therefore, it's important that you understand the distinction. To demonstrate, record a macro that inserts a new column. If you record the macro in absolute mode, the new column will always be inserted at the same position. If you record the macro in relative mode, the column will be inserted relative to the active cell.

Subroutines vs Functions

It is important to understand a key distinction. A VBA macro can be one of two types: a subroutine or a function.

VBA Subroutines

You can think of a *subroutine macro* as a new command that can be executed by either the user or by another macro. You can have any number of subroutines in an Excel workbook.

Subroutines always start with the keyword *Sub*, the macro's name (every macro must have a unique name), and then a pair of parentheses. The parentheses are required; they are empty unless the procedure uses one or more arguments. The *End Sub* statement signals the end of a subroutine. The lines in between comprise the procedure's code.

When you record a macro, the result is always a new subroutine.

VBA Functions

The second type of VBA procedure is a function. A *function* always returns a single value (just as a worksheet function returns a single value). A VBA function can be executed by other VBA procedures or used in worksheet formulas, just as you would use Excel's built-in worksheet functions. Creating VBA functions that you use in worksheet formulas can simplify your formulas and let you perform calculations that otherwise may be impossible. You cannot use the macro recorder to create function procedures.

A function looks much like a subroutine. Notice, however, that function procedures begin with the keyword *Function* and end with an *End Function* statement. Listed below is a custom function named CubeRoot which calculates the cube root of its single argument. Notice that the value returned by the function is the variable that corresponds to the function's name.

```
Function CubeRoot(num)
     CubeRoot = num ^ (1 / 3)
End Function
```

To use this function in a worksheet, enter a formula such as:

```
=CubeRoot(A1)
```

The function below is a more generalized function that takes two arguments (a number and its root). It returns the nth root of a number.

```
Function Root(num, n)
     Root = num ^ (1 / n)
End Function
```

A Subroutine Example

In this section I provide step-by-step instructions for developing a useful macro. This macro works with the selected range of cells. When the macro is executed, an input box asks the user for a value. Each cell in the selected range is then multiplied by the value supplied by the user. Note that both macros described below are included on the diskette in the file VBA.XLS.

Creating the Macro

1. Start with an empty workbook.
2. Enter some values into a range of cells. The actual values don't matter. They are simply to provide something to work with.
3. If you're using Excel 97, press Alt+F11 to activate the VBE. Select the current workbook in the Project window and use *Insert Module* to insert a VBA module. If you're using Excel 5 or Excel 97, select *Insert Module* to insert a VBA module.

4. Enter the following VBA code into the module (statements preceded by an apostrophe are comments):

```
Sub MultiplyCells()
'     Get the value for the multiplier
      Multiplier = InputBox("Enter multiplier:", "Multiply Cells")

'     Process each cell in the selection
      For Each cell In Selection
          cell.Value = cell.Value * Multiplier
      Next cell
End Sub
```

5. If you're using Excel 97, press Alt+F11 to reactivate the workbook. If you're using Excel 5 or Excel 95, click the worksheet tab to activate the sheet that contains the values you entered in Step 2.
6. Select the range that contains the values.
7. Choose *Tools Macro Macros* (in Excel 5 or Excel 95, choose *Tools Macro*) to display the Macro dialog box. Select MultiplyCells from the list and click the Run button.
8. The macro will display an InputBox. Enter a value and click OK.

Each cell in the selected range will be multiplied by the value you supplied.

How the Macro Works

The MultiplyCells macro is rather simple. It starts by displaying an input box (using VBA's InputBox function), and assigns the value entered by the user to a variable named Multiplier. Then a For Each loop cycles through each cell in the selection and modifies the cell's Value property by multiplying the cell's current value by the Multiplier. When each cell in the selection is processed, the macro ends.

Improving the Macro

The MultiplyCells macro works fine, but there's definitely room for improvement. Following is a list of problems with this macro:

- If something other than a range is selected, an error occurs. For example, if a chart is selected when the subroutine is executed, an error message appears.
- Clicking the Cancel button when the input box is displayed causes an error.
- Specifying a non-numeric value in the input box also causes an error.
- Blank cells in the selection are not ignored. After the macro runs, blank cells contain a value of zero. If the user makes a large selection (such as an entire column), the macro seems to take forever since it process every cell.
- If any cells in the selection contain text, an error occurs.

- Formula cells are not ignored. Running the macro wipes out any formulas in the selected range.
- The macro is not as fast as it could be, since screen updating is not turned off. In other words, you see the change occuring to each cell.

The MultiplyCells Macro – Improved

An improved version of the MultiplyCells macro is listed below.

```
Sub MultiplyCells()
'       Exit if a range is not selected
        If TypeName(Selection) <> "Range" Then Exit Sub

'       Get the value for the multiplier
        Multiplier = InputBox("Enter multiplier:", "Multiply Cells")

'       Exit if canceled
        If Multiplier = "" Then Exit Sub

'       Make sure value is numeric
        If Not IsNumeric(Multiplier) Then
                MsgBox "The multiplier must be a number."
                Exit Sub
        End If

'       Turn off screen updating
        Application.ScreenUpdating = False

'       Process only the cells that contain constants
        For Each cell In Selection.SpecialCells(xlConstants, xlNumbers)
                cell.Value = cell.Value * Multiplier
        Next cell
End Sub
```

This version of MultiplyCells has the following modifications:

- I use VBA's TypeName function to determine the type of the selection. If the selection is not a Range, the subroutine ends with no further processing.
- If the user clicks the Cancel button when the input box is displayed, the Multiplier variable contains an empty string. I use an If statement to test this condition. If Multiplier is equal to "" (which represents an empty string), the subroutine ends.
- I use VBA's IsNumeric function to test the Multiplier variable. This function returns True if its argument is numeric. If the Multiplier variable is not numeric, a message box appears and the subroutine ends.
- I set the ScreenUpdating property to False – which avoids seeing each cell change. This makes the macro run faster.

- I use the SpecialCells method to work with a subset of the selected range. This subset contains only the non-blank, non-formula cells that contain a value. To see how this works, record your action while you select the *Edit GoTo* command, click the Special button, and make various selections.

Both versions of the macro are provided on the disk as VBA.XLS.

Concluding Comments

The revised version of the MultiplyCells subroutine demonstrates several useful techniques that can make your macros work more efficiently and avoid Excel's error messages. This macro handles the input box properly and by-passes Excel's error messages. It is very efficient, since only cells that contain values are processed, and screen updating is turned off. If you are the only person who will use the MultiplyCells macro, you may be satisfied with the initial version and not want to take the time to improve it. However, if others will be using your macro you'll probably want to make these changes to the macro.

If you don't understand exactly how this macro works use the online help system to read about the various keywords.

Compatibility

As indicated previously, VBA is essentially a way to manipulate objects. Excel's object model continues to evolve. Therefore, objects that are available in Excel 97 may or may not be available in Excel 95 or Excel 5. Consequently, if you develop a macro there is no guarantee that it will work with previous versions of Excel. If you need a macro to work across various versions of Excel you should develop it using the earliest version of Excel that will be supported. Even then, it's important that you test it with all versions. Of course this will mean that you may have to dispense with some potentially useful advanced features of later versions.

Excel 97 introduced an entirely new way of working with custom dialog boxes. Excel 5 and Excel 95 use dialog sheets. In Excel 8 you create a custom dialog box on a UserForm, accessed in the Visual Basic Editor. However, Excel 97 continues to support dialog sheets.

Learning More About VBA

If this is your first exposure to VBA, you're probably a bit overwhelmed by objects, properties, and methods. Fortunately, there are several good ways to learn about objects, properties, and methods.

274

J. Walkenbach

Record Your Actions

The best way – without question – to become familiar with VBA is to turn on the macro recorder and record actions that you make in Excel. Try inserting new rows and column, creating a chart, formatting cells, etc. It's even better if the VBA module in which the code is being recorded is visible while you're recording.

Use the Online Help System

The main source of detailed information about Excel's objects, methods, and procedures is the online help system. Help is very thorough and easy to access. When you're in a VBA module, just move the cursor to a property or method and press F1. You get help that describes the word that is under the cursor. As in all Windows Help screens you can print the topic or save it, via the clipboard, to disk.

Consult a Book

There are a number of Excel books available that deal with VBA. Two that are widely cited are:

- John Walkenbach: Excel For Windows 95 Power Programming With VBA, 2nd edn. IDG Books Worldwide, Inc.
- Eric Wells: Developing Excel 95 Solutions. Microsoft Press.

Others are listed in the Appendix B.

Study Other People's Code

Several other chapters (2, 4, 5, 6, 8) in this book contain examples of VBA code. Chapter 8 has the most comprehensive applications. A good way of learning is to use VBA's debugger to trace through the code step-by-step at the same time trying to understand its intentions. The Appendix: Further Sources of Information on Spreadsheets contains more details.

You can also find many examples of VBA programming on the World Wide Web. A good starting location is The Spreadsheet Page, at

http://www.j-walk.com/ss/

Appendix B: Further Sources of Information on Spreadsheets

W.G. Filby

In the following we present some easily accessible sources of further information on spreadsheets. These lists make no claim to be complete. They are intended only to demonstrate the versatility of spreadsheet usage in the most recent literature.

General Books on Spreadsheets

Microsoft Excel 97 Step by Step, Complete Course
Catapult, Inc. Staff
Microsoft (1997)

Teach Yourself Excel 97 for Windows
Weingarten, John
MIS Press, (1996)

Microsoft Excel 97 for Windows Quickstart
Que, (1996)

Excel 97 Bible
Walkenbach, John
IDG Books, (1996)

Excel X Macro & VBA Handbook
Moseley, Lonnie E.
Sybex, (1996)

Microsoft Excel 97 Field Guide
Nelson, Stephen L.
Microsoft, (1996)

Running Microsoft Excel 97
Dodge, Mark; Kinata, Chris; Stinson, Craig
Microsoft, (1996)

Mastering Excel: A Problem Solving Approach
Gips, James
Wiley, (1996)

Excel for Windows 95 Power Programming Tech
Walkenbach, John
IDG Books, (1996)

The Beginner's Guide to MS Excel 5.0
McKay, Dave
INST Publishing, (1995)

Books on the Scientific Application of Spreadsheets

Chemistry and Biochemistry

Dynamic Models in Chemistry: A Workbook of Computer Simulations Using Electronic Spreadsheets
Atkinson, Daniel E.; Brower, Douglas C.; McClard, Ronald;
Simonson & Co, (1990)

Dynamic Models in Biochemistry: A Workbook of Computer Simulations Using Electronic Spreadsheets
Atkinson, David E.; Clarke, Steven G.; Rees, Douglas C.;
Simonson & Co (1989)

Spreadsheet Applications in Chemistry Using Microsoft Excel
Diamond, Dermot; Hanratty, Venita,
John Wiley and Sons, (1997)

Spreadsheets for Chemists
Filby, Gordon
VCH, (1994)

Spreadsheet Chemistry
Breneman, Gary L.; Parker, O. Jeffrey
Prentice-Hall, (1990)

Library Science

Electronic Spreadsheets for Libraries
Auld, Lawrence W.
Oryx Press, (1986)

Management Science

Applied Management Science & Spreadsheet Modeling
Clauss, Francis J.
PWS Pubs., (1996)

Management Science: A Spreadsheet Approach For Windows
Plane, Donald R.
Course Tech., (1996)

Practical Management Science: Spreadsheet Modeling & Applications
Winston, Wayne L.; Albright, S. Christian
Wadsworth Publishers, (1997)

Management Science Using Spreadsheets: Preliminary Edition
Hesse, Rick
Addison-Wesley, (1995)

Physics

Dynamic Models in Physics: A Workbook of Computer Simulations Using Electronic Spreadsheets
Potter, Frank; Peck, Charles;
Simonson & Co., (1989)

Spreadsheet Physics
Misner, Charles
Addison-Wesley, (1991)

Science and Engineering

Lotus in the Lab: Spreadsheet Applications for Scientists & Engineers
Ouchi, Glenn I.
Addison-Wesley, (1988)

Spreadsheet Modeling for Engineers & Scientists Using 1–2–3
Cress, David; Murtha, James A.
Prentice-Hall, (1997)

Spreadsheet Analysis for Engineers & Scientists,
Bloch, Sylvan C.
John Wiley and Sons, (1995)

Quattro Pro: For Scientific & Engineering Spreadsheets
Parks, R. G.
Springer-Verlag, (1991)

The Excel Spreadsheet for Engineers & Scientists
Kral, Irvin H.
Prentice-Hall, (1991)

Practical Spreadsheet Statistics & Curve Fitting For Scientists & Engineers
Mazei, Louis M.
Prentice-Hall, (1990)

Primary Scientific Literature

Analytical Chemistry

Use of spreadsheets in analytical chemistry. Part 2. Titrations of polyprotic acids
Reich, Leo S., Am. Lab. (Shelton, Conn.) (1996), 28(14), 42–45

Use of spreadsheets in analytical chemistry. Part 3. Titrations using soluble metal complexes
Reich, Leo S.; Brown, Pamela A., Am. Lab. (Shelton, Conn.) (1996), 28(17), 42–45

Chemistry

A convenient spreadsheet approach to the calculation of stability constants and the simulation of kinetics
Huskens, Jurriaan; van Bekken, Herman; Peters, Joop A., Comput. Chem. (1995), 19(4), 409–16

Use of spreadsheets in the kinetic analysis of two consecutive first-order reactions
Reich, Leo, Thermochim. Acta (1996), 273, 113–18

Useful spreadsheet for updating multistep organic synthesis
Ortega, Pedro A.; Guzman, Miguel E.; Vera, Leonel, J. Chem. Educ. (1996), 73(8), 726–728

Geosciences

PROBE-AMPH – a spreadsheet program to classify microprobe-derived amphibole analyses
Tindle, A. G.; Webb, P. C., Comput. Geosci. (1994), 20(7–8), 1201–28

An interactive spreadsheet for graphing mineral stability diagrams
Biddle, Dean L.; Percival, Harry J.; Chittleborough, David J., Comput. Geosci. (1995), 21(1), 175–85

Numerical solutions for the one-dimensional heat-conduction equation using a spreadsheet
Gvirtzman, Zohar; Garfunkel, Zvi, Computers & Geosciences, (1996), 22(10), 1147–1158.

GPT; an EXCEL spreadsheet for thermobarometric calculations in metapelitic rocks
Reche, Joan; Martinez, Francisco J., Computers & Geosciences, 22(7). (1996), 775–784.

Teaching geomorphology through spreadsheet modelling
Locke, William W., Geomorphology, (1996), 16(3), 251–258.

TERNPLOT; an Excel spreadsheet for ternary diagrams
Marshall, Daniel, Computers & Geosciences, 22(6) (1996), 697–699.

EQMIN, a Microsoft Excel spreadsheet to perform thermodynamic calculations; a didactic approach
Martin, Jordi Delgado, Computers & Geosciences, 22(6). (1996), 639–650.

BGT; the Macros driven spreadsheet program for biotite-garnet thermometry
Rameshwar Rao, D., Computers & Geosciences, 21(4)..: 1995 . p. 593–604.

Spreadsheet interpretation of seismic refraction data
Fourie, C. J. S.; Odgers, A. T. R., Computers & Geosciences, 21(2), (1995), 273–277.

Life Sciences

Measurement of Bile Salt Aggregation Equilibria Using Kinetic Dialysis and Spreadsheet Modeling.
Duane W C; Gilboe D P, Analytical Biochemistry 229 (1). 1995. 15–19.

Simple computer spreadsheet for standardized interpretation of oral glucose tolerance tests.
Chesher D; Burnett L, Pathology 27 (2). 1995. 140–141.

Simple spreadsheet models to study population dynamics, as illustrated by a mountain reedbuck model.
Norton P M, South African Journal of Wildlife Research 24 (4). 1994. 73–81.

A non-linear fitting program in pharmacokinetics with Microsoft Excel spreadsheet .
Delboy H, International Journal of Bio-Medical Computing 37 (1). 1994. 1–14.

Use of a spreadsheet program for circadian analysis of biological-physiological data.
Bourdon L; Buguet A; Cucherat M; Radomski M W, Aviation Space and Environmental Medicine 66 (8). 1995. 787–791.

An Excel spreadsheet computer program combining algorithms for prediction of protein structural characteristics.
Clotet J; Cedano J; Querol E, Computer Applications in the Biosciences 10 (5). 1994. 495–500.

A general method of curve fitting and error analysis using a spreadsheet : Determination of the binding constants of tight binding ligands in variable volume assays
Delahunty, Martha D.; Mack, Joseph P. G., Comput. Appl. Biosci. (1993), 9(2), 127–31

Monte Carlo spreadsheet modeling of stable isotope biosynthesis.
Masterson T M; Kelleher J K, Computers in Biology and Medicine 26 (5). 1996. 429–437

Web Sites and FTP sites

For readers with an Internet connection this provides far the most efficient way of getting up to speed. Not only is the latest product information placed on developers homepages often in advance of official release but also FAQs (frequently answered questions), tips and tricks, free software are often available to download. A couple of the more active ones are described briefly in Table 1. They will be more than enough to get you going.

Table 1. URLs and comments

URL	Comments
com/msexcel/[http://microsoft.]	The page of course. Everything you need to know about Excel
http://www.j-walk.com/ss/	Free Power Utility pak for Excel, spreadsheet jokes, info on author's books
http://www.vex.net/~negandhi/excel/	MSN chat sessions, FAQs, very good programming tips, book recommendations, useful links to other Excel sites
http://www.lacher.com	Mainly financial applications. Lots of free VBA macros to download
http://sunsite.univie.ac.at/Spreadsite/	Very comprehensive listing of applications of spreadsheets in science and education
http://ourworld.compuserve.com/homepages/Stephen_Bullen/	Excellent source of downloadable Excel utilities.
http://www.joanneum.ac.at/services/vbaexcel	A complete VBA tutorial and seminar

Appendix C: The Major New Features of Microsoft Excel 97

W. G. Filby

During the final stages of preparation of this book Microsoft released Office 97, a comprehensive office software package including Word 97, Excel 97 and several other programs useful in an Office environment.

All example material contained in this book is compatible with the new version. Note however that on some occasions informational messages may be displayed on reading our worksheets into Excel 97. One frequent such occasion is when macros are involved. Firstly a message dealing with the possibility of a computer virus infection by way of Excel macros appears. Secondly, owing to the complete restructuring which has been carried out on the VBA environment a warning that VBA modules are no longer used appears. This message can be switched off in succeeding work. It does not influence the correct running of our macros.

In the following we will outline those changes most expected to be of interest to readers of this book. Sources for more complete information are provided at the end of the section.

Increased Capacity

Excel 97 has been released with an increase in capacity commensurate with today's typical hardware configurations. Now a cell can contain 32,767 characters instead of the 255 allowed in previous versions. Also the number of rows per worksheet has been increased from 16,384 to 65,535 and the number of chart points per series from 4,000 to 32,000.

The Office Assistants

A love-hate one, this one. Any one of nine animated characters (a deforming paper-clip, Shakespeare, Einstein etc.) with optional sound pop up in a floating window and ask "What would you like to do ?".

The real novelty here is the Assistants' ability to cope with natural language querying. The success rate is quite high and the system can even deal with minor spelling mistakes. For example it dealt with the query "new workseet" with five suggestions, one of which was the correct one. Sometimes, however the suggestions are way off and for that eventuality the Help-seeking user will be grateful for the old fashioned Index Search. As ever to be had by pressing F1 or ? on the menu bar.

You can control various properties of the Assistant and scroll through the "characters" by clicking on the Options button which appears on activating it.

Entering Formulas

The Formula Palette

This new feature replaces the Function Wizard featured in earlier versions. Using it makes it easier to create and enter formulas. You can call it up by clicking the new Edit Formula button (=) on the edit line or by double clicking on a formula-containing cell for in-cell editing. Applied on an existing formula another new feature, the so-called **Range Finders**, causes each range used by the formula to be outlined in a different color. For example if we are editing the formula =AVERAGE(B4:C4) the bracketed B4:C4 range argument will be colored and the actual range on the worksheet receives a border of the same color. In turn, this border can be dragged and dropped to alter the range the function operates on.

The Formula Palette itself appears in place of the named range whenever the creation or editing of a formula is indicated by clicking on the Edit Formula button. It provides immediate access to several of Excel's most frequently used functions (SUM, AVERAGE, IF etc). Clicking on the last entry of the pull-down brings up a categorized list of available functions (mostly recently used, finance mathematical, informational etc.) as in earlier Function Wizard editions. A further bonus – the Formula Palette helps insert non-contiguous numbers into the formula's arguments, and displays them with the running result of the function as it is being completed.

Natural Language Formulas

A long-awaited novelty. Using this new feature it becomes possible to "teach" Excel to recognize the labels of cells to be inserted as function arguments. Thus you needn't formally define cell names, using this new feature you can use existing row or column labels to do it for you. It enables the user to write meaningful formulas in everyday language like =UnitPrice Screws*Order Screws. The tabbed worksheet **NatLang Formulas** in the file FEATXL97.XLS on the CD-ROM shows some simple examples for practice. You can access this facility using either the Wizard provided or via the Insert menu. With the latter select the Labels option and follow the on-screen instructions from there. On the way you'll find another new feature, **Collapse/Expand** dialog boxes. These dialog boxes collapse to a single line, making it easier to see and select ranges by point and click. After completion of data entry the dialog can be restored to normal size by a mouse click. Be careful to carefully define your row and column labels correctly or else it won't work and you'll be left with familiar sights (=B3*B5 and the like).

Another potentially useful aspect of this new feature is the way the intermediate results of selected terms of formulas can be calculated and displayed. For an example turn to the worksheet **NatLang Formulas** in the file FEATXL97.XLS on the CD-ROM. Select the cell N11 to display the formula in the edit panel. There select a monthly ozon value (Nov Ozon, Jan Ozon), press F9 to recalculate the worksheet. The numeric value of the chosen month's ozon will appear amidst the remaining synonyms. At this stage the worksheet will look something like this:

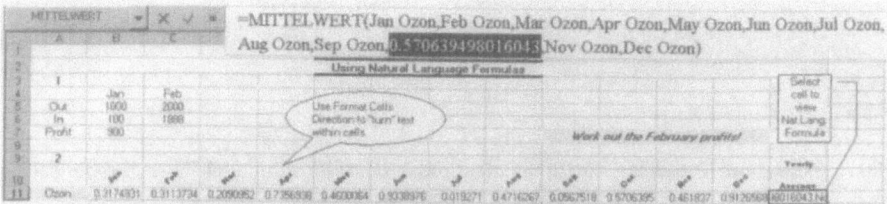

MITTELWERT is the German name for the Excel function AVERAGE.

Another example (number 4 on the worksheet NatLang Formulas uses the rather long equation describing the Gaussian dispersion of air pollutants. The full formula displayed in the edit panel looks like this:

$$=((Q/(2*PI()*SY*SZ*U))*EXP((-Y^2)/2*SIGy^2))*(EXP(-(Z\_-h^2)/2*SIGz^2)+EXP(-(Z\_+h^2)/2*SIGz^2))$$

But after selecting the first term (before the EXP terms) and recalculating it appears thus:

$$=(0.0161042959211949*EXP((-Y^2)/2*SIGy^2))*(EXP(-(Z\_-h^2)/2*SIGz^2)+EXP(-(Z\_+h^2)/2*SIGz^2))$$

The original formula is restored by pressing Esc. Clearly this feature will be of great use whenever long formulas look as if they're not working correctly. By the way the Gaussian equation was entered extremely quickly by point and click, the named cells option and copying in the edit panel.

Formula AutoCorrect

This feature makes limited suggestions for fifteen of the most common errors occurring during formula-creation. Thus it will find mistakes, sometimes multiple ones, like double operators (like ^^ or) and unmatched parentheses. Simple mistakes like the latter are corrected without confirmation. With more complicated mistakes Excel may make one or more suggestions, see Table 1. Examples (A1, A2 are cells containing numbers, A3 contains "SomeText") see next page.

Table 1. Suggestions made by Formula AutoCorrect for more complicated errors

Expression	AutoCorrect's Suggestion
=A1))	=A1
=A1*A2^^2	A1*A2^2
=(A1*A2))^^2)	no suggestion, error notified
=LENGT(A3)	no suggestion, error notified
=A3&&A3	=A3&A3

Table 2. Charting features

Chart Wizard and Toolbar	Facilitates creation and modification of charts by keeping all the charting options in one place
Chart Menu	Users no longer need to work with different dialogs displaying the same information
Chart Tips	Identify chart elements and data series values immediately when the mouse is moved over a chart
Single Click Selection	Users simply click once on the item they want to change to activate the chart and modify it.
Additional Chart Types	Additional 3-D and 2-D chart types like **Cylinder**, **Pyramid**, and **Cone** charts have been added, Modified types like **Bubble charts**, **Pie of Pie** and **Bar of Pie** too.
Time-scale axes	Dates are always displayed in the proper order and users can group their data without changing the underlying data.
Chart Data Tables	Data can be displayed in tabular as well as graphical form by appending a table directly on the chart.
Picture, Texture, and Gradient fills in chart	Users can format the fill of chart walls, floors, and the faces of 2D and 3D charts with pictures, textures, and gradient fills. (OfficeArt).

Charting

This is shown in Table 2.

Customizing Cells and Data Validation

Automatic Formatting Options

"Beyond the Grid" formatting options is what Microsoft calls these. One new feature allows you to automatically format specific cells based on certain values on the worksheet. Thus, rules enabling a simple and efficient form of data validation can be defined without writing any special code. Additionally, you can tie input and error messages to these cells to help users enter correct values. Using anoth-

Table 3. Further options

Visual Printing	• Page Preview: is now fully editable, super-imposes page numbers and page breaks so that users see exactly what will be printed
	• Draggable Page Breaks: dragging page-breaks and printer area borders allows easy repagination of a worksheet

er option, text within a cell can be rotated or indented according to hierarchical or other relationships in the worksheet. Merging cell contents to exceed one row or column has also been introduced in this version. Other options are shown in Table 3.

Data Validation

With this new feature you can stipulate which data or data types are to be allowed in a defined cell or range. For example, you can allow only integer or decimal numbers or only data, time and date values included, within a certain range. And it doesn't end there! You can also restrict the number of characters allowed in a cell, make an entry dependent on a valid calculation and check the complete worksheet for invalid input. For the latter there is also an option to circle offending cells red! All these options can be set via the Data/Validation menu and/or the Auditing toolbar. The worksheet **Cell Formatting I** in the file FEAT97.XLS contains a few examples of each of the main new formatting features.

Conditional Formatting, Improved Commenting and the New Drawing Toolbar

Conditional Formatting allows users to automatically highlight outliers by setting up rules that change a cell's formatting depending on its value. The worksheet **Cell Formatting II** in the file FEAT97.XLS contains a few examples of this feature together with some uses of the improved comments option and the reconstructed Drawing Toolbar. There, especially the new feature known as *AutoShapes* is a veritable treasure trove of over one hundred new drawing objects including various types of arrow symbols, legends, connectors and flow diagram icons. You can call these up either from the View Toolbars Drawing menu option or by clicking on the Drawing icon on the main toolbar (yellow cube, blue A, green cylinder) next to the zoom dropdown. From there you have also have access to WordArt, shadowing, 3-D effects, lines and arrows and more. The Drawing toolbar itself contains an interesting novelty. Submenus like those for accessing the lines, arrows, stars and stripes etc can be "pulled out" or the menu and left on the desktop, making them available as long as the user likes. This works even without the "parent" Drawing Toolbar being open. A solid stripe along the top of the dropdown menu indicates that this option is active.

Other Changes in Excel 97

These can be clearly seen in Table 4.

Readers with access to the World Wide Web can obtain the latest information on the Microsoft Office packet from *http://www.microsoft.com/office*. Two other Web sites are important: *http://www.baarns.com/Office97/excel* and *http://www.j-walk.com/ss/*. This appendix leans heavily on all of these sources.

Table 4. Tabular presentation of Excel 97's new features

Productivity	Excel 97 provides several new features to ease day-to-day use: *Multilevel Undo and Redo: with this improved feature you can reconstruct the original state of your worksheet at every stage of your last sixteen deletions* *Revised Menubar: the standard menubar has now become a toolbar; draggable, floatable and reconfigurable. Menus and buttons can be mixed on the same bar* *Active Web Searches for Excel or HTML pages are possible without a separate* *Open from URL* allows users to open Excel or HTML files from servers using the File Open command in Excel 97
Internet features	Excel 97 has several new features dealing with Web matters: <u>*Navigation and Searching*</u> *Hyperlinks can now be embedded in all Excel 97 documents* *Web Toolbar* allows easy navigation of hyperlinked documents *Active Web Searches* for Excel or HTML pages are possible without a separate search engine *Open from URL* allows users to open Excel or HTML files from servers using the File Open command in Excel 97 <u>*Online Publishing*</u> *Save as URL* allows spreadsheets to be saved as HTML documents *Save to URL* allows spreadsheets to be posted onto HTTP or intranet servers *Web Form Wizard* facilitates connecting an Excel form to a Web server *HTML tags* allow spreadsheet formatting in HTML tables
Microsoft Map	Allows geographic maps to be placed on a worksheet. These can be overlain with highways, towns, demographic data etc as required. The Map-Manager provides comprehensive tools for manipulation and user-defined activities.

Table 4. (continued)

Pivot Tables	**Structured Selection:** allows selection of related information within a PivotTable for formatting or analysis **Persistent Formatting:** Retains custom formatting through each pivot and data refresh **Calculated Fields and Items:** Users can now create new data fields and items without modifying source data by building formulas using existing PivotTable fields or items **Server based Paged Fields:** page fields can be stored on a server with external data **Dates displayed as numbers:** PivotTables handle dates correctly. **Page Field Layout:** PivotTable page fields can be arranged across columns and in multiple rows. **Options Dialog:** more control over elements like format, display, data source, and external data links when using the PivotTable Wizard. **AutoSort:** Now, the same sorting capabilities as in Microsoft Excel are possible in PivotTables. Sorting by two or more criteria in the same command. **AutoShow:** Allows displays of only the most relevant information in a PivotTable.
Solver	Improved speed and robustness, improved linearity testing, better control of non-linear convergence
Visual Basic Environment (VBE) for VBA version 5.0	The VBE is entirely new and will be common to all applications supporting VBA 5.0. Has improved code editor, object browser, multipane debugger, a property sheet, project explorer, and a statement builder

Springer
and the
environment

At Springer we firmly believe that an international science publisher has a special obligation to the environment, and our corporate policies consistently reflect this conviction.
We also expect our business partners – paper mills, printers, packaging manufacturers, etc. – to commit themselves to using materials and production processes that do not harm the environment. The paper in this book is made from low- or no-chlorine pulp and is acid free, in conformance with international standards for paper permanency.

H. Lohninger

INSPECT

A Program System for Scientific and Engineering Data Analysis

1996. 2 MS-DOS diskettes with handbook IX, 211 pp. 46 figs. Hardcover DM 598
ISBN 3-540-14530-3

INSPECT is a DOS-based system for exploratory analysis of multivariate data. It provides the practitioner with a set of mathematical tools for the interpretation of multivariate data, including principal component analysis, multiple linear regression, cluster analysis, and neural network modeling. **INSPECT** has been designed to be an easy-to-use tool in everyday work of scientists and engineers. The software package includes a 215 pp handbook. Over 250 commands provide the basis for editing, displaying, analyzing and modeling of data. All major types of charts can be generated, including an on-screen 3D-rotation of data. A parser for mathematical formulas allows the processing of the data by entering almost arbitrary formulas.

 Springer

Springer-Verlag, P. O. Box 31 13 40, D-10643 Berlin, Germany

R. Henrion, G. Henrion

Multivariate Datenanalyse

Methodik und Anwendung in der Chemie
und verwandten Gebieten

1995. XI, 261 S. 60 Abb., 3 1/2" MS-DOS Diskette
Geb. DM 128
ISBN 3-540-58188-X

Die Anwendung multivariater statistischer
Verfahren auf umfangreiche Datensätze vornehm-
lich aus der analytischen Chemie ist das zentrale
Thema des Buches. Das Autorenteam – Chemiker
und Mathematiker – stellt die klassischen und
modernen Methoden und deren Kombination zur
Lösung analytischer und physikalisch-chemischer
Problemstellungen praxisnah dar. Das Buch ist für
Anfänger und erfahrene Praktiker gleichermaßen
geeignet, weil es die komplizierten Sachverhalte
durchgehend deskriptiv und mathematisch-
theoretisch darstellt. Zusätzlich bietet das Buch die
Möglichkeit, viele der vorgestellten Verfahren
anhand der auf Diskette im Sourcecode mitgeliefer-
ten Computerprogramme (Turbo-Pascal 5.0) und
ebenfalls mitgelieferter bzw. eigener Datensätze zu
erproben.

Please order from
Springer-Verlag Berlin
Fax: + 49 / 30 / 8 27 87- 301
e-mail: orders@springer.de
or through your bookseller

Price subject to change without notice.
In EU countries the local VAT is effective.

Springer

Springer-Verlag, P. O. Box 31 13 40, D-10643 Berlin, Germany

Copyright information

License Agreement and Warranty